D0161880

Vertebrate Zoology is a major new textbook that is intended to lead students away from purely descriptive zoology courses into an experimental approach that emphasizes asking and answering questions about nature. The book gives a panoramic view of vertebrate life, classification, ecology, and behavior.

In Part I of the book the major groups of vertebrates and their origins are described, and a special chapter discusses the ways in which vertebrates sense their environments. The familiar human senses of sight, hearing, taste, smell, and touch are compared with those of other vertebrates that exceed human capabilities, as well as the magnetic and electric senses that are outside of our sensory world. Part II covers classification and its methodology and points to the ways in which some vertebrates deviate from the routinely accepted system. Part III describes the ecology of vertebrates from two standpoints: how individuals cope with environmental extremes, and principles of population and community ecology, as illustrated by experiments carried out in the field. Part IV describes the geographic distributions of vertebrates and considers theories of how they came to be in their current locations. Part V discusses migration and describes experiments that have elucidated the mechanisms of navigation over vast distances. Vertebrate behavior is the subject of Part VI, which deals with observations and the theories and experiments they have inspired.

This book will find use as a text for second- or third-year university or college students who have completed one or more courses in zoology or biology.

Vertebrate Zoology

Vertebrate Zoology

An experimental field approach

NELSON G. HAIRSTON, SR.
The University of North Carolina
Chapel Hill

Published by the Press Syndicate of the University of Cambridge
The Pitt Building, Trumpington Street, Cambridge CB2 1RP
40 West 20th Street, New York, NY 10011-4211, USA
10 Stamford Road, Oakleigh, Melbourne 3166, Australia

First published 1994

Printed in the United States of America

Library of Congress Cataloging-in-Publication data
Hairston, Nelson G.
Vertebrate zoology : an experimental field approach / Nelson G. Hairston, Sr.
p. cm.
ISBN 0-521-41703-1 (hardback).
1. Vertebrates. I. Title.
QL605.H34 1994
596—dc20 93–27878
CIP

A catalog record for this book is available from the British Library

ISBN 0-521-41703-1 hardback

To the 520 students
who, over thirteen years, made it a joy
to teach vertebrate field zoology

Contents

Preface *page* xiii
Acknowledgments xv

Part I. The major groups of living vertebrates: Their origins and special adaptations 1

The geologic time scale 1
Continental drift 5

1. The living vertebrates 11
Agnatha 11
Chondrichthyes and Osteichthyes 14
Chondrichthyes 15
Osteichthyes 17
Sarcopterygii 17
Actinopterygii 18
Teleostei 21
Unique abilities of Chondrichthyes and Osteichthyes 22
Amphibia 23
The origin of terrestrial vertebrates 23
Living Amphibia 27
Reptiles 30
Mammals 41
The range of eutherian adaptations 53
Birds 54

2. Vertebrate sensing of the environment 65
Sight 65
Visual fields of vertebrates 68
Relative visual abilities among vertebrates 70

Hearing 71
Structure 71
Echolocation 80
Smell and taste: Chemical senses 82
Touch and pressure 86
Heat sensing 87
Sensitivity to weak electric fields 87

Part II. Classification 91

3. **Classification: Principles and problems** 93
Introduction 93
Categories 94
Use of the system in practice 95
Arbitrary definition of all categories above species 96
An objective category: The species 99
Problems in applying the species concept 100
Practical problems 100
Unrecognized geographic variation 101
Species so similar as to be difficult to distinguish 102
Color phases within species 103
Speciation that has reversed itself 104
The open ring: One species that can seem to be two 105
Conceptual problems 107
Numerical taxonomy and "electrophoretic species" 107
Parthenogenetic species 113
Summary 116

Part III. Vertebrate ecology 117

4. **Adaptations to different environments** 121
Seasonal activity patterns 121
Winter 121
Spring 126
Summer 128
Autumn 129
Coping with physical environments 131
Cold 131
Heat 137

5. **A field study in population ecology: The rusty lizard** 145

6. **Field experiments on competing vertebrates** 151
Terrestrial salamanders 151
Desert rodents 158
Forest birds 161
Great tits and blue tits 161
Magpies and jackdaws 162
Lizards 163

7. **Groups of vertebrate species (communities)** 167
Niche partitioning among sunfishs 167
Tadpoles and newts 169
Streambank salamanders 172

8. **An unsolved problem: Population cycles** 177

Part IV. Geographic distributions of vertebrates 185

9. **Distributions of North Carolina vertebrates** 187
Summary 191

10. **Global pattern and distribution of climates** 193
Worldwide pattern 193
Patterns within continents 195

11. **Vertebrate distribution: The importance of other species** 205
Competition 205
Predation and parasitism 210
Food species 212
Summary 217

12. **Principles of global distribution** 219
The theories 224

Part V. Migration 235

13. **Migrations of vertebrates other than birds** 239
Summary 249

14. **Bird migration: Patterns and capabilities** 251
Reasons for migrating 251
Observational methods 252
  Banding 252
  Direct watching 253
  Listening 253
  Radio tracking 253
  Radar 253
Experimental methods 254
Migration routes 254
  North American flyways 254
  European flyways 255
  Central Eurasian routes 256
  Oceanic routes 257
Achievements of migrating birds 260
  Nonstop distances 260
  Time consumed 261

Flight speeds 261
Height of migratory flight 262
Accuracy of finding their way 264
Accuracy of timing of migration 265
Questions about migrating birds 266
Direction finding 266
Timing 266
Energy demands 266

15. Direction finding and navigation by birds 269
Landmarks 270
Sun 271
Stars 273
Magnetism 274
Smell 277
Sound 278
Navigation 278
Summary 279

16. Physiological solutions to problems of migration 281
Timing of migration 281
Energetics of migration 286

Part VI. Vertebrate behavior 289

17. Independent behavior of individuals 293
Feeding behavior 293
A representative study: The European bee-eater 297
Predator avoidance 300
Running away 300
Taking cover 301
Freezing 301
Using special defenses 301
Confronting the predator 302
Two special cases 302
Cowbird victims 302
The notorious hawk–goose model 303
Summary 304

18. Behaviors of two individuals interacting 307
Territorial behavior 307
Auditory displays 307
Visual displays 308
Chemical advertisement 309
Kinds of territories 311
Benefits and costs of territoriality 312
Mating behavior 314

Parental care 316
  Incubation 316
  Feeding the young 317
  Protecting the young 319
Summary 319

19. Behavioral interactions in groups of vertebrates 321
Predator avoidance 321
Energetic benefits to vertebrates in groups 324
Origins of group behavior 326
Group selection 326
Requirements for group selection 329
Summary 333

Index 335

Preface

The title of this book looks like taxonomic chauvinism (favoritism toward those animals most like us), as do a number of other similar titles. Without denying that vertebrates are a numerical minority among animal species, I claim that the fact that we are vertebrates makes the subject inherently interesting, and worthy of special attention. I also argue that because of that relationship, our understanding of the behavior and ecology of vertebrates is likely to be more realistic than our understanding of invertebrates, and that should contribute to a more accurate interpretation of the natural world.

These impressions should not be attributed to any specialization on my part, although I have been called a urodelan specialist (meaning that I have firsthand experience with salamanders). I have spent half of my scientific career in research on invertebrates and have published on six phyla and conducted research on a seventh.

To some people, the term "field zoology" carries the meaning "natural history," with a vaguely unscientific connotation. The connotation is a holdover from an earlier period when it was thought impossible to conduct rigorous observations and experiments out-of-doors. The fields of behavior and ecology have demonstrated the contrary since the 1920s, but the dogma of laboratory biology, like other dogmas, dies slowly. In common with the course from which it has been developed, this book is directed to students who have a background in general zoology or biology and is intended to introduce them to the fascination of animals in the field, and at the same time to show that field study can be made scientifically rigorous, with examples of the kinds of careful field observations that can lead to formulation of proper hypotheses, as well as examples of the experimental testing of those hypotheses.

Several books that overlap this one in subject matter cover such topics as anatomy, physiology, and genetics. I believe that separate books are needed to give those topics the amount of coverage that they require, and they are touched upon here only to the extent that a good course in general zoology or general biology will already have supplied the necessary background. The field of biology is so broad that it would be unreasonable for a book intended for students who have already been introduced to the subject to attempt coverage of more than small parts of the esoteric and complex subareas that constitute the next level.

Students should be aware of the all-pervading influence of evolution by natural selection on all aspects of biology, especially as manifested in the natural world. It is impossible to understand nature without accepting the concept of evolution. The alternative, as proposed by the creationists, is an attempt to promote the dogma of a particular religious sect in the guise of science and should be rejected by anyone attempting to understand natural phenomena.

As this book developed out of my lecture notes for a course at the University of North Carolina, there is inevitably a geographic bias toward the southeastern United States. In admitting that, I also admit my preference for the authority that personal experience conveys, as compared with a secondary presentation given in the interest of a geographically balanced coverage. Anyone presenting such a course in the southwestern United States would surely give more space to rodents and reptiles, as they provide the varied and abundant examples of vertebrates for that area, just as amphibians and overwintering waterfowl do in the Southeast. Nevertheless, where appropriate to the topic under discussion, I include examples from all of the major groups of vertebrates and from other parts of the world, such as the Southwest, the Northwest, Europe, and the Philippines. I largely avoid discussion of marine situations, both because I have no experience with them and because there would be few opportunities to show marine vertebrates to students. An author who attempted to give complete coverage geographically and taxonomically would soon be writing an encyclopedia. I believe that this book will be useful in most parts of North America, especially when supplemented with lectures describing examples from studies of the local fauna.

The book is divided into six separate topics: Part I, an introductory section describing the classes of living vertebrates and their special adaptations, especially the ways in which they sense the environment; Part II, a section on classification; Part III, a section describing a number of detailed ecological studies of different groups of vertebrates; Part IV, geographic distribution; Part V, migration; Part VI, an introduction to vertebrate behavior. These topics are mostly independent of each other, and each part begins with its own introduction. Where appropriate, I have included references at the ends of the chapters so that students can pursue further those subjects of special interest to them.

Acknowledgments

A number of people have helped me by reading various parts of the manuscript. They include my former colleagues at the Museum of Zoology of the University of Michigan: William R. Dawson, Gerald R. Smith, Ronald A. Nussbaum, Robert B. Payne, and Philip Myers. Their comments and constructive suggestions prevented me from making a number of errors and brought me up to date on recent advances in physiological ecology, theories of classification, and nomenclature. I fear that what I have written will not fully satisfy all of them, partly because I have reservations about some of the more esoteric aspects of the subjects. Peter J. Morin of Rutgers University, R. Haven Wiley of the University of North Carolina, and Earl Werner of the University of Michigan were equally thorough in their comments on the parts dealing with ecology, behavior, and classification, respectively. To all of these colleagues, I am truly grateful.

I have redrawn many previously published figures, and I appreciate the generous spirit with which the original authors have given permission to copy their work. Listed alphabetically, they are Andrew J. Berger, Peter Berthold, James H. Brown, William R. Dawson, Stephen T. Emlen, Carl Ernst, J. Alan Feduccia, Richard Highton, Erik Jarvik, Charles J. Krebs, Jason A. Lillegraven, H. G. Lloyd, G. A. Manley, Robert May, Peter J. Morin, James C. Munger, Ronald Nowak, Eugene P. Odum, Robert T. Orr, E. T. Pengelley, Frank A. Pitelka, Knut Schmidt-Nielsen, R. Jack Schultz, Carl K. Seyfert, Robert C. Stebbins, Joseph R. Tomelleri, Timothy C. Williams, and Edward O. Wilson. Professor R. Haven Wiley kindly provided his original prints of the sage grouse displays. In most cases, permissions were freely given by the publishers, for which I am also grateful.

Susan S. Whitfield made the drawings for Figures 1.15 and 2.2 and helped by photographing most of the others. She is not responsible for any flaws.

PART

I

The major groups of living vertebrates

Their origins and special adaptations

Chapter 1 introduces the vertebrates and describes them and the special adaptations of each group. Also described is what we know from the fossil record of the origin of each group. The conditions of vertebrate life are such that the perceptual abilities of the animals are extremely important. Some of them we can only imagine, as they greatly exceed our human capabilities. Because many of these abilities are not confined to particular classes of vertebrates, or even to vertebrates in some cases, they are covered separately in Chapter 2.

We can understand vertebrates and their adaptations better if we learn the sequence of their appearance on earth and evaluate their geographic distribution. To do so, we must first appreciate the nature of geologic time and the impact of continental drift on the biosphere.

The geologic time scale

An animal's structure is most important if it represents an improvement on any structure of similar function that was possessed by organisms that lived earlier. We can learn the sequence of appearance of the major groups of vertebrates only by consulting the geologic record, as shown by the fossils that have been discovered. The fossils, in turn, make sense only if we know their respective ages.

There are several ways of determining the ages of fossils. The first, and easiest to understand, is to establish the relative ages of the rocks in which they are found. The most obvious way is to follow the law of superposition, which simply states that among rock layers found at the earth's surface, those laid down first are at the bottom of the series, and the youngest are at the top. Barring extreme deformation, which is readily recognizable, younger strata always overlie older strata. This is seen dramatically in the Grand Canyon, where the oldest rocks are at the bottom, beside the Colorado River, and the youngest are at the rim of the canyon, a mile higher. Even there, the total number of layers is only a tiny fraction of all the sedimentary and volcanic layers that are known. There were long periods when no sediment was being laid down at that location. The layers in the Grand Canyon can be correlated with those of other areas by matching one or more layers in the two areas, and then filling in the gaps in one area by studying the layers that are present in the other. In referring to these relative ages, it is conventional to use the units of geologic time that were established early in the history of geologic exploration. It was observed that there were conspicuous gaps in the sequence, reflecting times when apparently little or no sediment was being laid down, and when erosion was removing some of the rocks that had already been formed. Such gaps marked the end of one period and the beginning of another, and the various periods were given the names by which they are still known. The gaps were not worldwide, and their relative magnitudes allowed geologists to set up major and lesser divisions of geologic time. The most conspicuous boundaries in the record are recognized on the basis of major changes in animal and plant life. Subsequent research has shown that in many cases the gaps can be completely filled in by studying a variety of locations, but it is still convenient to use the names. The most conspicuous gaps separate the "eras." We are concerned only with the three eras during which vertebrates have been present – a small fraction of all geologic time. The youngest or most recent era, and the one in which we live, is the Cenozoic, a name meaning "recent animals." It has lasted about 65 million years. Notice that geologists think of time in units of millions of years. Major changes in the fauna marked the boundary between the Cenozoic and the next older era, the Mesozoic, the era of "middle" animals. The Mesozoic began about 225 million years ago, and thus lasted longer that the Cenozoic has – about 160 million years. It was preceded by the Paleozoic, the era of "old" animals. The Paleozoic began about 600 million years ago and hence lasted for around 375 million years. The eras are divided into periods, and each of the two periods of the Cenozoic era is again divided into epochs. Geologists find further subdivisions convenient, but they need not concern us in this brief account. Table I.1 gives the eras, periods, and epochs in chronological order, with their durations in millions of years, and the origins of their names, many of which reflect the influence of British scientists in the pioneering studies of geologic history.

As an aside in presenting the geologic record, it seems strange that there should be a "third" (Tertiary) period and a "fourth" (Quaternary) period, but no "first" or "second." That anomaly can be explained on basis of the original classification of the geology of the Alps. The law of superposition placed the oldest rocks on the "inside," and they became known as the *primary* formation.

Table I.1. *The subdivisions of geologic time*

Era	Period	Epoch
Cenozoic (recent animals) 65 (65)	Quaternary (fourth) 2.5 (2.5)	Holocene (completely Recent) 0.005 (0.005)
		Pleistocene (most Recent) 2.5 (2.5)
	Tertiary (third) 62.5 (65)	Pliocene (more Recent) 4.5 (7)
		Miocene (less Recent) 19 (26)
		Oligocene (little Recent) 12 (38)
		Eocene (dawn of the Recent) 16 (54)
		Paleocene (old dawn of Recent) 11 (65)
Mesozoic (middle animals) 160 (225)	Cretaceous (chalk) 71 (136)	
	Jurassic (Jura Mtns., Switzerland) 54 (190)	
	Triassic (three parts) 35 (225)	
Paleozoic (old animals) 345 (570)	Permian (Perm, eastern Russia) 55 (280)	
	Carboniferous (coal) 65 (345)	
	Devonian (Devonshire, England) 50 (395)	
	Silurian (Silures, a Welsh tribe) 35 (430)	
	Ordovician (Ordovices, another Welsh tribe) 70 (500)	
	Cambrian (Cambria, Roman name for Wales) 70 (570)	

Note: Only those parts during which vertebrates have been present are shown. Duration of each (and time since the start of each) shown in millions of years.

They are igneous rocks, containing no fossils. Outside (above) the primary rocks were the obviously oldest sedimentary rocks, and they were given the name *secondary.* Above them were the *tertiary* formations, a name that is still used. The secondary formations were later found to cover an enormous span of time, one that had a conspicuous break within it, and so the name had to be replaced by two names: Paleozoic and Mesozoic. "Primary" is no longer used, because there are igneous rocks from all periods, and because there are sedimentary rocks older than the original "secondary" rocks.

The obvious next question is how the real time is to be estimated. For most of the Holocene, there is the method of using tree rings, which, because they vary in width with good years and poor years, can be coordinated over long periods of time, for which fossils of successive overlapping ages can be found. At best, this is a short-term method. A second, similar method is the use of *varves,* alternating bands of light and dark clay that represent seasonal changes, with the layers of dark organic matter representing what was dropped to the bottom of a pond in summer and autumn, and the lighter layers of pure clay being deposited in winter. These are obvious in fairly recent clay deposits, such as those at the bottoms of ponds. They can be observed by drilling deep cores and carefully removing them from the coring device. They have proved useful for deposits well into the Pleistocene. In other rocks, the same kind of structure can be identified, sometimes with fossils preserved well enough to identify seeds (autumn) and flowers (spring). Even where such details cannot be found, individual years can be detected, and such formations can be shown to represent very long periods. The Eocene Green River Shales of Wyoming and Colorado are 793 m (2,600 ft) thick. They are composed of identifiable varves averaging less than 0.2 mm (<0.008 in.) thick. A simple calculation shows that the formation represents 6 million years of earth history, and yet it covers only about one-third of the Eocene epoch.

Although tree rings and varves are revealing, they are of limited use because they are found in very limited parts of the record. A much more general method for determining the age of a formation involves the use of the rate of radioactive decay of isotopes of certain elements. Among those in use is carbon-14 (^{14}C), which is formed by cosmic-ray bombardment of nitrogen-14 in the upper atmosphere. Chemically, ^{14}C behaves exactly like stable forms of carbon (the other isotopes) and is incorporated as CO_2 into plants, and from them into animals. Carbon-14 decays at a rate that will reduce it to one-half its original amount in 5,730 years. Once a plant or animal dies, new carbon is no longer incorporated into its body, and the ^{14}C is gradually changed into the stable ^{12}C. Thus, by measuring the proportion of ^{14}C present, one can determine the time since the death of the organism. Incidentally, that is how the age of the Shroud of Turin was determined. Three independent laboratories were given samples of the shroud, and also samples of the same kind of cloth of historically known ages. Without being told which was which, the three laboratories agreed that the plants from which the Shroud of Turin was made had been alive no earlier than about the end of the thirteenth century. It is thus less than half as old as it would have to be to have been the shroud of Jesus.

Carbon-14 dating is useful up to an age of about 40,000 years, when the amount is no longer accurately measurable. Other radioactive atoms have much longer half-lives and can be used in the same way, by measuring the amount left and the amount of the decay product. The first such isotope used was uranium-235, which has a half-life of 713 million years and is useful wherever it has crystallized in a position such that its relative age can be established, as, for example, where molten magma was forced up through sedimentary rocks and then cooled among the strata, allowing the contained ^{235}U to crystallize. The decay of potassium-40 (^{40}K) to argon, with a half-life of 1.42×10^8 years, has been especially useful for dating fossil-bearing strata associated with igneous rocks. There is a certain amount of error inherent in these methods, but for most ages determined by the radioactive method, the potential error is no more than 2%.

With this background, the first appearances of the fossils bearing on the history of the different vertebrate classes can be placed in their historical sequence. As each group is discussed, we shall examine the special adaptations that made that group better able to cope with life under its existing conditions.

Continental drift

Knowledge of the order in which the eras, periods, and epochs came is not sufficient to allow us to understand the conditions under which the various groups of vertebrates arose. Had the continents always been in their current positions on the surface of the earth, the sequence alone would be more helpful than it is, but even so, such seemingly unexplainable facts as the finding of the oldest known fossil amphibians in Greenland led to assumptions about dramatic shifts in worldwide climates. Despite such anomalies, until the 1960s the stability of the continents was generally assumed, although the theory of continental drift had been propounded by the geologist Alfred Wegener before 1920. His theory was rejected by most geologists because there was no known force that could be invoked to move the continents as Wegener proposed. An examination of a map of the world leads the imaginative to see that if the South Atlantic Ocean were closed, Africa and South America would fit together rather nicely, and that was the starting point for the theory. The force that could move continents was discovered when the nature of the Mid-Atlantic Ridge was

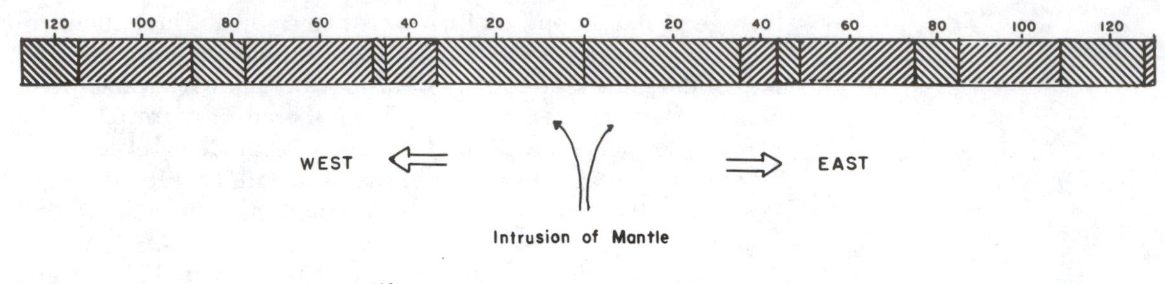

Figure I.1. *Correspondence of magnetic reversals on both sides of a ridge of sea floor spreading. Time given in millions of years ago. (Redrawn from Seyfert & Sirkin 1979.)*

investigated. There, a crack in the ocean floor is literally being forced open, with the result that huge quantities of lava are being produced and spread over the ocean floor. The buildup of this lava is not as rapid as might be expected, because the sides of the crack are continually moving apart. This has been shown through observations of the magnetism in the solidified lava. The earth's magnetic field exerts an influence on any substance that has a magnetic property, such as iron, as we see in a compass. If the substance is molten, the particles orient themselves with the magnetic field of the earth; this orientation is retained when the lava solidifies, and it can be detected by appropriate instruments. Now, it is an astonishing fact that the polarity of the earth reverses itself at irregular intervals, and these reversals are detected in successive bands of solidified lava on the ocean floor on either side of the Mid-Atlantic Ridge. As the widths of the bands vary, and those on either side of the ridge are symmetrical, we have conclusive evidence that the ocean floor is spreading apart (Figure I.1).

The Mid-Atlantic Ridge is part of a worldwide system of such features. At present, most of them are submarine, but one terrestrial example is the Great Rift Valley system in Africa. The huge parts of the earth's surface lying between such cracks (the "plates") are thus moving. The part of the earth lying below the crust is the mantle, which is hot enough not to be considered solid, and it is thought that currents within this "liquid" mantle are dragging the plates along as they float on its surface. Under these conditions, the plates are sure to collide, and where they do, earthquakes and active volcanos are numerous, and often major mountain ranges are formed. When two plates collide, one of the plates is forced under the other.

This movement has resulted in major rearrangements of the continents through geologic time, and the reconstructions of what maps would have looked like in those ancient times have provided some resolutions to puzzles such as the amphibian fossils in Greenland.

Such maps are constructed from two sources of evidence: correlation of the orientations of magnetism in the rocks, which orients the continental blocks with the poles, and correlation of major geologic formations, particularly those representing belts of sedimentary rocks that reveal the former locations of major elongate seas (geosynclines) – it being thought that such formations, when very similar, reveal once-continuous shallow seas where now the formations are separated on different continents. These correlations have shown that what are now single continents were once divided into two or more, and what are now separate continents were once combined into single landmasses. Figures I.2–I.5 show the positions of these landmasses and their locations with respect to the poles and the equator. For example, during the Cambrian and Ordovician periods (Figure I.2), at the time of the earliest known vertebrates, there were four separate continents lying close to the equator: North America, Europe, Siberia, and China; separated from that group by the no-longer-existing Tethys Sea was a single supercontinent, consisting of South America, Africa, Arabia, Antarctica, Australia, and India. That supercontinent has been named Gondwanaland. Magnetic correlations have placed the South Magnetic Pole in northern Africa during the Ordovician, explaining the evidence of glaciation there at that time. During the Silurian, the separate continents across the Tethys Sea from Gondwanaland began to drift together into a single supercontinent called Laurasia, and in the Devonian the Tethys Sea began to close, a process completed by the Carboniferous. The result was that all of the continents had coalesced into a single continent, the name of which is Pangaea (Figure I.3). During the Devonian, the period when the Amphibia arose, Greenland was on the equator, and North America, Europe, and western Asia were all between 30° north and south latitude, or virtually tropical. Except for Australia, Gondwanaland was in a region of cold, centered on the South Pole, which was located in southern Africa. Eastern Siberia, China, and Japan were also cold, being north of 30° north latitude. The North Pole was in the Pacific, opposite Japan.

Pangaea lasted from the beginning of the Carboniferous (Figure I.4) through the Triassic. There was a general northward drift of Pangaea from Devonian to Triassic time. As noted, the locations of different parts of Pangaea with respect to the poles and the tropics are most helpful in understanding the existence and locations of different kinds of fossils. In the Jurassic, the present-day continents began drifting apart (Figure I.5). The sequence of steps in the breakup of Pangaea is important for an understanding of the current geographic distributions of the families of vertebrates, as discussed in Part IV.

Reference

Seyfert, C. K., & L. A. Sirkin. 1979. *Earth History and Plate Tectonics.* New York, Harper & Row.

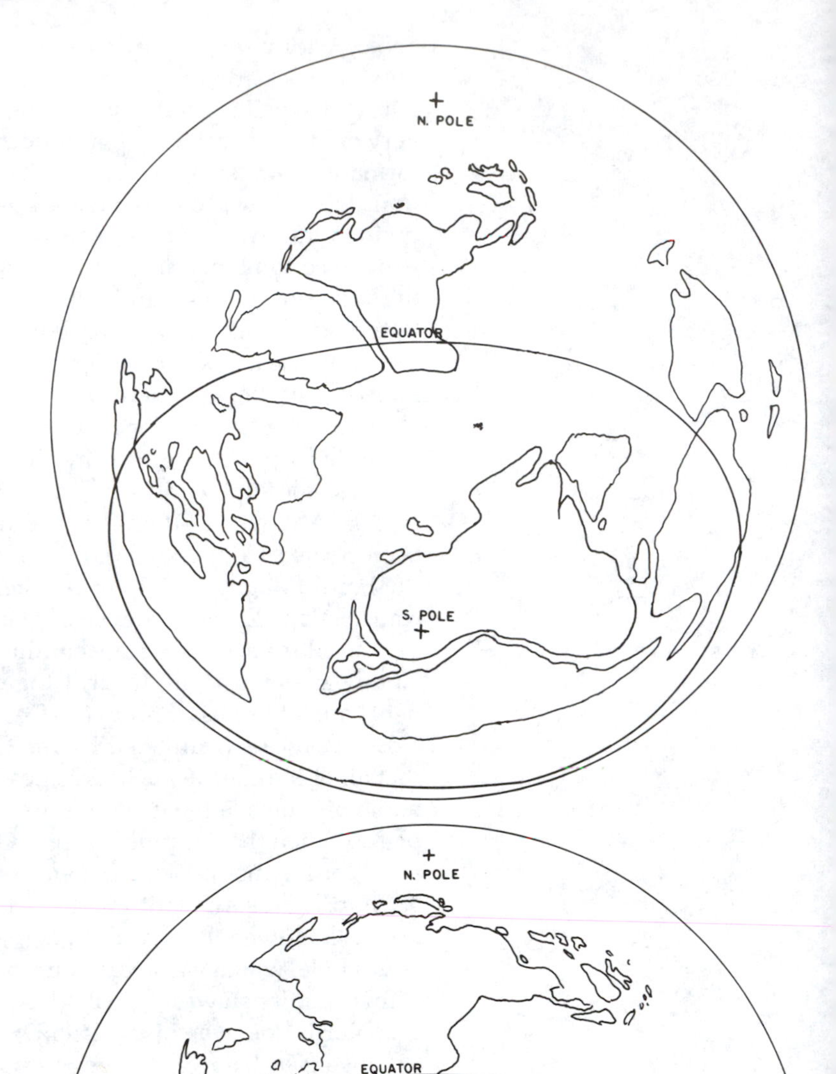

Figure I.2. *Positions of the continents (present-day outlines) during the Ordovician period. (Redrawn from Seyfert & Sirkin 1979.)*

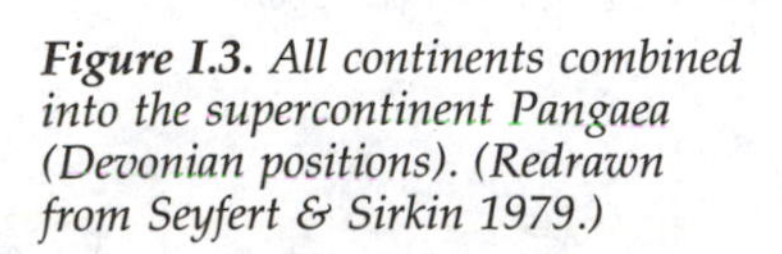

Figure I.3. *All continents combined into the supercontinent Pangaea (Devonian positions). (Redrawn from Seyfert & Sirkin 1979.)*

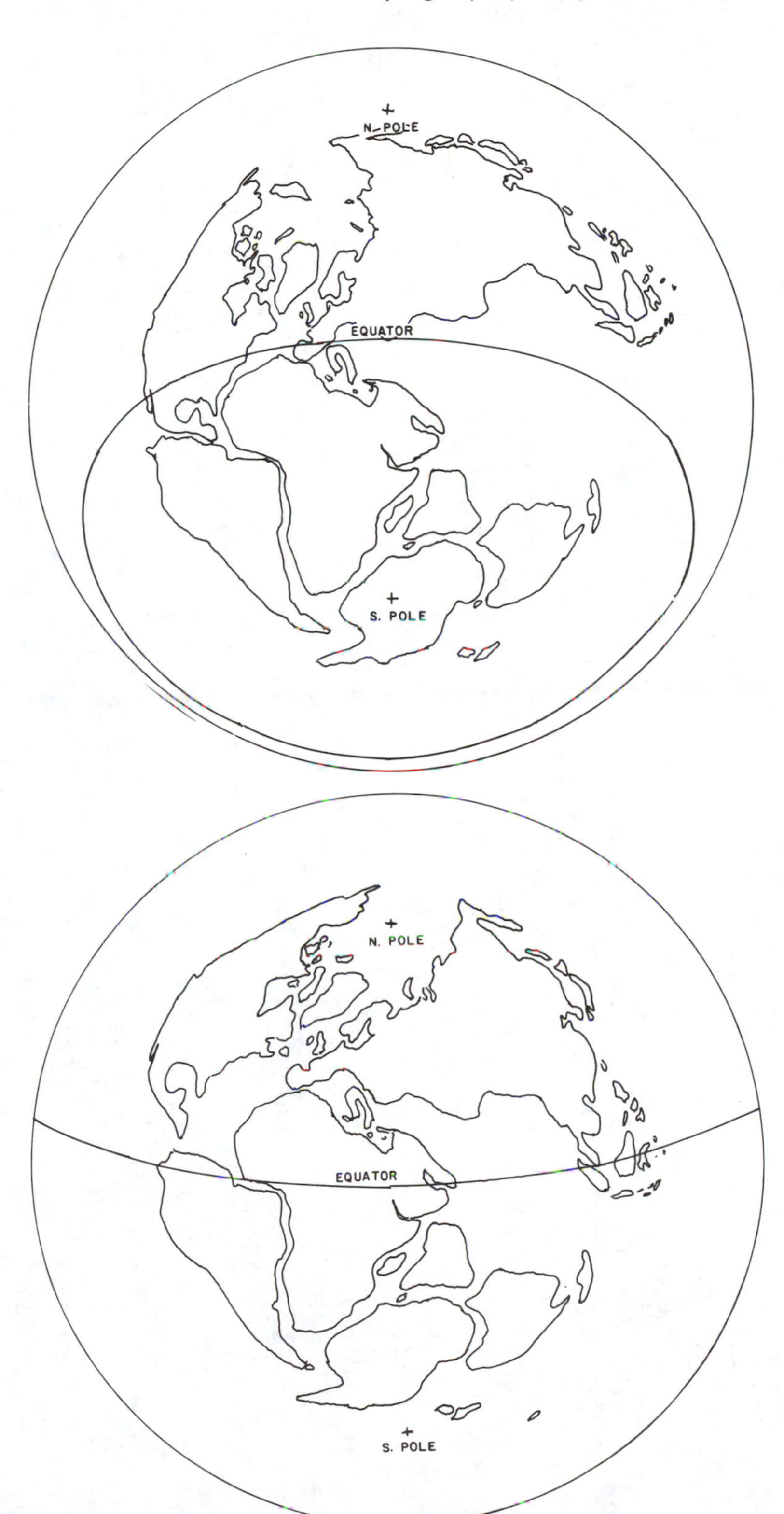

Figure I.4. *Position of Pangaea in the Carboniferous. (Redrawn from Seyfert & Sirkin 1979.)*

Figure I.5. *The beginning of the breakup of Pangaea in the Jurassic. The northern continents had become Laurasia, and the southern ones Gondwanaland. (Redrawn from Seyfert & Sirkin 1979.)*

1

The living vertebrates

There are seven major subdivisions of the vertebrates. They are no longer called the "classes," because in most cases they represent a series of successively descendant groups, rather than the equivalent groups that the name "class" implies. Moreover, it is recognized that some of them, especially the Reptilia, are heterogeneous, because logically they should include other "classes," and hence are described as *paraphyletic* (representing different lineages).

Each group had a major advantage over its ancestral group, and this chapter is concerned with the ages of the first fossils that can be assigned to each group, and with the major adaptations that gave each an advantage. No fossils of any sure ancestor of the vertebrates have been found, but it is possible that something like a recently discovered fossil containing the toothlike structures called *conodonts* may qualify. The fossil in question is Carboniferous in age, not early enough to qualify as a vertebrate ancestor. What we know of the structures of the earliest vertebrates, the structures of living animals that are most similar to vertebrates, and the features observed in the development of living vertebrates allows us to make reasonable reconstructions of the ancestor.

Agnatha

The name Agnatha means "without jaws," contrasting the living hagfishes and lampreys with all other vertebrates, collectively known as Gnathostomata, meaning "jaw-mouthed." In part because of recently discovered fossils, the placing of the two living groups together has been challenged on the grounds that lampreys share a more recent ancestor with the gnathostomes than they do with the more primitive hagfishes. If that contention proves to be correct, the name Agnatha will no longer be usable, because the contained groups would be paraphyletic. The question is not easily resolvable, because there are several different pieces of information that must be judged more important or less important, depending on the side of the argument, and both anatomical and molecular data are involved. If the name Agnatha can no longer be used, our classification should start with the most primitive "vertebrates" – the hagfishes, or Myxiniformes. The term "vertebrates" is in quotation marks because they lack vertebrae and are primitive in other regards and hence are properly called Craniata, a term that includes the lampreys and gnathostomes, and thus all vertebrates.

The first vertebrate fossil fragments come from the Late Cambrian and Early Ordovician periods. These fragments could not be assigned to the vertebrates were it not for their microscopic structure, which closely resembles that of the bony plates covering more complete fossils that have recently been found in both Lower and Upper Ordovician deposits and are much more abundant in the Silurian and Devonian. A Late Cambrian appearance places vertebrates on the scene much later than most of the invertebrate groups, which appeared more or less simultaneously at the beginning of the Cambrian or slightly earlier.

The earliest complete fossil vertebrates were formerly called ostracoderms, but actually belong to several distinct groups (Aranaspida, Pteraspida, and Cephalaspida). They lacked jaws and had holes at the sides through the bony covering of the anterior end. It appears that they fed and respired by drawing water through the mouth, filtering out the food and absorbing oxygen, before expelling the water out through the holes leading from the pharyngeal cavity to the outside.

There remains today at least one kind of nonvertebrate animal that feeds and respires in that manner: amphioxus (*Branchiostoma lanceolatum*), belonging to the Cephalochordata. A small animal, only a few centimeters long, it has a notochord, in common with vertebrates and the related Urochordata, and its central nervous system is an elongate, tubular structure dorsal to the notochord. Unlike vertebrates of similar general structure, it keeps water flowing through its pharyngeal cavity by the action of cilia. That is an effective means of moving small currents of water, especially inside small cavities, and it is used by many invertebrates. Amphioxus lives on the bottom in bays and estuaries and burrows into sand. It has no hard parts that would fossilize easily, but that is not a sufficient reason for the absence of similar forms in the fossil record, as worms and even jellyfish have been preserved in Early Cambrian formations, and also in Precambrian rocks.

It is essential, if a pharyngeal cavity as large as those of vertebrates is to be kept operating, that the sides be supported by more than epithelium and gills. Modern jawless forms have a "branchial basket" of cartilage supporting the gills and the openings at the sides of the pharynx through which the water is ex-

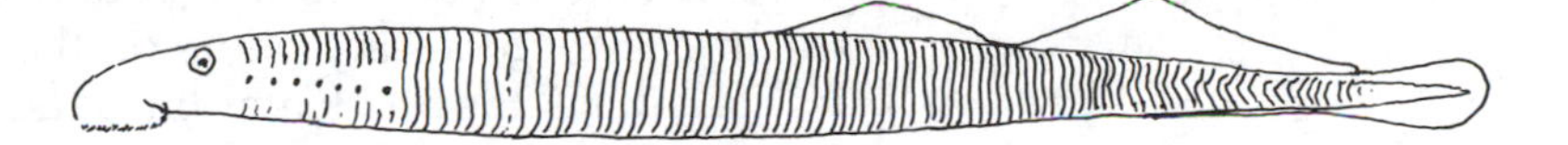

Figure 1.1. *The parasitic sea lamprey,* Petromyzon marinus. *(Redrawn from Smith 1979.)*

pelled. A major advance over their presumed invertebrate ancestors was the development of a set of muscles by which water was pumped through the pharyngeal cavity, allowing a much larger size than was possible for the cilia-bearing ancestors. The branchial basket and the cartilaginous cranium and axial skeleton would not fossilize easily, and there are no certain fossils connecting them to the ancient "ostracoderms," which became extinct in the Devonian. It is thought that a group of "agnathans," the Anaspida, known from the Silurian and Devonian, represent the ancestors of modern lampreys, but fossils have not been found that might connect them and the more recent Carboniferous forms that are identifiable as true ancestors of those living now.

The living forms are generally inconspicuous and except for a single species would be known only to specialists. That species is *Petromyzon marinus,* the parasitic sea lamprey (Figure 1.1). It is held to be responsible for the destruction of most of the lake-trout fishing industry in the Great Lakes. As its name implies, it was originally an anadromous species – one that spawns in freshwater streams and grows to maturity in the ocean. Originally it was confined to streams entering the Atlantic Ocean. After the Welland Canal to bypass Niagara Falls was constructed, *Petromyzon* was able to enter Lake Erie and then the rest of the upper lakes. After its eggs are laid in gravel banks in streams, the lampreys hatch and drift with the current until they reach a mudbank, where they make burrows and feed by filtering the water. This continues for several years. Eventually they make their way down to a large body of water and become parasitic. The sea is not necessary. In the Great Lakes, the primary hosts were lake trout and whitefish, to which the lampreys attached by their suckerlike mouths. The inside of the mouth and the tongue are armed with horny, spikelike teeth, which are used to rasp away the skin of the host. The lamprey feeds on the host's body fluids and continues to keep the wound open. The host ordinarily is not killed, but when the fully grown lamprey leaves it to spawn, it leaves a weakened fish, and frequently such fish die before reproducing. The enormous reproductive capacity of the lampreys and the absence of any defense mechanism in their hosts resulted in greatly reduced catches of whitefish, and elimination of lake-trout populations except in Lake Superior. Large sums were spent to bring the lamprey population under control. But by the time that had been accomplished the trout populations in all of the lakes except Lake Superior were gone. Recently, the trout numbers have been increased by planting millions of hatchery fish, but the rate of natural reproduction has not returned to normal, apparently because of the absence of necessary odors on the former spawning sites. In the meantime, the accidental introduction of the alewife and the intentional introduction of several species of salmon have prevented a return to the original conditions.

There are other native parasitic species and around thirty species of lampreys that are not parasitic living in the streams of eastern and far western North America. Their life cycles are similar to those of the parasitic forms, and after leaving their burrows in mud banks, the adults, like those of *Petromyzon*, spawn and die.

A second group of extant jawless species, more primitive than the lampreys, comprises the hagfishes, which live on the sea bottom and are predators on invertebrates. Like many predators, they are also scavengers, able to tear off pieces of a fish carcass and eventually work their way into the body cavity, where feeding on the soft tissues is easy.

Chondrichthyes and Osteichthyes

True fishes belong to one of the two groups Chondrichthyes and Osteichthyes. The names refer to the fact that Chondrichthyes have only cartilaginous skeletons, whereas most Osteichthyes have true bone. They share the most important advance over the hagfishes and lampreys, one that the late anatomist and paleontologist A. S. Romer considered the most important anatomical advance in the history of the vertebrates: the acquisition of jaws. All vertebrates that have jaws compose the Gnathostomata. It is thought by many that the jaws were derived from one of the nine pairs of jointed rods, now always called arches, that supported the pharyngeal cavity in the early "ostracoderms." The fossils do not show the arches clearly, and extant lampreys and hagfishes have cartilaginous branchial baskets that are complex single units that cannot be clearly related to the branchial arches supporting the gill region in true fishes. Therefore, we must rely on studies of the anatomy of development in extant fishes and other vertebrates to support the theory. Some of the rods in this series are still in place as supports for the gill structure in modern fishes, but one of the first four apparently has been lost, and the next three have become modified to other functions over the course of evolution. During development, one pair of rods (apparently the second in the original series) forms part of the beginning of the floor of the skull; the third in the original series develops into the jaws. The upper part of the fourth arch develops into a support for the jaws in fishes, connecting them to the skull and providing a lever on which to open the jaws widely. The original gill-supporting function of this fourth or hyoid arch is shown by the fact that in Chondrichthyes and some primitive Osteichthyes there is an opening between the arch and the jaw: the spiracle, corresponding to the gill slits between the remaining arches. Water for respiration is taken into the pharyngeal cavity through the spiracle and forced out over the gills.

In the mid-Silurian there appeared the first fossil vertebrates that had jaws: the Placodermi and the Acanthodia. The ability to bite and grasp was evidently an enormous advantage. The ancient jawless forms continued in great variety and numbers through the Silurian, but in the Devonian the more efficient Chondrichthyes and Osteichthyes appeared and became diverse and abundant.

In addition to jaws, each of the members of these groups had two pairs of fins: pectoral and pelvic. Previously existing vertebrates lacked paired fins, having several different structures that performed similar functions less efficiently. Like them, lampreys and hagfishes lack paired fins. Regulation of position and forward movement, essential to predatory animals, is made far more efficient by the possession of paired appendages. The paired fins of a fish allow the animal to prevent the unwanted motions of yawing (side-to-side movement), rolling, and pitching (up-and-down movement). The paired fins of some early fishes evolved into the limbs of terrestrial vertebrates.

It is tempting to theorize that as a result of competition from these fishes, with their superior jaws and their paired fins, most of the Agnatha became extinct, but there is a problem with that theory and with others that postulate extinctions as resulting from competition. The problem is that some Agnatha continue to exist, just as amphioxus continues to exist, as do the Amphibia, which were replaced by the reptiles. Part of the solution to the problem may reside in the restricted and unusual habitats of the survivors, in which habitats they remain competitive; and although that explanation may suffice for the living Agnatha, the habitats of living amphibians are too diverse, and overlap with those of at least some reptiles, for competition to account for the whole story.

Chondrichthyes

The cartilaginous fishes are represented now by the sharks, rays, and an obscure deep-oceanic group, the chimaeras. The last are commonly placed in a different group, the Holocephali, from the rest, the Elasmobranchii. The popular image of sharks is that they are all dangerous killers, and it is true that several species have attacked swimmers in many parts of the world. It is also true that most of the approximately 225 known species of sharks are predators, but the largest species are harmless giants that filter minute organisms from the water. Most species live near the surface of the open oceans, and swimming over deep water is more dangerous than in the surf, but there are places, such as eastern Australia, where there have been attacks in shallow water. There are a few freshwater sharks, including the ferocious *Carcharrhinus leucas* of Lake Nicaragua and its drainage, the San Juan River. This species is said to attack in knee-deep water.

The characteristic "head-shaking" by sharks feeding on large prey has been described in both technical and popular literature. It represents the completion of the bite by the ventrally oriented mouth. When the jaws are opened, muscles pull the supporting part of the hyoid arch (the hyomandibular) outward and forward, forcing the upper jaw (the palatoquadrate) forward and allowing the maximum opening of the mouth. With the bite, the upper jaw, with its strong, serrated teeth, is forced deep into the flesh of the prey. When this process is at its maximum, the body is thrown into strong lateral undulations, resulting in the removal of a large piece of flesh.

Most sharks (Figure 1.2a, left) are unusually large for vertebrates, with an average length of 2 m, but there are some small species only 0.25 m in length. Most of these are found at considerable depths. They have luminescent organs, the function of which is discussed later. Some species hunt in packs and are formidable enemies of larger squids and octopuses.

The second and larger group of the Elasmobranchii, the Rajiformes, are the skates and rays (Figure 1.2a, right), with some 400 or more species. They are flattened dorsoventrally, with widely spreading pectoral fins. Most species occur in shallow marine habitats, where they remain on the bottom, often mostly covered with sand, with only the eyes and spiracles showing. Unlike most sharks, the bottom-living species have hard, flat teeth for crushing shelled invertebrates. Unlike other fishes, skates and rays propel themselves forward by undulations of the pectoral fins, rather than by lateral undulations of the tail. This requires flexibility, and most species lack scales on the pectorals. It has been suggested that covering the body with sand is a necessary means of protection from predators. In addition, some species (the stingrays) are armed with modified scales at the base of the tail in the form of spikes equipped with poison, and others have organs that generate electric current, as described later. The largest rays, unlike most species, feed in open water by filtering small organisms, as do the largest sharks. Some of these giants have "wingspans" up to 6 m.

A few species of rays have invaded fresh water, most notably the members of the genus *Potamotrygon*, which have inhabited the Amazon Basin for a long time, perhaps since the Tertiary period.

All members of the Chondrichthyes have internal fertilization, a trait that is shared by only a small minority of the Osteichthyes. Internal fertilization is a much more certain means of ensuring contact between sperm and egg than is the common means, employed by agnathans and many fishes and frogs, whereby the female and male jointly eject eggs and sperm into the surrounding water. Males of the Chondrichthyes have parts of their pelvic fins modified as claspers, one of which is inserted into the cloaca of the female during copulation. Sperm are passed down a groove in the clasper, in some cases being propelled by water and secretions from a special bulb, the siphon sac, located under the skin of the pelvic fin.

The development of the fertilized eggs may take place in shells, some of which have elaborate filaments that become entangled with objects on the ocean floor, or the eggs may be retained in the oviduct, where they develop until hatching or even later, then being released as miniature replicates of the adults. In some species, the yolk is supplemented by growths from the wall of the oviduct that enter the mouths of the embryos and add nutrients; in other species, the yolk is dispensed with, and the yolk sac develops elaborate vascularization, which contacts the wall of the oviduct and receives nutrients from and exchanges dissolved gases with the circulatory system of the female. From this description of the variety of developmental systems, it is clear that the old terms "oviparous" and "viviparous" ("egg-laying" and "live-bearing") are insufficient. "Ovoviviparous" is a term that was coined for species that retain the eggs in the oviduct until hatching, but it is clear that "viviparous" covers more phenomena than was once thought. Whatever the terminology, the retention of fertilized eggs in the body of the female provides protection for the developing embryos. There is, however, a price in the limited space available, and hence in the number of young that can be produced at one time. Most sharks have only about ten to fifteen young at once, in contrast to the thousands of eggs that are laid by some bony fishes.

Osteichthyes

There are more extant species of bony fishes (more than 20,000) than of all other vertebrates combined. Most of them belong to one of the two major subdivisions: the Actinopterygii, a name referring to the actinosts, small bones that make up the fin rays, in contrast to the muscular extensions into the bases of the fins of the Sarcopterygii, a group that has only a few species, but is of great egolutionary interest because they are most like the fishes of the Devonian period that gave rise to the terrestrial vertebrates.

SARCOPTERYGII. The name of this group means "fleshy fins" in Greek, and the appendages are quite different from those of other fishes, being like paddles in two species, and long and tapering in the remaining five. The differences in the shapes of fins do not define the two major groups, however. What is important is the presence of functional lungs in one of the two groups, the Dipnoi (Figure 1.2b, right). Although all dipnoians also have gills, those of the single South American species, *Lepidosiren paradoxa*, and the four African species of the genus *Protopterus* have gills that are so reduced that they will drown if held under water. The large (1–2 m long) *Protopterus* thrives in areas that flood during the rainy season, feeding in the flooded lowlands. When the lowlands begin to dry up after the rains, the lungfish makes a burrow 1 m deep, ending in a larger chamber. As even the water in the burrow dries up, the fish becomes inactive and secretes a mucoid "cocoon" around its coiled body. These lungfishes have been shown to survive four years of drought in this condition of aestivation. The Australian lungfish, *Neoceratodus forsteri*, is markedly different from the other dipnoians, being covered with large scales, which are lacking in the others, and having fleshy, paddlelike fins, which it uses to "walk" along the bottom of its stream habitat. Like the other Sarcopterygii, it is a large fish, reaching 1.5 m in length. Most of its respiration is through its gills.

Although functional lungs are found in relatively few species of Osteichthyes, most of them have a structure that is *homologous* (having the same embryological origin) to lungs. In most Actinopterygii, the structure usually does not function in respiration, but is a closed bladder that in different species may contain any of a variety of gases. In many species, the gas is oxygen. The volume of gas is regulated by a special structure, the oval organ, from which the oxygen can diffuse back into the circulatory system. This gas bladder acts as a hydrostatic organ, giving the fish neutral buoyancy. In fishes that move through large vertical distances in water, the volume of gas is kept constant by an accessory bladder from which oxygen can diffuse back into the blood. As described in Chapter 2, gas bladders also function in sound reception and sound production. No structure equivalent to lungs or gas bladder is found in Agnatha or Chondrichthyes.

The second division of sarcopterygians consists of the single species *Latimeria chalumnae* (Figure 1.2b, left), which belongs to a group, the Crossopterygii, that was thought to have become extinct at the end of the Cretaceous period, until a living specimen was caught near South Africa in 1938. Since then, more than seventy-five specimens have been taken in deep water (around 1,000 ft, or 300 m) off the Comoro Islands, northwest of Madagascar. They are large fish, usually more than 1 m long. They have large fleshy fins with skeletons very like those of the fossil crossopterygians, and they are covered with prominent scales. They are aggressive predators and are ovoviviparous. The principal interest in crossopterygians derives from the strong probability that the similar Devonian forms were the ancestors of the Amphibia. The evidence supporting that view is given later, when the Amphibia are discussed.

ACTINOPTERYGII. The ray-finned fishes have become adapted to virtually every aquatic situation, the exceptions being those waters that are too hot and those, like the Great Salt Lake and the Dead Sea, that are too saline. Groups of animals that have evolved

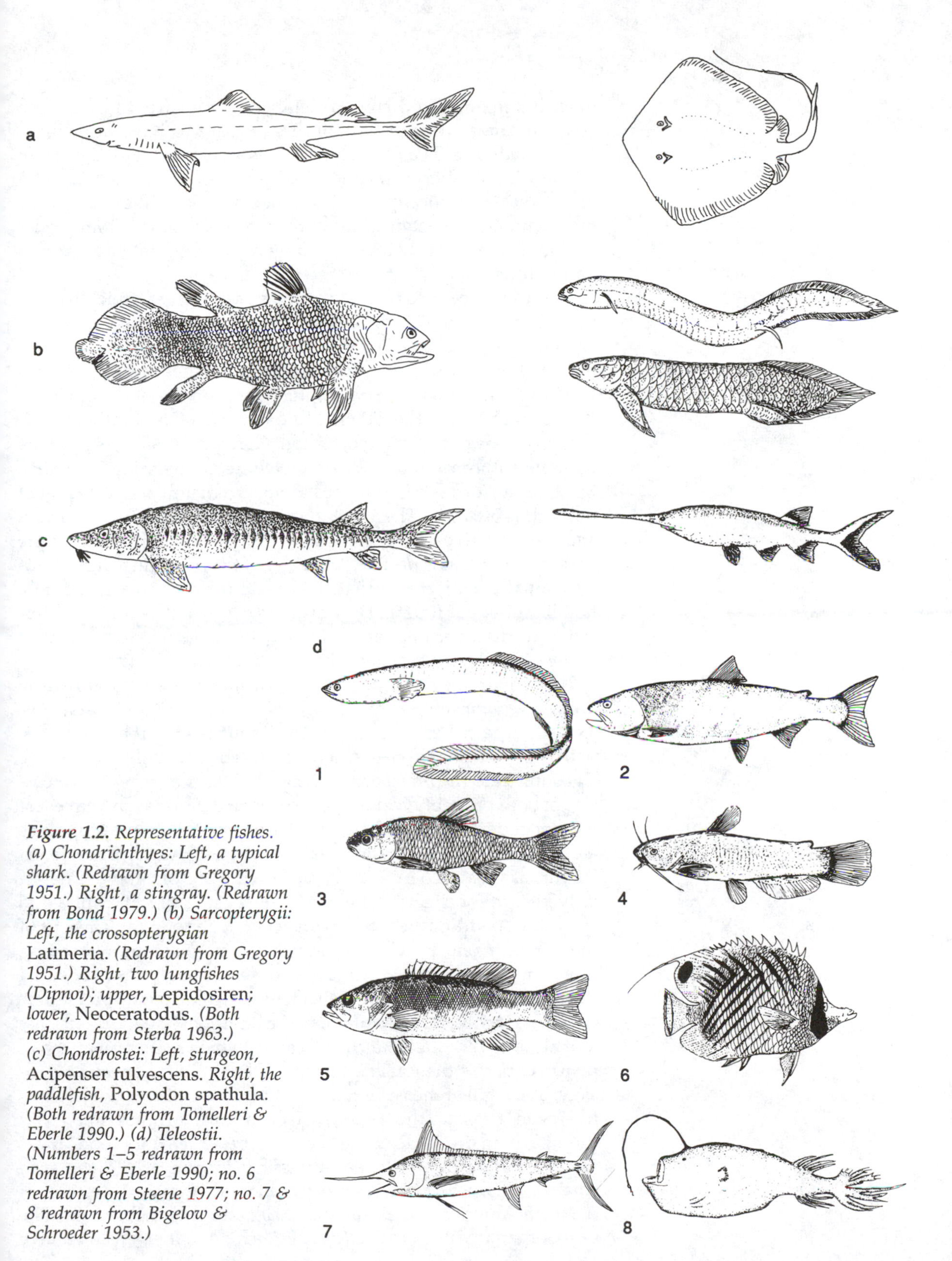

Figure 1.2. *Representative fishes. (a) Chondrichthyes: Left, a typical shark. (Redrawn from Gregory 1951.) Right, a stingray. (Redrawn from Bond 1979.) (b) Sarcopterygii: Left, the crossopterygian* Latimeria. *(Redrawn from Gregory 1951.) Right, two lungfishes (Dipnoi); upper,* Lepidosiren; *lower,* Neoceratodus. *(Both redrawn from Sterba 1963.) (c) Chondrostei: Left, sturgeon,* Acipenser fulvescens. *Right, the paddlefish,* Polyodon spathula. *(Both redrawn from Tomelleri & Eberle 1990.) (d) Teleostii. (Numbers 1–5 redrawn from Tomelleri & Eberle 1990; no. 6 redrawn from Steene 1977; no. 7 & 8 redrawn from Bigelow & Schroeder 1953.)*

in such a manner and have developed many varied forms, each adapted to a particular way of life, are said to have undergone *adaptive radiation*. The process usually is associated with the origin of a new and superior means of existence, or with the entry of a group into an unoccupied geographic area. In either case, they meet with little resistance, either from competitors or from predators, and are relatively free to occupy many different habitats and adopt different ways of life.

That there are recognizable groups within the Osteichthyes is agreed by all authorities on fish classification, but there is little agreement on the relationships among these groups. It is not clear that the lungfishes and crossopterygians are any more closely related than either is to the actinopterygians, on the one hand, or to the Amphibia, on the other. Moreover, there is another small group of species that have been considered to be sarcopterygians by some authors, but to be a completely separate group by others. Most authorities seem to regard them as a distinct subdivision of the Actinopterygii. These are the Polypteri, the African bichirs and their relatives. They have fleshy bases for their fins, but the skeleton of the fin is unlike that of any of the possible relatives. An additional peculiarity is that the dorsal fin is subdivided into a number of equal finlets. They also have lungs. It would be interesting to discover enough of their natural history to be able to understand the functions of their unique structural features.

The remaining actinopterygians fall into more or less natural groups. One of the groups, the chondrosteans (Figure 1.2c), consists of large fishes with cartilaginous skeletons: the sturgeons, with twenty-three species, and the paddlefishes, with two species, all confined to the Northern Hemisphere. As the Devonian ancestors of both probably had true bone, the two groups may have lost it independently and may not be closely related. Some sturgeons are anadromous, and others remain in fresh water. They reach lengths as great as 6 m. They forage on the bottom and are popularly known for their production of the best caviar. In contrast to sturgeons, paddlefishes feed by filtering floating organisms out of the water. If one watches these fish in an aquarium at feeding time, it is obvious that they are able to sense the presence of small floating crustaceans, collectively included in the term "plankton." On encountering a group of them, the paddlefish will drop its jaw to make a cavernous pharyngeal cavity, which makes an effective plankton net as the crustaceans are filtered out by special extensions of the gill arches (gill rakers). When the cloud of crustaceans has been passed, the paddlefish will turn around and pass through it again. It clearly senses the presence of its food. In its natural habitat, it would be impossible for the fish to see such small objects, as the two species live in muddy rivers (Mississippi in North America and Yangtze in Asia) and their tributaries. The sense organ involved is the "paddle," a long, flat extension above the mouth. This rostrum is richly innervated, and it detects minute electric currents generated by the swimming prey.

The remaining actinopterygians have more efficient mechanisms for opening and closing the jaws than do the chondrosteans. Fishes with these systems are first found as fossils from the Permian period. The more primitive of the two groups, the Holostei, consist of the gars (seven species) and the bowfin (one species), both confined to North America. The bowfin is exclusively a freshwater fish; the gars are found in both fresh water and brackish water. Unlike the other Actinopterygii, the gars have heavy, nonoverlapping scales, thus resembling their Mesozoic ancestors.

TELEOSTEI. The teleosts represent the pinnacle of fish evolution. The extreme variety of their adaptations to virtually all aquatic habitats and ways of life within those habitats apparently resulted from basic structural changes in their jaws and locomotor apparatus. Such changes made both feeding and movement more efficient than those functions had been in their ancestors.

Unlike feeding in air, lunging after prey in water is likely to push the prey away, because of the density of the medium. For those feeding on large prey, as is common for sharks, this effect is, of course, negligible, but smaller prey can easily be missed. The structure of the teleostean jaw and orobranchial (mouth-and-gill) chamber is such that sudden opening of the mouth produces suction that draws the prey inside. Interestingly, some turtles also use suction to capture prey. Not all teleosts use the suction method to feed. Some, like barracuda and tarpon, feed on prey large enough not to be affected by the pushing phenomenon. Others, such as the cleaner fishes that pick parasites from the surfaces and orobranchial chambers of large fishes, use delicate and precise feeding methods. Still others, like the parrot fish, actually bite off pieces of living coral.

The varied feeding methods are joined by highly effective means of locomotion (Figure 1.2d). For forward swimming, most fishes use undulations of the body and caudal fin. That creates a sculling motion, driving the fish forward. The sustained speed that can be achieved will vary inversely with increasing flexibility of the body as the fish executes its undulations. In pelagic fishes, such as mackerel, tuna, and swordfish, the body is nearly rigid, and the undulations are confined to the large crescent-shaped tail. The result is a powerful thrust that can be sustained. Short bursts of speed can be achieved by most fishes, the speed depending on the rapidity with which the undulations are produced. Fishes such as trout can accelerate quickly, but most of the time they only maintain position in the current. Eels are slow swimmers because the thrust from the undulations is maximally effective only during a small part of the undulatory cycle.

Although nearly all species can use their undulations to move, some fishes use them only for escape, most of the time using either their paired fins or their dorsal and anal fins to move. Ordinarily, such fish live where precise movements rather than rapid ones are most effective. Finally, a few species are armored and are unable to undulate their bodies, relying on their tails to move them forward. Thus, their combinations of feeding and swimming specializations have made the teleosts the most varied and widespread vertebrates.

Unique abilities of Chondrichthyes and Osteichthyes

Both major groups of living fishes have two adaptations that are not shared by other vertebrates: the possession of light-emitting organs and the ability to produce electric current in amounts greater than those created in the normal life processes.

The production of light by fishes is always associated with life in the deep sea, below the level to which sunlight penetrates, which is a maximum of 1,000 m. The chemical means of light production is the same in all organisms: independent secretion of the enzyme luciferase and its substrate luciferin. When they are brought together, the oxidation of the substrate produces the bioluminescence. This ability is possessed by many invertebrates and protists, including bacteria. When one walks along a beach at night and sees the breakers glowing, it is nearly always from luminescent bacteria, but sometimes, as a wave rolls up the beach, individual points of light can be seen. Frequently they are the protozoan *Noctiluca*. Similarly, moving at night in a boat, one can sometimes see transient movements of glowing patches. These are caused when fishes move and stimulate the bacteria to emit light, just as in the motion produced by breakers. Fishes that have light-emitting organs achieve the effect in two different ways. Some species have the ability to make their own light in photophores; others have organs that act as culture containers for luminescent bacteria. In either case, the light emission is concentrated in individual spots. These spots are arranged in a specific pattern for each species of fish, an arrangement that at least allows for the recognition and location of conspecifics in the depths of the ocean, where the only other means of communication would be the production of sound or a specific chemical, neither of which would give a precise direction for finding the other individual.

Like the ability to produce light, the ability to generate electric current is an exclusive property of fishes. At least 500 species can generate electric current. They fall into two distinct groups, which use the ability in two very different ways. The strongly electric fishes produce potentials of 300–600 volts (electric catfish of Africa and electric eel of South America, at the two ends of that scale) to stun large prey. The torpedo ray of the Mediterranean produces a weaker potential of 50 volts, but of 50 amperes, the difference being due to the shape of the electric organ. It also uses the organ as a means to stun its prey.

The cells producing electricity, the electrocytes, are modified muscle cells. In most electric fishes, they are arranged in stacks, a series arrangement that allows a great increase in voltage when the different layers are discharged simultaneously. The simultaneous discharge is accomplished by synchronization of the nerve cells that individually innervate each electrocyte. The electric catfish and the electric eel have the electric organs in long series covering most of the length of the body; the torpedo ray has them in the pectoral fins, requiring a flat shape, with less length to build up a high voltage.

The weakly electric fishes create an electric field around them and can detect the presence of a nonconducting object, such as a rock, or another organism that creates a differently conducting field. As these distort only a part of the field generated by the electric fish, it can detect their direction and even distance out to a meter or more. As these fishes are largely confined to murky waters in Africa and South America, the advantage of this extra sense is obvious. More interesting are the differences in character of the electric signals produced by the different species. Some produce regular, brief pulses, at different intervals for different species; others produce discharges in the form of prolonged waves. Both kinds can be modulated by the individual, a capability that suggests that the discharges are used in intraspecific communication, perhaps to identify the species and sex of the sender.

Amphibia

The origin of terrestrial vertebrates

Before vertebrates could evolve into terrestrial forms, two conditions had to be met, in addition to their possession of appropriate structural and physiological modifications. Those two conditions were a habitat suitable for occupancy and ecological conditions that would make it advantageous to leave their existing habitat. An organism that is superior to its competitors or is unmenaced by predators where it lives is unlikely to take advantage of mutations that might make it well adapted to another habitat. These considerations have caused biologists to develop theories about why some Osteichthyes evolved into Amphibia. The time is reasonably well known to have been the mid-Devonian, and attention has been directed to the ecological conditions deduced to have been prevalent. North America and Europe were located over the equator (see Figure I.3) and were therefore warm. Land plants had appeared in the Silurian, and by the Devonian there were large horsetails and scale trees. The nature of the deposits laid down indicates that the land was low and probably swampy. Sluggish streams apparently were common, connecting ponds and pools, but the deposits known as the Devonian Red Beds indicate that periodic droughts occurred.

In those streams there were abundant freshwater fishes, and they were subject to two kinds of stress. From the physiological standpoint, such habitats, which were warm and shallow, had abundant dead organic matter. They must have had low concentrations of oxygen, but the fishes living there had lungs, as do the living dipnoians, or they gulped air, as do many living fishes in similar habitats, or they were able to respire through their skins, as do many living amphibians. It was advantageous to remain near the surface, or in very shallow water. As such habitats would have contained dense plant growth, moving through it would have placed a premium on strong appendages, rather than swimming ability. In this regard, the sarcopterygians would have been at an advantage, with bony skeletons in the fins, rather than the thin rays of actinopterygians. The ability to move through dense aquatic vegetation and to respire under conditions of low oxygen would have made a transition to terrestrial life possible, but without some other pressure they would have remained aquatic, as have the lungfishes. The droughts may have made it advantageous to be able to move out of drying ponds toward those holding more water, thus beginning terrestrial life.

The second stress must have been ecological. With warm, shallow waterways containing abundant life, two possibilities suggest themselves: competition from the numerous fishes existing in the same habitat, or predation by larger sarcopterygians or other fishes that were present. Under those conditions, the streambank muddy habitat, with its terrestrial plants and existing arthropods for food, might well have been favorable, relative to the fresh waters of the period.

Whatever the pressures, there were many adaptive changes in anatomy and physiology that were necessary to the evolving Amphibia. To penetrate a membrane, oxygen must be in solution. Therefore, in air it is necessary that the respiratory membranes be kept moist. As the lungs of even present-day amphibians are simple bags (or have been lost, as is the case for the most abundant and diverse family of salamanders), the skin became important in respiration. Keeping the skin moist means supplying water from internal sources, and that is easier for a large animal than for a small one of the same shape, because the volume of the source is greater, relative to the surface area to be moistened. The problem for evolutionary theorists here is the fact that the Paleozoic Amphibia had scales, which would have reduced their ability to respire through the skin. It should not be surprising that the earliest Amphibia that have been found (*Ichthyostega*) were large animals, more than 2 ft (64 cm) long. There were difficulties in that size, however. Terrestrial animals face a problem that does not exist for aquatic animals: supporting the body in a medium that provides negligible support (unlike water, which is close to the specific gravity of most animals). In air, the limbs must be strong enough to hold the body above the ground, and the axial skeleton must be rigid and strong enough for the body to be suspended below it. These are not large problems for small animals such as insects, but for heavy-bodied forms like the first amphibians, several structural changes were necessary. As has been described, the sarcopterygian limb skeleton was stout enough, but the musculature was not strong enough to support the body in air. The vertebral column and ribs were much larger and stronger in the first amphibians than in their immediate ancestors. The structural changes between the amphibians and their immediate ancestors did not confer direct advantages over the latter, as had the jaws of true fishes over the Agnatha. The progress was in the ability to occupy a novel environment.

Among the Devonian sarcopterygians there were two different groups, either of which could have been the ancestors of the first amphibians: the lungfishes (Dipnoi) or the rhipidistian Crossopterygii. Both are first known as fossils from the early Devonian, and there are arguments favoring each as the likely ancestral form. The arguments concern whether parts of the skeletons of fossil and living lungfishes are more like the corresponding parts of fossil and living amphibians than the latter are like the skeletal parts of fossil crossopterygians. Writers favoring the rhipidistians place stress on the similarities between the limb skeletons of rhipidistians and those of all terrestrial vertebrates (Figure 1.3). These similarities concern a single bone attached to the pectoral or pelvic girdle, with two bones attached to the single one, followed by a series of smaller bones. The single bone corresponds to the humerus or femur; the next two correspond to the radius and ulna or the tibia and fibula; the smaller bones correspond to the carpals or tarsals. In lungfishes, there is a single series of bones attached to the girdle, with smaller bones radiating from them. Authors favoring the lungfishes as ancestors of the amphibians stress the anatomy of the head, especially the openings involved in breathing. More recently, comparisons of mitochondrial DNA from modern amphibians, actinopterygians, lungfishes, and the living crossopterygian *Latimeria* have revealed the greatest similarity between the frog and lungfish, but the data have been shown to be inadequate for the statistical analysis used. Detailed analysis of the amino acid sequences of the hemoglobin chains indicates a closer relationship between *Latimeria* and frog tadpoles than between the lungfish *Lepidosiren* and any amphibian. One should exercise caution in evaluating these data, because biochemical comparisons are most reliable among groups that have diverged much more recently than the Devonian. It should be

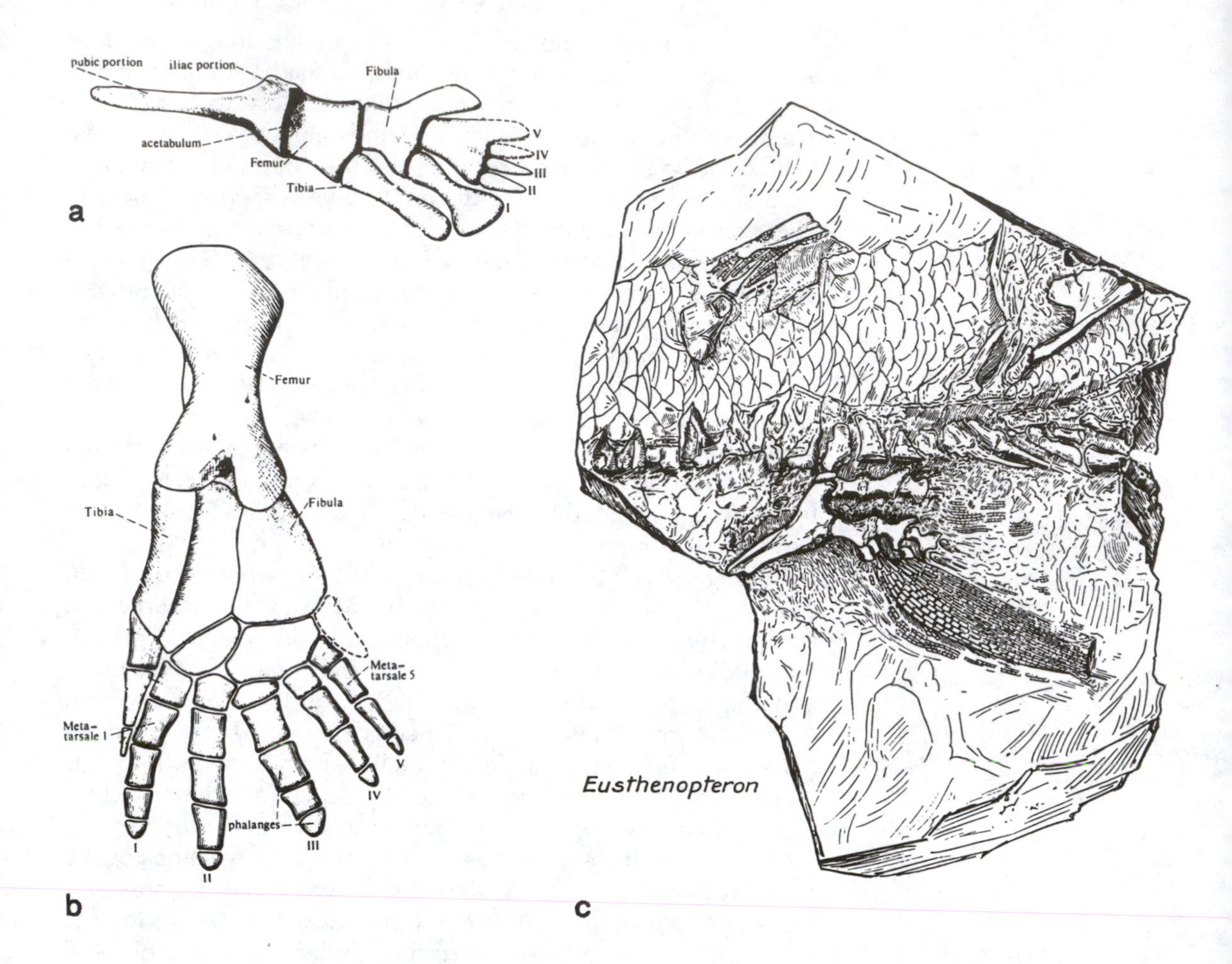

Figure 1.3. *Evidence for the origin of Amphibia. (a) Reconstructed hindlimb and pelvic girdle of* Eusthenopteron, *a rhipidistian from the late Devonian. (b) Reconstructed hindlimb from* Ichthyostega, *the earliest amphibian, also from the late Devonian. (c) Actual appearance of the pelvic girdle and hindlimbs of the fossil* Eusthenopteron. *(Parts a and b redrawn from Jarvik 1980; part c redrawn from Gregory 1951.)*

realized that without well-preserved intermediate fossils from the appropriate part of the Devonian, the argument is not likely to be settled.

The earliest amphibians were large, heavy-bodied animals that must have had difficulty walking on land, and most species probably lived in swampy places. Their name, Labyrinthodontia, comes from the peculiar structure of their teeth, in which there were complex foldings that produced a grooved outside surface. It may be significant that the rhipidistians also had such teeth. One of the orders of labyrinthodonts, the anthracosaurs, gave rise to the reptiles in the Carboniferous. In fact, because our current distinction between amphibians and reptiles depends on the mode of reproduction, a trait not fossilizable, it is not clear for some of these fossil vertebrates whether they were Amphibia or ancestral reptiles.

It is an interesting fact that there are no fossils connecting the ancient amphibians with any of the three living groups. All of the three groups of Paleozoic amphibians had become extinct early in

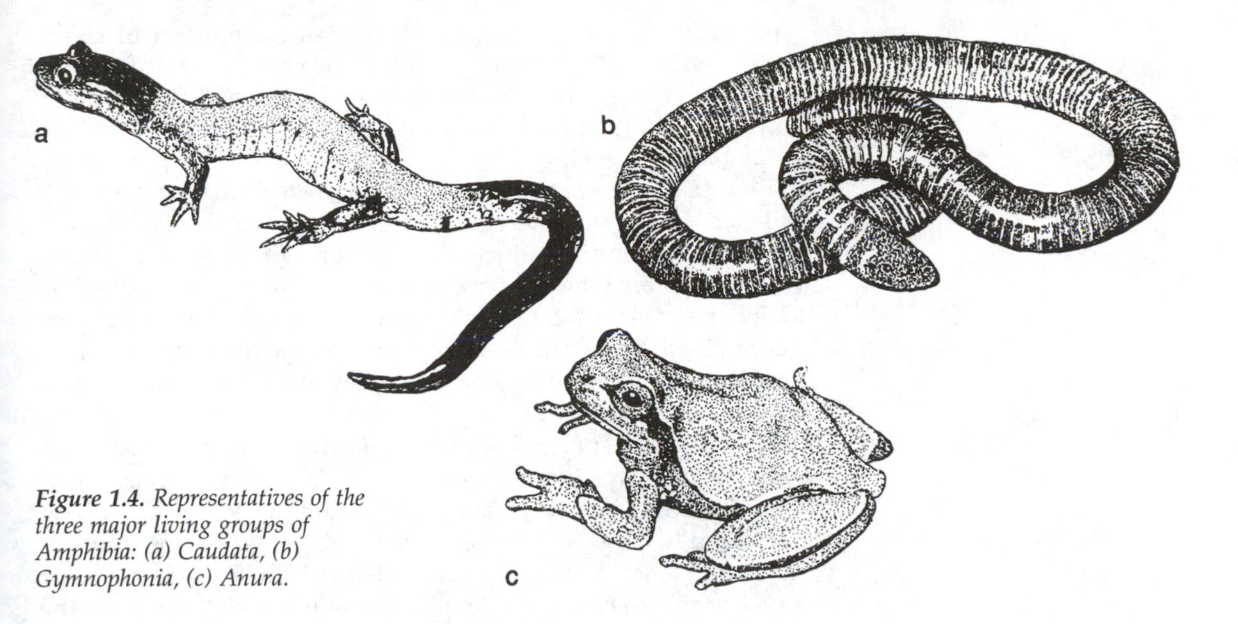

Figure 1.4. *Representatives of the three major living groups of Amphibia: (a) Caudata, (b) Gymnophonia, (c) Anura.*

the Mesozoic, perhaps because of competition from the reptiles (but keep in mind the earlier warning about this interpretation).

Living Amphibia

The three distinct groups of amphibians alive today are placed in the inclusive group Lissamphibia, referring to the trait (a smooth skin) that suggests that they have a common origin. They are the Caudata (salamanders), with more than 350 species, the Anura (frogs), with about 3,440 species, and the Gymnophonia (caecilians, a name that means no more to most people than Gymnophonia), with more than 160 species (Figure 1.4). New species are being added to all three groups. No fossils link them together. Except for a Triassic fossil, *Triadobatrachus*, that may be ancestral to the frogs, all fossil Lissamphibia are clearly assignable to one of the three living groups – frogs from the Early Jurassic, salamanders and caecilians from the Late Jurassic. The three have two traits in common anatomically: their pedicellate teeth, in which the upper and lower parts are separated by a narrow zone of uncalcified tissue, and their smooth, wet skins. They have two other common attributes: Most frogs and some salamanders and caecilians have life histories involving an aquatic larval phase and a terrestrial adult phase, and all species of living Amphibia are predaceous. There are no herbivores among them.

The life histories of amphibians have been much studied. Of the more than 350 known species of salamanders, about 200 are completely terrestrial, having direct development on land, and lack a larval stage. Of the remainder, about 25 are either *paedomorphic,* reproducing as larvae, or *neotenic*, retaining a number of larval traits as adults; 23 more species have some populations that reproduce as larvae. All of these, of course, are permanently aquatic. Fewer than 100 species of salamanders always have the "typical" amphibious life history.

Two species retain the eggs in the oviducts until they are ready to hatch, but in most terrestrial species the females lay eggs in humid places and remain with them until they hatch. In places with reasonably long winters, this does not allow the female time to obtain enough food before the mating season to be able to produce a clutch of eggs in the next year. They are thus alternate-year breeders. The protection given the eggs more than compensates for the loss in fecundity.

Many aquatic species lay their eggs either singly on pond bottoms or vegetation (newts) or in large jelly-covered masses attached to sticks in a pond (most ambystomatids). The females do not attend the eggs in these species. The stream-dwelling lungless salamanders lay a mass of eggs under rocks in streams or under wood or moss nearby, the female remaining with them as in the case of their terrestrial relatives.

Some salamanders, like many other ectothermic vertebrates (obtaining heat from the environment, rather than generating it metabolically), are surprisingly long-lived. Herpetologists, who are specialists working on amphibians and reptiles, like to keep captive specimens, partly because they require little attention, and often they have recorded the lengths of time that captive specimens have lived. The shortest record that I could find was 10 years for the dwarf mudpuppy, *Necturus punctatus*, and the longest was 55 years for the giant salamander *Andrias japonicus* of Japan. Such records may be too short, because it is not known how old they were when captured, or they may be too long, because captive animals are not exposed to the usual hazards of life. A more realistic but difficult estimate is the mean generation time, or the age at which the average female has laid half of the eggs that she will ever lay. Such estimates have been made for populations under field conditions by obtaining information on survival rates of females at each age and on the number of female eggs laid in each successive breeding season. Table 1.1 gives the available information for nine species. The values range from 2.6 to 9.8 years. It should be noted that for the second value, a minority of the females must live for 20–25 years.

Six of the nine families of salamanders have internal fertilization. There is no intromittent organ, and introduction of the sperm into the cloaca of the female is accomplished by spermatophores. The male secretes a transparent jellylike structure about 3 mm high, on top of which is placed a mass of sperm, the sperm cap. The female must be stimulated to pick up the sperm cap with the lips of her cloaca. The behavior accompanying this stimulus varies

Table 1.1. *Mean generation times for nine species of salanders, arranged from shortest to longest*

Species	Mean generation time (yr)
Ambystoma talpodeum	2.6
Eurycea bislineata	4.4
Desmognathus fuscus	5.14
D. wrighti	5.31
D. quadramaculatus	5.35
D. monticola	5.56
D. ochrophaeus	5.6
Triturus vulgaris	7.6
Plethodon jordani	9.79

Source: Data from Hairston (1987).

among genera. In *Ambystoma maculatum,* several males assemble in early spring in a pond. The presence of one or more females causes the males to pack together, swimming over each other, and nosing each other and the females. As the activity progresses, individual males leave the group and deposit spermatophores, many spermatophores being deposited within a small area. A female, when sufficiently stimulated by the actions and possibly by secretions of the males, moves to the group of spermatophores and picks one up with the cloacal lips. A preferential location for depositing the spermatophore is on top of an existing one – an obvious way of gaining an advantage over a rival's sperm cap. In *Plethodon jordani, P. glutinosus,* and related species, mating is by individual pairs. The male approaches the female and touches her side, back, and tail with his nose, following which he attempts to touch her back and tail with his chin, the location of a conspicuous disk-shaped gland. Unless she runs away, the male then works his head under her chin and moves forward, placing her chin over his back above his hind legs or the base of his tail. This is accompanied by undulations of his tail. If properly stimulated, the female then moves astride his tail, and they move slowly forward in a straight line. The male stops and rocks his sacral region, the female moving her chin in opposite directions. The male then presses his vent to the ground and deposits a spermatophore. Then both move forward, still in a straight line, until the female's vent is over the spermatophore. She then lowers her vent and picks up the sperm cap, leaving the jelly stalk attached to the substrate. The whole complex behavior ensures that the sperm cap will be picked up accurately. As there is no pond to bring numbers together, and these species are territorial, more than two individuals do not normally encounter each other in the breeding season, and the accuracy is important.

Male Gymnophonia have a copulatory organ that can be extruded through the vent, and as far as is known all species use this structure for internal fertilization, whether they are aquatic or terrestrial. Some of the viviparous species secrete a nutritive substance from the wall of the oviduct after the yolk in the developing eggs is exhausted. This may be after less than half of the embryonic growth has been achieved. The oviparous species remain coiled around the clutch of eggs until they hatch, like many salamanders.

In contrast to the Gymnophonia and most salamanders, relatively few frog species have internal fertilization, sperm and eggs being passed simultaneously into the water by the clasping male and the female, or in a few cases into a foam mass produced by the pair. The foam mass protects the developing eggs from predators; it may be floated on the water or attached to plants hanging over the larval habitat. There are several other modifications of the behavior of frogs in amplexus, as the clasping position is named. The exceptions to external fertilization are the "tailed" frog, *Ascaphus truei,* of the Pacific Northwest, in which the "tail" is an extension from the cloaca and is used by the male as an intromittent organ, and several ovoviviparous species whose methods of copulation are unknown.

Most species of anurans leave their eggs to develop on their own, but a number of them guard nests or egg masses, and some species carry the eggs attached to the back or in a pouch of the skin. The eggs of one species develop in the vocal sacs of the male. The most remarkable is the Australian *Rheobatrachus,* which broods the eggs in its stomach, which of course must suspend its digestive function during the development of the eggs.

The Anura are worldwide in distribution, but more families of frogs are confined to the tropics than elsewhere. There are peculiarities in the distribution of some families, and they are discussed in Part IV, which deals with biogeography. The Gymnophonia are all tropical in distribution. The Caudata are peculiar in being almost exclusively a temperate-zone order. One family, the Plethodontidae, has many representatives in Central and South America, but most of the species are montane, and hence not in what is normally regarded as a tropical climate. Their spread into the tropics apparently has been recent.

Reptiles

The name Reptilia should no longer be used in classification, because some of its members are what is known as *paraphyletic,* which means that they represent different phylogenetic lineages. To use the term Reptilia, we would have to include birds and mammals as subdivisions. In the following sections, "reptile" is used in the vernacular sense to include turtles, snakes, lizards, and crocodilians, which have many traits in common.

Despite the opportunity on land, the amphibians made relatively short use of it as the dominant group of vertebrates. That was because in evolving under the conditions of terrestrial life, some of them became the ancestors of the reptiles, and hence of all other terrestrial vertebrates. The descendants of those that did not are the living Amphibia. The two types of changes that usually are accepted as characterizing the reptile grade of adaptation concern their eggs and their jaws.

The changes in their jaws involved shifts in the positions of the muscles used in biting. The new arrangement made them more efficient than the early amphibians in feeding on the many new kinds of insects that were appearing in the Carboniferous.

The amphibians faced, and still face, severe problems in getting away from the vicinity of water. Their eggs must be kept in water or in saturated air, and most of them have moist skins from which water evaporates rapidly in conditions of low humidity. A few species of frogs have managed to colonize some deserts by remaining buried deep enough to survive between the infrequent rains, when they appear above ground, breed in the temporary pools, and then retreat below ground. Development is rapid, and the newly metamorphosed froglets dig into the ground before they become desiccated. The eggs are never exposed to dry conditions. The completely terrestrial salamanders lay their eggs below ground or in rotting logs, where the humidity is never below saturation.

The reptile solutions to these problems were to protect the skin by retaining the scales of the ancestral amphibians and to evolve an egg that was much more resistant to desiccation. To diminish water loss by encasing the egg in a shell would seem a simple solution, but in fact it creates a new problem that is equally difficult: disposal of nitrogen-containing wastes. When protein is digested, nitrogen is left over in the form of ammonia (NH_3). This compound is both very toxic and very soluble in water. Therefore, aquatic animals have no problem with it, as the ammonia can be disposed of through the gills or through a permeable skin before it reaches toxic levels. For terrestrial animals, and their terrestrial eggs, no such easy mechanism is available, and two different solutions have been evolved. Only one is feasible for an embryo inside an eggshell. The reptiles, which have such eggs, are able to convert ammonia into uric acid, which is virtually insoluble in water and is nontoxic. As the embryo grows, an outpocket arises from the hindgut, and as it increases in size it acquires a blood supply and comes to occupy a large part of the space between the shell outside and the embryo and yolk inside (Figure 1.5). This structure, named the allantois, is a depository for uric acid in the form of an insoluble paste, and because of its position next to the shell and its rich supply of blood vessels, it also functions in the respiration of the developing embryo, as the shell permits the exchange

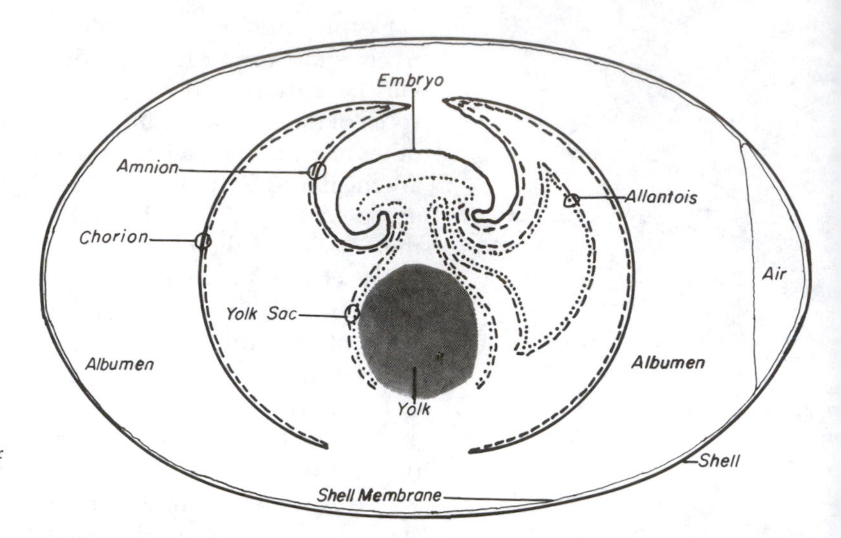

Figure 1.5. *Longitudinal section of the egg of a reptile, showing the extraembryonic membranes.*

of gases (oxygen, carbon dioxide, and water vapor). When the reptile hatches, the allantois is left outside the animal in the same way as the placenta (the homologous structure) is left outside the newborn mammal. Bird eggs have the same structures as reptile eggs. This kind of egg, named the amniotic egg because of another membrane (the amnion) that surrounds the embryo and provides an aqueous medium for it, cannot endure complete desiccation, but when buried in sand or earth it is much more resistant to water loss than is any amphibian egg. With it, the reptiles occupied the land and were the dominant land vertebrates for more than 250 million years, about five times as long as the Amphibia. It must be pointed out that ammonia is converted into urea by mammals. It is nontoxic and is soluble in water. It can thus be retained in the body until enough water is available for excretion. Urea is also present in Sarcopterygii and Chondrichthyes. It raises the osmotic pressure of the body fluids and prevents dehydration in seawater.

Given their advantages, the reptiles achieved an important adaptive radiation in the Late Carboniferous and Early Permian. It was then that the main branches of the reptiles became established, and in the Mesozoic era there was a great diversification, culminating in the two great groups of dinosaurs. It is tempting to digress on the subject of these animals, which must have been truly wonderful, but as they are extinct they can scarcely come under the topic of field zoology.

As with the separation from the Amphibia, the distinctions were in the ways in which the skull became modified for the attachments of the enlarging jaw muscles. The modifications involved gaps in the part of the skull that extends beyond the braincase. The gaps left bony arches (apses) in three lines. Originally there were no gaps, and the ancestral reptiles were the Anapsida. The skull type is represented in the present-day turtles. The Synapsida had a single gap not touching the skull cap (parietal bone). This group eventually gave rise to the mammals. A third group and a fourth group are sometimes combined under the name Diapsida because both of them had two gaps and two bony arches behind the orbit (eye socket). One of these, the Lepidosauria, has left descendants in the form of lizards (Lacertilia), snakes (Ophidia), and a lizardlike single species (Sphenodontia) living on islands near New Zealand. The second diapsid group, the Archosauria, gave rise to the dinosaurs, as well as to the Crocodilia and to the birds. In Chapter 3, the subject of cladistics is discussed. This is an approach that tries to estimate ancestor–descendant relationships, and an important principle is that classification should follow these relationships strictly. Were that system followed for the reptiles, the turtles would have to be recognized as a separate class to match the mammals, and in order to make the birds a class, there would have to be one separate class for the crocodiles and another for the lizards and snakes. Many herpetologists still find it convenient to recognize three major groups of reptiles: Testudinata (turtles), Crocodilia (crocodiles and relatives), and Squamata (lizards, snakes, amphisbaenians, and sphenodonts).

The Testudinata (Figure 1.6) evolved an early solution to the perils of existence (at least as early as the Triassic), and their heavy shells have made them a persistent if not dominant group ever since. The 240 species are grouped into twelve families, of which nine are classed as freshwater animals, although the allocation of time between land and fresh water varies among species, with box turtles being exclusively terrestrial. The Testudinae, or land tortoises, are represented by forty completely terrestrial species, and there are two distinct groups of sea turtles.

All turtles lay their eggs on land, digging a cavity for them in the earth or sand, and covering them with the same material. For most vertebrates, sex is determined genetically, with one pair of chromosomes being responsible. One member of the pair is different from the other in one sex; in the opposite sex, the two chromosomes are alike. Some turtles are quite different. Their sex is determined by the temperature at which the eggs develop, low temperatures producing males, and high temperatures producing females. The critical difference occurs within a range of 3–4°C. The basic trait is not confined to turtles. The American alligator and a few species of lizards also have temperature determination of sex, but with the opposite effect: low temperatures for females, and high for males.

In common with most ectothermal vertebrates, turtles suffer high mortality in the first year after the eggs are laid. The nests are not guarded, and many predators find them. Young individuals

Figure 1.6. Representatives of five distinct groups of turtles. (Parts a–d redrawn from King & Burke 1989; part e redrawn from Ernst & Barbour 1989.)

are rarely seen, partly because in freshwater species they do not leave the aquatic habitat and are not caught by the fencing methods used to trap the adults. Adults have high survival rates, commonly exceeding 80% per year.

Freshwater turtles, unlike most vertebrates, change their diets radically as they grow. Early in life they are predaceous and will eat almost any animal that they can catch. As they grow, their diet shifts to herbivory.

The longevity of some members of the order is legendary. Goin, Goin, and Zug, in their *Introduction to Herpetology* (1978), list maximum known ages for various reptiles. The champion is the giant tortoise, at 152 years, with other species ranging from 25 to 59 years.

The navigational ability of sea turtles is phenomenal. From the coast of Brazil, they migrate to Ascension Island in the middle of the Atlantic. As the island is only 34 square miles in area, or about 6 miles in diameter, their accuracy is remarkable. That kind of ability is the subject of Part V of this book.

Like the turtles, crocodilians have changed little since the Triassic. All twenty-one species are predaceous and semiaquatic, living on the shores of swamps, rivers, and estuaries. Most of the species are large, and some have reached a length of 7 m. Their most interesting trait is the parental care that they show. Nests are guarded, and the American alligator helps the hatching young to free themselves from the nest and then carries them to the nearest water. The female remains with the young for up to 2 years. The young have an alarm call, to which the parents respond by coming rapidly to the rescue.

The dominant reptiles now are the Squamata, with more than 3,000 species of lizards and more than 3,500 species of snakes. The single unique species *Sphenodon punctatus* was formerly placed in a separate order, but it is now considered to belong with the squamates. It inhabits small islands near New Zealand and has remarkable longevity, at least 77 years. Its most notable trait as a reptile is the ability to remain active at the low temperature of 11°C.

Lizards (Lacertilia) have occupied many kinds of habitats, and their adaptations to some of them are worth describing. Most species are ground-dwelling, but most climb readily, and nearly all are capable of rapid locomotion (Figure 1.7). They are especially abundant and diverse in deserts and semideserts. Unlike desert mammals, lizards are able to be active during daytime in such locations. They have quite precise control over their body temperatures, despite being ectotherms. This is accomplished by positioning their bodies so as to obtain maximum exposure to the sun when their bodies are below the optimum temperature, and by gradually changing the direction they face and the amount of body surface exposed as the temperature rises. Eventually they move into shade, avoiding further heating. They conserve water physiologically. Uric acid is secreted as a paste, and the water and salts are reabsorbed. Like crocodilians and sea turtles, several groups of lizards have special glands that secrete salt in a higher concentration than is possible from the kidneys.

A few lizards have evolved some interesting unique abilities. In the American tropics, it is a fascinating sight to see *Basiliscus* skitter across streams. It accomplishes this with extremely long toes, which spread its weight sufficiently that the lizard does not break through the surface film. As the individuals grow, it is necessary for them to move faster to remain on the surface; the large ones are unable to do so, and when frightened they simply dive in and swim away under water.

A second unique adaptation is the ability of *Draco* to glide. These small lizards of Southeast Asia and the East Indies are arboreal, and except for nest-making they never have to be on the ground. They glide from one tree to another, usually with a surprisingly small angle downward. The "wings" are flaps of skin covering long ribs. The ribs can be extended outward to spread the gliding surface, or folded against the body. Thus, unlike the situation for birds or bats, the foreleg is not committed to the flight surface, leaving it free for normal lizard functions.

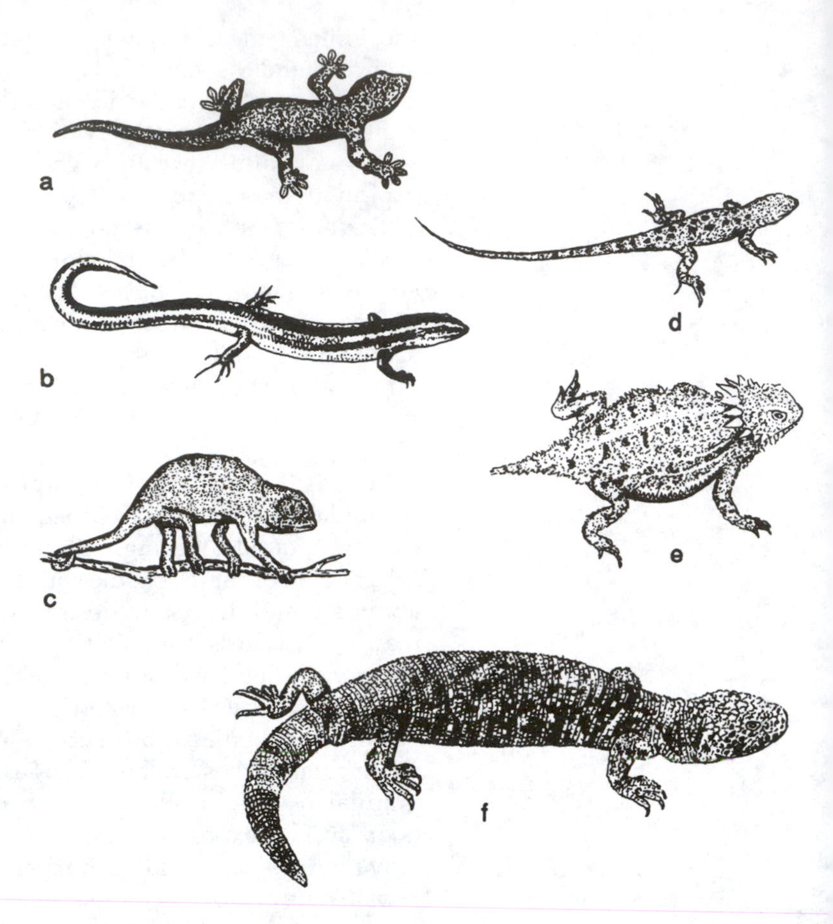

Figure 1.7. *Different body forms in lizards. (Parts a and c redrawn from Cansdale 1955; parts b and d–f redrawn from Stebbins 1966.)*

Most lizards are diurnal, but nearly all of the many species of geckos are active only at night. Their eyes, in common with those of other nocturnal vertebrates, have slit-shaped pupils that are capable of extreme opening (Figure 1.8). Most species of geckos are specially adapted to climbing on smooth surfaces, including vertical walls, and even on ceilings. Those species have expanded toes, each with a series of flat lamellae. The lamellae have numerous setae that provide attachment by surface tension.

Most lizards are territorial, and many of them advertise their territories with bright colors, like the yellow-headed lizard of Africa, *Agama agama*, several species of *Draco*, and many others. The color comes from several kinds of pigment cells in the skin, and some species, especially the chameleons, are able to change colors by nervous control of the degree to which the pigment granules in individual cells are concentrated in the center of the cell (in which case that color is diminished) or are spread throughout the extensive pseudopodia (bringing out that color). The stimulus is in the physiological state of the animal, not the color of the background, as is commonly supposed.

a

b

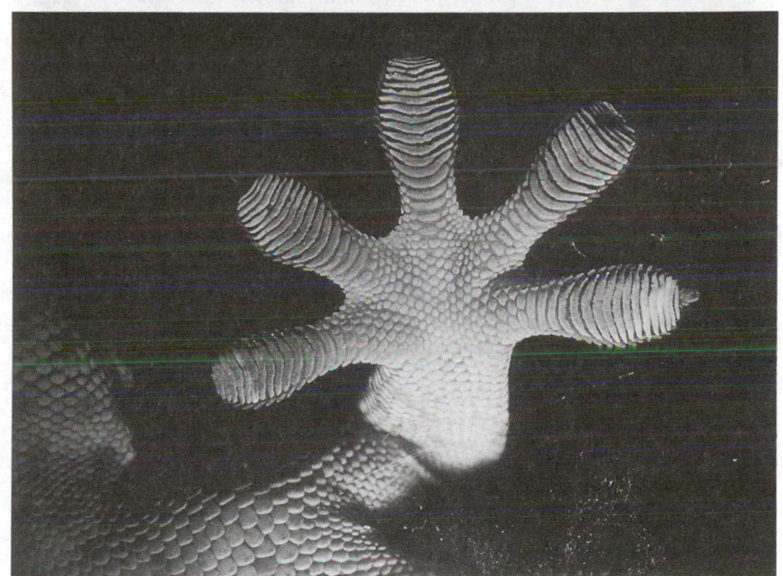

Figure 1.8. *(a) Eye and (b) underside of foot of* Gekko gekko, *Philippines.*

In contrast to other reptiles, most lizards are short-lived. The life histories of a number of species have been studied in sufficient detail to yield life tables. One of the early examples is described in Chapter 5. As with salamanders, the calculated mean generation time is neither the minimal time from egg to adulthood nor the maximum known longevity. Table 1.2 shows the mean generation times for twenty-four species. There is a strong American bias in the total number of records, but there are enough from other parts of the world to confirm the range of mean generation times.

The four species with mean generation times of 1 year are annuals. In two cases, at least one life-history stage cannot survive

Table 1.2. *Mean generation times for a number of species of lizards*

Species	Mean generation time (yr)	Location
Sceloporus undulatus	1.0	Texas
Uta stansburianus	1.0	Texas
Mabuya buettneri	1.0	W. Africa
Panaspis nimbaensis	1.1	W. Africa
Draco volans	1.2	Philippines
Emoia atrocostata	1.4	Philippines
Sceloporus virgatus	1.6	Arizona
Takydromus takydromoides	1.7	Japan
Scincella laterale	1.9	Florida
Crotaphytus wizlizenii	1.9	Nevada
Sceloporus jarrovi	2.0	Arizona
Mabuya maculilabris	2.0	W. Africa
Sceloporus olivaceous	2.0	Texas
Cnemidophorus sexlineatus	2.3	Kansas
Crotaphytus collaris	2.4	Kansas
Scelopporus graciosus	2.5	Utah
Eumeces fasciatus	2.6	Kansas
Sceloporus magister	2.7	Utah
Cnemidophorus tigris	2.9	Nevada
Sceloporus poinsetti	3.1	Texas
Leiolopisma zealandica	3.6	New Zealand
Xantusia vigilis	4.4	Arizona
Eumeces obsoletus	4.7	Kansas
Anguis fragilis	8.0	France

Source: Data amassed from many sources; calculations made in cooperation with D. W. Tinkle.

through the harshest season, so that only one age class is represented most of the time. Adults in the West African savanna do not survive the dry season, though the species survive because their eggs have been placed underground. The mean generation times for the remaining species reflect different adaptations by different species, sometimes to the same physical environments. This can be seen by comparing the calculations for different species listed from a given general area. Texas and Arizona are notable examples. The table also shows that the traits determining the details of life tables do not follow taxonomic lines. This is true even within species. *Sceloporus undulatus* in Texas has a mean generation time of 1.0 year; in Ohio it is 1.5 years, in South Carolina 1.9 years, and in both Colorado and Utah 2.7 years. Clearly, the local environment has selected for the traits that determine the life history of *S. undulatus*.

One comparison can be made between a mean generation time and a maximum longevity, in the case of the slow worm, *Anguis fragilis*, a legless lizard of Europe. It has survived 54 years in captivity, compared with its 8-year mean generation time; the individual involved was a female that at a known age of at least 51 years mated successfully with a male known to be at least 21 years old. The great discrepancy between mean generation time and maximum known life comes from two sources: the protection afforded captive animals, and especially the great influence of the early part of life on the calculations, regarding both rate of population growth and mean generation time.

Leglessness raises the question of an interesting trend among lizards. It has evolved independently in nine different groups of Lacertilia. Many legless lizards are *fossorial,* burrowing in sand or loose soil. Two possible reasons for the repeated tendency are, first, that they are safer from predators when below ground and, second, that legs are a disadvantage when a lizard is moving through dense vegetation, such as tall grass.

Snakes, the Serpentes, show that one case of the evolution of leglessness led to a major success. Whereas legless lizards compose a rather small fraction of the total number of lizard species, there are more species of snakes than of all other living reptiles combined. The success of snakes is due to their ability to engulf prey much larger than their heads. They can thus obtain food in large pieces and feed less frequently, as compared with legless lizards. That means that they can remain secluded most of the time and are thus exposed to larger predators only infrequently. It is the great flexibility and mobility of their jaws that make the snakes capable of swallowing large prey. Whereas the upper jaws of lizards are immovably attached to each other and to the cranium, and the different elements of their lower jaws are fixed to each other, the several parts of both the upper and lower jaws of snakes are joined by ligaments both to each other and to the cranium. The movable quadrate bone, to which both upper and lower jaws are connected, also aids in opening the jaws widely. The snake is able to move each part of each jaw independently and "walk" the jaws around its prey, which is thus brought into the mouth and swallowed.

Despite the fact that poisonous species are a minority of snakes everywhere except in Australia, the ability of some species to kill large animals, including humans, has earned all snakes the enmity of the general public, most of whom are sure that any snake they encounter is poisonous. Doubtless the ability of snakes to move quickly without legs has added to that enmity, reflected so vividly in the Book of Genesis. Three different groups of snakes have quite different kinds of teeth for delivering their poison. The simplest are grooved teeth located in the back of the upper jaw. These "rear-fanged" snakes were considered not to be dangerous until a bite from one of them caused the death of the distinguished zoologist Karl P. Schmidt of the Field Museum of Natural History. The fangs of cobras, kraits, coral snakes, and all of the poisonous species of Australia, as well as the sea snakes, are located at the front of the upper jaw and are hollow. The base of the fang is connected

a

b

Figure 1.9. South American boa on a uniform background (a) and on the forest floor (b). (Redrawn from Ditmars 1937.)

to a sac containing the poison, which is injected into the prey (or attacker) by contraction of the jaw muscles when the snake bites. The most elaborate fangs are those of the vipers, which include all poisonous species in the Western Hemisphere except the coral snakes, plus many species in Africa, Asia, and Europe. These fangs are also hollow and are much longer than are those of a cobra of comparable size. The possible difficulty in closing the mouth or swallowing with such a large tooth is avoided by having the maxillary bone, to which the tooth is attached, hinged via a ligament to the rest of the upper jaw. Opening the mouth automatically brings the fang forward, in position for the strike. When the mouth is closed, the hinge brings the fang into a folded position against the upper jaw.

The strike of a viper is much faster than that of a cobra. No doubt many have seen a motion picture of the legendary mongoose killing a cobra, but when a group of filmmakers attempted to film the mongoose killing a fer-de-lance, a pit viper of tropical America, a number of mongooses were killed by the snake. Mongooses are quick enough to avoid the strike of a cobra, but not quick enough to cause a pit viper to miss.

Like other animals, snakes have responded to the various forms of natural selection imposed by their habitats. Seen against a uniform background, the markings of many vipers and boas are bold, even spectacular, but when placed against a background of dead leaves on a forest floor, these snakes are quite difficult to see (Figure 1.9).

One of the more interesting aspects of the adaptations of snakes to individual environments is the effect on the shape of a cylindrical, legless animal. Species that live below ground most of the time, such as worm snakes, *Carphophis amoenus,* have a uniform diameter for most of their lengths. Most vipers are heavy-bodied for their lengths and do not move much, being "sit-and-wait" predators, and relying on their bite for both hunting and defense (Figure 1.10). The majority of species are prey to raptorial birds and some mammals, and they rely on speed to escape, a feature that has selected for an elongate shape. This is seen in an extreme form in some of the tree snakes (Figure 1.10), their slender shape allowing them to reach across from one supporting branch to another. The remaining special shape is that of the sea snakes, which have flattened, oar-shaped tails.

Mammals

The mammals are considered before the birds in recognition of their earlier appearance in the fossil record, mammals generally being considered to have developed in the Late Triassic. The three features used by most zoologists to distinguish mammals from most other vertebrates (possession of hair, milk production, and bearing living young) are not fossilizable. Paleontologists (students of fossils) and anatomists have therefore used the characteristics of skulls and teeth to distinguish the early mammals from their synapsid contemporaries and predecessors. That is practical, because skulls and teeth are the parts of a vertebrate that are most likely to be fossilized. Both the upper and lower jaws in reptiles are composed of more bones than are present in the jaws of adult mammals, in which some bones have been modified to serve other functions, or have been lost. The quadrate, which in present-day Squamata is movable and hinges the jaws to each other and to the cranium, has become modified in mammals to become the incus, one of the auditory ossicles involved in hearing. The upper jaw of a mammal consists of only two bones, the maxilla and premaxilla, and the lower jaw is a single bone, the dentary, which is hinged directly to the skull.

There is an additional difference in the smooth surfaces, the occipital condyles, that join the back of the skull to the vertebral column. Reptiles have a single condyle; mammals have a pair.

Most mammals have different kinds of teeth in an individual

a

b

Figure 1.10. *Extreme shapes in snakes. (a) Vine snake,* Dryophis acuminatus; *(b) Gaboon viper,* Bitis gabonica. *(Redrawn from Ditmars 1937.)*

jaw, serving different functions, a condition that is known as *heterodont;* such a differentiation is rare in other vertebrates and is poorly developed in the few where it is found, as in crocodilians. Mammals also have only two sets of teeth as they grow: the milk teeth, which are shed, and the permanent teeth of adults, which replace the milk teeth. Other vertebrates replace individual teeth throughout life.

As these changes were gradual, the mammals were closely similar to their ancestors and contemporaries, the Therapsida, and it is not easy to decide whether some of these Triassic fossils should be identified as reptiles or as mammals. Throughout most of the Mesozoic, mammals were small animals, insignificant in size and abundance compared with the dominant reptiles. Marsupial fossils are known from the Cretaceous, some of them much like the present-day opossum. A few Insectivora, now represented by shrews and moles, are known from the Late Cretaceous, but it was not until the Tertiary that the mammals became the dominant vertebrates. It is a matter of controversy whether the mammals in some way caused the decline and extinction of the dinosaurs toward the end of the Cretaceous, or whether the dinosaurs became extinct through other causes, and the mammals were able to take advantage of their absence. A currently popular hypothesis supporting the latter view is that the impact of an asteroid brought on a worldwide extreme deterioration in climate for long enough to cause the extinction of the dinosaurs and other prominent groups of animals. There is good geologic evidence for a major asteroid impact at the end of the Cretaceous, but the difficulty with the hypothesis is that the effect of that impact would have had to have been selective, as there is no evidence of reductions in the abundances of mammals, birds, fishes, turtles, lizards, snakes, or crocodilians.

Some characteristics of mammals frequently are cited as demonstrating superiority over other vertebrates, and that appears to be true with respect to present-day reptiles and amphibians and, for some characteristics, birds as well. Assuming that the mammals of the Mesozoic had these characteristics, they did not make those mammals superior to those reptiles. Three of the traits are unknown in other classes: providing milk for the newborn (an exception is pigeon "milk"), the possession of hair, and the possession of teeth that are specialized for different functions in an individual jaw. Three traits shared with some other vertebrates are the four-chambered heart (also possessed by crocodilians and birds), viviparity (shared with some fishes, amphibians, and reptiles), and endothermy (shared with birds).

Milk is produced in mammary glands, which give the class its name. (Etymological purists claim that "mammalogists" should be specialists on breasts, and that to be specialists on mammals, they should be "mammalologists." An escape is the term "theriologist," a student of beasts.) The milk of each species is specifically adapted to nourish the newborn young of that species, thus providing for optimal growth. The proportions of the different constituents of milk vary among different mammals. The constituen s are proteins, sugar (lactose), fats, and inorganic salts.

Milk high in albumin or fats makes for rapid growth of the young. Seal pups, whose milk is 45–55% fat, quadruple their weight in 2–4 weeks; human babies, living on milk that is only 2–4% fat and also is low in proteins but high in sugar content, grow much more slowly.

For the newborn, milk is more digestible than any other food. This gives the mammals a large advantage, as it frees the mother from searching for special food suitable for very young animals. Some parent birds must find food for their newly hatched young that is unlike their own food. Sparrows, for example, must provide insects, although the adults feed on seeds, because the beaks of the young are not hard enough to crack seeds. Newly hatched reptiles must forage for themselves.

For most mammals, hair serves as insulation, retaining the body heat generated by their endothermic physiology. There is a close inverse relationship between the thickness of the coat of hair and the temperature of the environment in which the mammal lives, with mammals from boreal regions having thicker coats than those from warmer climates. Most mammals have three kinds of hair: long, coarse guard hairs; soft, fine underfur; and long, thick vibrissae that are connected to nerve cells, making them sensitive to touch. It is the dense underfur that provides insulation by trapping a layer of air, which is a poor conductor of heat.

As hair is nonliving, it wears out and is replaced during molting, with summer coats being thinner than winter coats. It is the latter that have always been prized as furs. This is most conspicuous in the weasels, whose winter fur in northern latitudes is white and is called ermine (once reserved for royalty). The white color reflects the fact that in those areas where ermine is taken, the ground is snow-covered nearly all winter (Figure 1.11), making the white color adaptive camouflage, as it is for the hares, Arctic foxes, and even some birds (ptarmigan) inhabiting equally snow-covered regions. The summer coats of all of these animals are nearly completely brown, white being present only on the underside. The pigment in the hair gives the color, which usually matches the animal's habitat. The widespread deermouse, *Peromyscus maniculatus,* of North America has many local races, each of which matches its habitat. In the damp forests of the Pacific Northwest the mice are dark brown, as are those living in the high Appalachians. In the Midwest, those living in forests are darker brown than are those living in grassland. Desert races are notably pale, and one race living in an area with yellow, sandy soil is yellowish in color.

Only a few mammals have coats of hair that make them conspicuous. There are two quite different reasons for a conspicuous appearance. For some primates, such as the colobus monkeys of Africa, the color and pattern appear to be related to advertisement, either to notify rival males of their presence or to attract

Figure 1.11. *Ermine, shorttail weasel,* Mustela erminea, *with winter and summer coats. (Redrawn from Burt & Grossenheider 1964.)*

Figure 1.12. *Striped skunk,* Mephitis mephitis.

females. In skunks, on the other hand, the conspicuous pattern of black and white (Figure 1.12) serves to warn potential predators of the skunks' offensive scent, which is secreted by glands at the base of the tail. One species of skunk stands on its forelegs and directs scent over its back into the face of a threatening adversary.

Most species of mammals have teeth that are specialized for dealing with specific kinds of foods (Figure 1.13). This is quite different from the situation for other living vertebrates, most of which have simple straight pegs for grasping prey (as in present-day amphibians and nearly all reptiles and fishes), or all crushing teeth (as in some sharks). The hollow, "hypodermic needle" teeth of most poisonous snakes constitute an exception in being different from the rest, which are simple pegs. The basic mammal pattern is four kinds of teeth on each side of each jaw. From front to back there are three incisors, one canine, four premolars, and three molars (Figure 1.13a). Marsupials have slightly different numbers: 5–1-3–4 in the upper jaw, 4–1-3–4 in the lower jaw. In most mammals, at least some of these kinds of teeth are missing, having been lost when their original function no longer was part

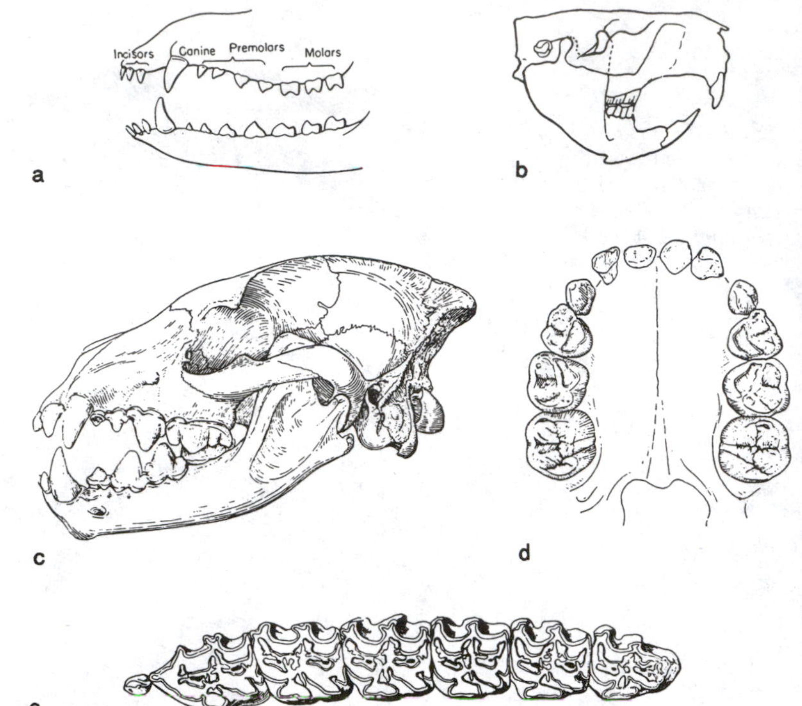

Figure 1.13. Mammal teeth. (a) Tooth complement of a generalized mammal. (Redrawn from Romer 1949.) (b) Side view of beaver skull, to show specialized incisors. (c) Side view of hyena skull, to show canine and slicing cheek teeth. (d) Surface view of teeth of Australopithecus, to show low, rounded cusps. (e) Surface view of cheek teeth of a horse, to show complex enamel pattern, specialized for grinding harsh grasses. (Parts b–e redrawn from Gregory 1951.)

of the particular feeding pattern of their possessor. The incisors are used for biting off parts of the food and also for grooming. They usually are simple points or flat blades. In rodents and lagomorphs (rabbits and relatives), the incisors are much elongated and function as chisels in gnawing (Figure 1.13b). This action causes wear, which is compensated for by continual growth from the root. In some hoofed mammals, such as deer and some domestic ruminants (cattle and sheep), the upper incisors are missing, the nipping function being taken over by the tongue and the lower incisors. Horses, however, have both upper and lower incisors. The most spectacular incisors are the tusks of elephants. The canines are used by carnivorous mammals for grasping or stabbing prey (Figure 1.13c). They are longer than the other teeth and are pointed and usually slightly curved. Most herbivores have lost their canines, there being a gap between the incisors and the premolars. The premolars and molars frequently are described together as the cheek teeth, because in most mammals they have much the same function. In their simplest form, they have low rounded bumps (cusps) and function to crush nuts, seeds, and fruits. Such teeth are found in omnivorous mammals, such as humans, pigs, squirrels, and cricetene mice (Figure 1.13d). Some of the rodents have no premolars. True herbivores, the leaf eaters, which must chew and frequently grind their food, have evolved specialized cheek teeth. A simple bumpy surface would not suffice, and the surfaces of the teeth are characterized by complex cusps that wear down to form elaborate ridges of enamel, alternating with dentine and cement (Figure 1.13e). The enamel is the hardest, and the result is a rough grinding surface that in some cases, such as horses and microtine rodents, is effective in reducing harsh grasses, which contain silica, to digestible fragments. Grazers have higher cheek teeth than do those species that browse on herbs, shrubs, and trees. Most meat-eating carnivores have reduced numbers of cheek teeth, and those that remain are highly specialized for shearing through tendon and bone. The last upper premolar and first lower molar are flat from side to side and have sharp edges (Figure 1.13c). In chewing, these carnassial teeth slide past each other, having a scissorslike action.

In fishes, amphibians, and reptiles, teeth that have worn out or have been lost are replaced from below, a process that continues throughout life. In mammals, however, only two sets of teeth ever form. The original incisors, canines, and premolars are lost spontaneously and are replaced by the "permanent" teeth; only one set of molars is ever formed, and they thus count as part of the milk-tooth set. If a fossil is sufficiently well preserved to determine that milk teeth were present, it is declared to be a mammal.

The remaining mammalian traits, though shared with some other vertebrates, do constitute an advantage for mammals. True viviparity, in which the developing embryo is nourished through the wall of the uterus (in mammals) or the oviduct (in other forms), provides protection for the developing embryo until it is ready for independent existence. The female does not have to leave the eggs to develop on their own (which would mean high mortality in many cases), nor does she have to guard them (which would expose her to increased risk of predation and would interfere with her ability to obtain food). The effect of this last problem has been described in the case of the plethodontid salamanders, where failure to obtain enough food has forced a system of reproducing in alternate years. In combination with the availability of milk, viviparity gives each newborn mammal a much better chance of surviving to adulthood than is enjoyed by oviparous forms.

As with many traits concerned with life histories, there is a trade-off involved with viviparity; it entails a cost in fecundity, or the number of offspring that can be produced. The fact that embryos are kept in the body of the mother limits the number that she can accommodate at one time. Thus, litter sizes for mammals tend to be smaller than clutch sizes for egg-laying amniotes. Table 1.3 compares the annual fecundity (litter size times the number of litters per year) for North American mammals with annual fecundity for oviparous lizards, most of them North American. The latter is, on average, about three times the magnitude of the for-

Table 1.3. *Comparison of annual fecundities of 239 North American mammals and 30 lizard species*

	Mammal spp.			Lizard spp.		
No./yr	No.		%	No.		%
1	48		19.6	0		0
2	22		9.0	1		3.3
3	17		7.0	0		0
4	31	median	12.7	2		6.7
5	27		11.1	1		3.3
6	17		7.0	0		0
7	13		5.3	1		3.3
8	20		8.2	1		3.3
9	9		3.7	3		10.0
10	7		2.9	2		6.7
11	1		0.4	1		3.3
12	12		4.9	2		6.7
13	2		0.8	0		0
14	3		1.2	1	median	3.3
15	1		0.4	4		13.3
16	4		1.6	0		0
17	1		0.4	2		6.7
18	2		0.8	1		3.3
19	0		0	0		0
20	0		0	2		6.7
21	1		0.4	0		0
22	0		0	1		3.3
23	0		0	0		0
24	0		0	0		0
25	0		0	1		3.3
26	0		0	1		3.3
27	0		0	1		3.3
28	0		0	0		0
29	0		0	0		0
30	1		0.4	0		0
39	0		0	1		3.3
98	0		0	1		3.3
Mean	5.49			17.16		

Source: Mammal data from Burt & Grossenheider (1964). Lizard data from Tinkle (1967) and D. W. Tinkle (pers. commun.).

mer, as is also true for the median and maximum values. The difference would be even more pronounced were it not for inclusion of the opossum. Marsupials give birth to tiny young that are then carried in the maternal pouch, and therefore their movements are not as restricted as are those of placentals. Many species of birds are exceptions, laying small clutches. That this is probably related to the necessity for incubating the eggs and caring for the young is suggested by the megapodes, which do neither, and lay up to twenty-four eggs in a clutch.

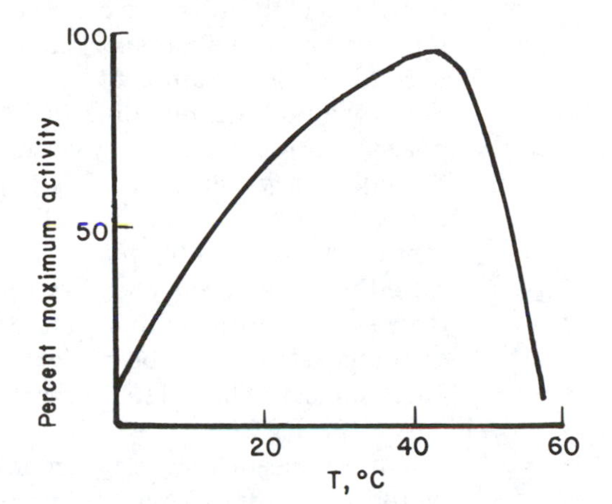

Figure 1.14. *Effect of temperature on the activity of an enzyme. The descending portion of the curve is due to thermal denaturation. (Redrawn from Lehninger 1975.)*

The four-chambered heart of the mammal is efficient in keeping the oxygenated blood from the lungs separated from the blood coming from the rest of the body. Amphibians and most reptiles have three-chambered hearts, with a single ventricle that pumps blood both to the lungs and to the rest of the body, at least risking mixing the two kinds. Such an arrangement is less efficient than the four-chambered heart.

Endothermy confers great advantages. By using metabolic heat to maintain a constant high body temperature, mammals (and birds) can remain active at times, such as at night, when ectothermic reptiles are not able to maintain a high temperature by basking. Most mammals, but only a few reptiles, are nocturnal. In addition to the extra time available for activity, endothermy allows mammals to be able to react to stimuli in a consistent and highly efficient way. A cool reptile is a sluggish reptile, but except for periods of torpor, as in hibernation, mammals are alert at all times, both to opportunities for food-getting and to the need for quick escape from enemies. Nearly all biochemical reactions are most rapid at fairly high temperatures, as the enzymes involved show well-defined peaks in the curves that describe the relations between their rates of activity and temperature, and these peaks are within the range of body temperatures (30–40°C) of endothermic vertebrates (Figure 1.14). A reptile will function well during a sunny day, but will have much less efficient metabolism when it can no longer raise its temperature by behavioral means.

There are substantial costs to maintaining a constant body temperature when that temperature is different from the temperature of the surrounding air, and the costs rise rapidly with the difference outside a fairly narrow range of temperatures. More than 80% of the energy from metabolism may be consumed in maintaining body temperature, and that energy must come from food, either directly or from stored energy in the form of fat. Thus mammals (and birds) must consume a great deal more food than reptiles or amphibians of comparable size, and the amount needed rises as the ambient temperature falls. Small mammals can avoid the problem by going into *torpor* (hibernating in protected places). Torpor involves lowering an animal's metabolic rate to the point at which the body temperature is only a few degrees above ambient, and the animal's reactions are reduced to a minimum. Arousal from torpor requires an expenditure of energy to raise the body temperature to normal, and mammals any larger than a woodchuck are unable to do that within a reasonable energy budget. Thus black bears, which remain inactive in winter, and are popularly believed to hibernate, lower their body temperature only to 31–35°C.

It might seem that many mammals, especially aquatic ones, would have difficulty in maintaining a temperature higher than their surroundings because their extremities, such as feet and ears, would quickly become chilled. Like seabirds with the same problem, they have heat exchangers. The heat exchanger is a series of blood vessels, going in both directions, located close together in a small region of the body next to the extremity. Heat is lost from blood flowing away from the body core, and gained by blood returning. Such structures are found in a number of mammals and birds, most typically warming blood that has been chilled in the flippers of seals or whales, or in the feet of seabirds, and cooling blood that is going to those organs, where it will have less heat to lose to the environment.

During the evolution of endothermy, the nearly constant temperatures have been set at levels higher than most of the ambient temperatures encountered, especially at night, but some environments, especially deserts and semideserts, become hot enough to cause severe problems in getting rid of metabolic heat. Small desert mammals escape the heat by burrowing to depths at which there are cooler temperatures and by being active only at night. That option is not available to camels, gazelles, and antelopes, which usually have no shade for protection. They have several mechanisms that utilize insulation, in the form of thick fur and a layer of fat on their backs (camels), or evaporative cooling. Some mammals have sweat glands over much of their skin; others, having them only on the pads of their feet, as in the case of canids, must resort to panting to evaporate body water. Some gazelles take advantage of the evaporation of water from their nasal passages during breathing and use a heat-exchange system to lose heat from blood going to the nose. Evaporation of water is a highly effective means of eliminating heat from the body, but a high price is paid in terms of water lost, especially in dry, hot climates, where water either is lacking for most of the year or occurs in very restricted locations. In a matter of hours, the loss of water through evaporative cooling would be fatal to small rodents in the open, and they must seek refuge in their burrows. For the camel, recovery of lost water depends on finding surface water, and it is unique in its ability to sustain a greater loss than other mammals, most of which will die if they lose water in excess of 10% of their body weight. Camels have remained healthy after losing 25% of their body weight. Some desert mammals obtain water from the sparse vegetation, especially succulents that pick up water during the night, but kangaroo rats are capable of obtaining all of their water from the metabolism of the seeds on which they feed.

One group of mammals, the Monotremata, do not share one of the traits just described. They are the egg-laying mammals, the duck-billed platypus of Australia and several species of spiny anteaters of Australia and New Guinea. Egg laying, assumed to be a primitive trait inherited from their reptilian ancestors, places the monotremes in a separate group, the Prototheria. Unfortunately, they are almost unknown in the fossil record earlier than the Pleistocene, and there is no direct connection with any of the Mesozoic mammals.

The remaining group, the Theria, consists of two divisions: the Metatheria (with a single subgroup, the Marsupiala) and the Eutheria (the placental mammals, with seventeen subgroups). The marsupials give birth to their young in an extremely undeveloped state. The newborn then crawls to a pouch in the skin that covers the teats and attaches itself to a teat. Most of its development takes place within the pouch. Marsupials closely similar to the opossum are known from the Cretaceous of North America and from the Early Tertiary of Europe, but at present, except for the opossum, marsupials are confined to the Australian region and Central and South America. The abundance and variety of adaptive types in Australia provide a classic example of adaptive radiation. Until the arrival of people and their introduced placental mammals, the marsupials (and the monotremes) had no competition from placentals except bats, mice, and seals. Under those conditions, different groups of marsupials evolved adaptations to many different ways of life, and it has frequently been stressed how similar some of these groups are to various placental mammals in the rest of the world (Figure 1.15). Thus, there are forms that are more or less equivalent to wolves (Tasmanian wolf), cats (Tasmanian devil and "native cats"), marmots (wombats), rabbits (rat kangaroos), squirrels (phalangers), flying squirrels (flying phalangers), moles (marsupial "moles"), weasels (phascogales), and anteaters (numbat). Some marsupials occupy niches similar to those occupied by placentals, but are unlike their placental equivalents. Kangaroos and wallabies are grazers and browsers, but are unlike antelope or deer. The koala is arboreal and feeds on leaves. It is unlike leaf-eating monkeys, but has some features in common with sloths. Bandicoots behave something like rabbits, but are omnivorous. There are conspicuous gaps in the parallelism. There

Figure 1.15. Convergent types of marsupials (left) and placental mammals (right): (a) mole types; (b) jumping types; (c) gliders; (d) anteater types; (e) marmot types; (f) cat types; (g) dog types. (Drawn from photographs in Nowak & Paradiso 1983.)

is nothing like an otter, beaver, or other aquatic placentals, and there is nothing like a bear or lion. However, an extinct huge herbivorous marsupial was much the equivalent of the extinct giant ground sloth. The parallelism, while fascinating, is a broad-brush affair.

Like Australia, South America was isolated from the rest of the terrestrial world for a long period. Only primitive placentals reached that continent, probably by island-hopping before a major break at Panama at the end of the Paleocene, and the marsupials there had their own adaptive radiation, which still includes more than seventy species in three families. Except for the opossum in North America, the descendants of the Mesozoic marsupials are confined to Central and South America and the Australian region. The evidence for the superiority of the placentals is most convincing.

The placental mammals (Eutheria, meaning "true animals") have become the dominant vertebrates on land, and they rival the large fishes for dominance in their parts of the seas. It took a single unique species of mammal (*Homo sapiens*) and our commensals to displace the overwhelmingly dominant mammal faunas of the Pleistocene, and they are still represented by most of the adaptive types.

The range of eutherian adaptations

In the following account, the names of the groups are given in parentheses. The placental mammals have adapted to and occupied every habitat, though they have not entered the realm of diurnal flight, which is dominated by birds. Terrestrial mammals, both herbivores and carnivores, large and small, are found from the equator to the Arctic ice pack, and from rain forests to deserts. The most spectacular examples are the great herds of hoofed herbivores (Perissodactyla and Artiodactyla) and their predators (Carnivora) that occupy the savannas of eastern and southern Africa. The bison of North America were comparable in abundance as late as the middle of the nineteenth century, and no earlier than the Late Pleistocene were accompanied by mammoths (Proboscidia), camels and antelopes (Artiodactyla), horses (Perissodactyla), and a large number of other species of large mammals.

The arboreal squirrels (Rodentia), monkeys (Primates), and sloths (Edentata) are conspicuous in their habitat, and squirrels and flying "lemurs" (Dermoptera) have added the gliding adaptation.

Many small terrestrial mammals have burrows as refuges, but members of two families of Insectivora (moles and golden moles) and a number of Rodentia (e.g., pocket gophers, mole rats, and African mole rats) remain in their subterranean burrow systems throughout their lives. The moles and pocket gophers have stout forelimbs that are used in digging, but both families of mole rats use their incisor teeth.

In taking up an aquatic life, there have been a number of degrees to which mammals have acquired adaptations. The carnivorous mink and the herbivorous marsh and swamp rabbits are not greatly different from completely terrestrial Mustelidae and Lagomorpha, respectively. Somewhat more specifically adapted to aquatic life are water shrews (Insectivora), polar bears (Carnivora), tapirs (Perissodactyla), and capybaras (Rodentia). Next in the progression come otters (Carnivora) and beavers, muskrats, and nutrias (all Rodentia). The most extreme aquatic adaptations in any species with close terrestrial relatives are those of the sea otter (Carnivora). Seals, sea lions, and walruses (Pinnipedia) do come ashore for breeding, but they are so modified for aquatic life that their limbs are flippers, unsuited for progression on land, where they are extremely clumsy. Finally, there are the whales and dolphins (Cetacea) and manatee and dugong (Sirenia), completely confined to an aquatic existence.

Unlike flying squirrels and flying "lemurs," bats (Chiroptera) execute true flight, and their ability to avoid obstacles and to capture flying insects in the dark depends on their use of echolocation, as described in Chapter 2. They are such competent flyers that it is remarkable that they have not succeeded in becoming diurnal in activity. As already suggested, they may not be able to compete with birds during the daylight hours.

Three orders are not mentioned in the foregoing discussion: Pholidota, the scaly anteaters; Tubulidentata, the aardvark; and Hyracoidea, "rock rabbits."

Birds

Birds were the last class of vertebrates to appear as fossils. The discovery of *Archaeopteryx* in 1861 was a severe blow to Darwin's opponents, who rejected the concept of evolution on the grounds that no fossils intermediate between any of the major groups of animals had been found, and were almost impossible to imagine. These six fossils (three complete or nearly complete specimens are known) are fully covered with feathers, including flight feathers, but in most other characters they are more reptilian than avian. In fact, one of the specimens was misidentified for 20 years as the small dinosaur *Compsognathus*, because the feather impressions are so faint that they were not noticed. The fossils are from the late Jurassic period and were all found in a small area in Bavaria. The most conspicuous reptilian feature is the long tail, which is unlike that of any bird, but has a pair of long feathers (rectrices) on each vertebra. The teeth are also reptilian, but a few bird fossils from the Cretaceous also have teeth; so they are not an exclusively reptile feature. The reptile origin of birds is unquestioned, but there are no fossils that are definite links between *Archaeopteryx* and preexisting forms. Feathers are such elaborate structures that their evolution from scales must have taken considerable time. They are also delicate structures, and it was only the fortunate circumstance of fossilization in the extremely fine grained Solnhofen Limestone that made the feather impressions of *Archaeopteryx* identifiable. It is unlikely that such good luck will reveal older impressions, and the similarity between the skeleton of the first bird and those of related reptiles makes it impossible to decide at present from which of two different lines of reptiles birds are descended, and also impossible to establish how much earlier in the Jurassic the crucial step of the origin of feathers took place.

Birds have only three traits that are unique to them: feathers; the retention of the right aortic arch, the artery from the heart to the rest of the body, excluding the lungs; and the possession of a furcula, the fused clavicles of the pectoral girdle, essential for the attachment of flight muscles. The four-chambered heart is shared with crocodilians and mammals, and endothermy is shared with mammals.

The advantages of a four-chambered heart and endothermy have been described, as they are also mammalian traits. The choice between the right aortic arch of birds and the left of mammals seems a matter of chance. There may be an advantage in having a single arch in endotherms, instead of both, as in reptiles and amphibians.

Some birds, albatrosses in particular, are able to drink seawater. Like some reptiles, they have salt glands that secrete salt in high concentration, thus avoiding the dehydration from drinking seawater that would kill a mammal.

Birds that live in cold climates, seabirds especially, lose heat to the environment rapidly from their chilled extremities. Unlike cold-climate mammals, they have no layer of blubber to insulate them, but like the mammals, they have heat-exchange systems that lose heat to the blood returning to the body core, and thus they lose less heat via the blood that is going to the feet or flippers.

There is a debate among ornithologists about the origin of feathers, some arguing that they first evolved as insulation for the maintenance of endothermy, and others arguing that such a complex structure was unnecessary for insulation and that they most likely arose in connection with the origin of flight. According to the second hypothesis, elongated scales on the rear surfaces of the forelimbs would have aided climbing reptiles in leaping from one branch to another, and each step in the origin of feathers would have provided an additional advantage. Without the presence of such scales on some fossil, the argument is not likely to be settled.

Whatever the origin of feathers, all birds that are capable of flight have the same arrangement of feathers. The lifting surface consists of long, stiff feathers (quills) arranged in two series at the trailing edge of the wing (Figure 1.16). The outer set is attached to the bones equivalent to the human hand and fingers. They are the primaries. The secondaries are attached to the ulna, which corresponds to the outer bone of the human forearm. The bases of these feathers are covered by two series of feathers, the upper and lower wing coverts, and the bases of these are also covered by coverts, the sets continuing forward to the leading edge of the wing. The shapes of the wing and of the primaries are adapted to the particular kinds of flight that different birds use, and the kinds of flight that they permit have given birds many uses.

From our earthbound viewpoint, the most obvious use is escape from predators. Ground-dwelling birds are most at risk from nonflying predators, and they tend to have short, rounded wings, adapted to a very rapid beat, giving a fast start, essential in the first moment of discovery of the predator. Bobwhites and pheasants are typical. Wings of this shape are also useful in quick maneuvering, as within forests. That is especially true if the outer part of each primary is narrower than the inner part, making the wing slotted, which provides for greater lift. Most small, familiar birds, such as chickadees and titmice, have this kind of wing. It is useful for rapid movements and quick changes of direction (Figure 1.17). Short, rounded wings, however, are not the most efficient for prolonged flight, and birds having them must stop frequently on long migrations, like those undertaken by many woodland birds such as warblers and vireos.

Birds that actively pursue their prey have wings that are most efficient in rapid flight. They are long and pointed. The wings of swallows, swifts, and falcons are examples. Some maneuverability is sacrificed for speed. In that connection, it is interesting to compare the wings of two kinds of bird-eating hawks. Falcons, which pursue their avian prey in the open, have the wings that were just described. Hawks of the genus *Accipiter*, like the sharp-shinned hawk, slip quietly through the forest and usually attack birds that are perched. They have wings like those of their prey.

Many hawks, eagles, and vultures search for their prey (or carrion) from high in the air. Soaring is their normal method of flight, and their wings are shaped accordingly. They are broad, giving a light wing loading, and the primaries are slotted for lift and short turns. Soaring involves a minimum expenditure of energy, and, when possible, they make use of thermals – regions in the air where it is rising because is has been heated on the ground. As thermals tend to be confined to small areas, the soaring birds must turn continually to remain in them. When progressing, they rise on one thermal and glide into the next, to repeat the process. On migration routes, they can be seen in groups in the morning, circling on the first thermals, or in front of cliffs where there is a strong updraft, to reach a sufficient height before departing in the correct direction.

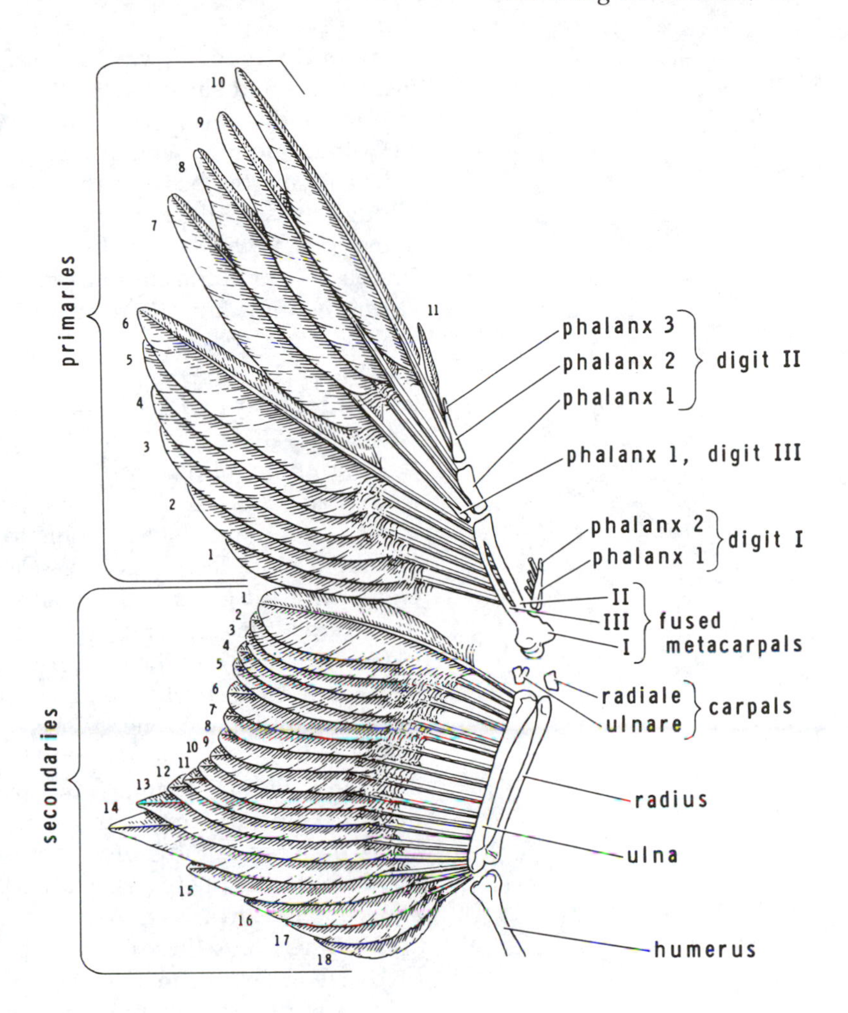

Figure 1.16. *Attachment of the flight feathers to wing bones. The number of each kind of feather varies among kinds of birds. (From Van Tyne & Berger 1959.)*

Figure 1.17. *Shapes of bird wings, adapted to different kinds of flight. (a) Small woodland bird. (Redrawn from Dorst 1974.) (b) Eagle and (c) albatross, for soaring on updrafts and strong winds, respectively. (Both redrawn from Aymar 1938.) (d) Falcon in flight.*

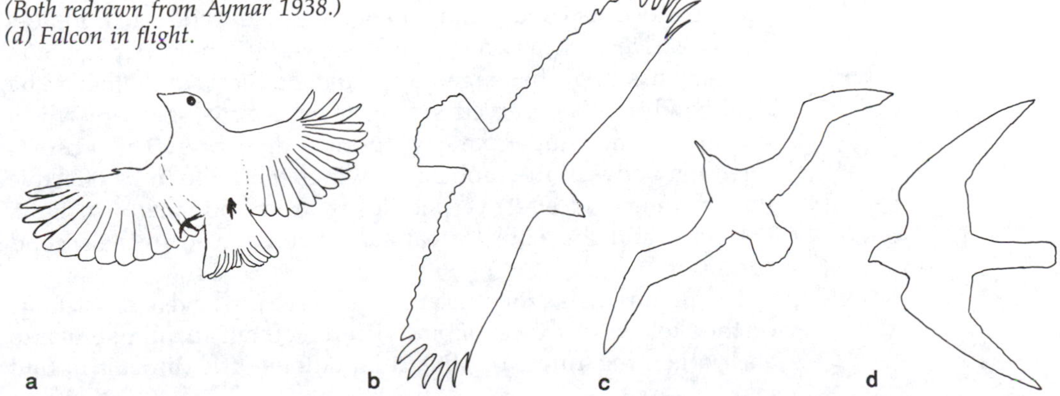

A fourth type of wing is characteristic of large seabirds – albatrosses and shearwaters. It is most effective where there is a strong wind, as over the southern circumpolar ocean. The wing is very long, pointed, and flat. Its success depends on using considerable inertia, because the bird gains speed from being high enough to avoid the drag of waves on the wind. From a height of some 15 m it will swoop down into the lower, slower air, and its inertia will be sufficient to carry it back up to where the wind is strong enough for it to soar. Unlike the wing of the eagle, the seabird wing does not generate enough lift in light winds to keep it aloft.

Hummingbird wings are in a class of their own. Unlike other small birds that are able to flutter and remain in place for a short time, hummingbirds hover upright, with wings beating back and forth, instead of up and down. The relative sizes of the different wing bones make the wing able to move in any direction at the shoulder joint. The result is a wingstroke that describes a figure eight, with lift generated on both forward and back strokes. In most birds, the cambered shape of the wing, like that of an airplane, provides lift when in forward flight. In hovering, that advantage is lost, and that entails a high cost in energy, which is recouped in the high-energy nectar acquired while hovering.

Notwithstanding the value of being able to fly, some birds have evolved to become flightless (Figure 1.18). This has happened under three quite different sets of circumstances. The most spectacular is the case of the large flightless birds now confined to the Southern Hemisphere: ostriches of Africa (but formerly of Asia), emus and cassowaries of Australia and New Guinea, and rheas of South America. To these can be added the extinct elephant birds of Madagascar and the moas of New Zealand. These birds apparently have undergone a common adaptation: an increase in size to the point that the process led to difficulty in flying, and eventually to the loss of flight. A second reason for the loss of flight is isolation on oceanic islands where predators are lacking. That has happened to a number of species, especially pigeons, rails, and ducks, and is extreme in the kiwi. Kiwis have long been considered to be flightless because of a common ancestry with the large flightless birds (ostriches, rheas, emus, and cassowaries), but their common ancestry has been challenged by J. Alan Feduccia and others. The final condition is observed among marine birds, many of which use their wings under water to pursue their prey. This poses a dilemma between the problem of wings too large to be effective under water and effective paddles that are too small for flight. Penguins and the extinct great auk have evolved to the second condition.

The requirements for efficient flight have put severe restrictions on the skeletal structures of birds, keeping them all quite similar to each other, regardless of other adaptations. It is difficult to find

Figure 1.18. *Three flightless birds that have become flightless under different kinds of selective pressures: (a) ostrich; (b) penguin; (c) kiwi. (Parts a and c redrawn from Feduccia 1980; part b redrawn from Van Tyne & Berger 1959.)*

adequate characters to distinguish different lineages, whereas lineages are easier to identify in mammals through teeth and skeletons. For that reason, there is disagreement among ornithologists about relationships, and the compromises have led to the acceptance of between twenty-five and thirty distinguishable groups, with the arguments being over the relationships among them and among the subgroups within them. Table 1.4 lists the groups recognized by one pair of authorities. Recognition of so many has led directly to the conclusion that there have been many examples of different lineages evolving identical or very similar adaptations to similar environments. Such phenomena are examples of convergent evolution, but with fewer groups there would be fewer examples. Accepting the current classification, six groups have evolved webbed feet (Gaviiformes, Procellariiformes, Pelecaniformes, Ciconiiformes, Anseriformes, and Charadriiformes), and three have evolved lobed toes for swimming (Podicipediformes, Gruiformes, and Charadriiformes). Long, pointed beaks have evolved among four groups of fish-eating birds (Gaviiformes, Pelecaniformes, Ciconiiformes, and Coraciiformes), and hooked, pointed beaks have evolved in predaceous birds in the Falconiformes, Strigiformes, and a few Passeriformes. Convergent wing shapes have already been described. The list of convergencies can be continued by consulting almost any book on ornithology.

The specializations of the teeth of mammals for processing different kinds of foods have been described. Although birds are constrained by the absence of teeth and the simple horny structure of their beaks, there have been many special adaptations for the

Table 1.4. *The major groups of birds*[a]

Group	Types
Sphenisciformes	Penguins
Struthioniformes	Ostrich
Rheiformes	Rheas
Casuariiformes	Emus and cassowaries
Apterygiformes	Kiwi
Tinamiformes	Tinamous
Gaviiformes	Loons
Podicipediformes	Grebes
Procellariiformes	Albatrosses and petrels
Pelecaniformes	Pelicans, gannets, and cormorants
Ciconiiformes	Herons, storks, ibises, and flamingos
Anseriformes	Ducks, geese, and swans
Falconiformes	Hawks and vultures
Galliformes	Grouse, pheasants, and turkeys
Gruiformes	Cranes and rails
Charadriiformes	Plovers, sandpipers, gulls, and auks
Columbiformes	Pigeons
Psittaciformes	Parrots
Musophagiformes	Turacos
Cuculiformes	Cuckoos
Strigiformes	Owls
Caprimulgiformes	Frogmouths, nightjars, and whippoorwills
Apodiformes	Swifts and hummingbirds
Coliiformes	Coly
Trogoniformes	Trogons
Coraciiformes	Kingfishers, bee-eaters, and hornbills
Piciformes	Woodpeckers, toucans, and honeyguides
Passeriformes	Flycatchers, lyrebirds, larks, swallows, titmice, wrens, thrushes, crows, shrikes, wood warblers, vireos, and finches (more than half of all bird species)

[a]Only the more familiar types are listed for some groups.
Source: Van Tyne & Barger (1959).

kinds of foods to be captured and/or processed (Figure 1.19). The hooked beaks of raptors (hawks and owls) have been mentioned. They are not used in the capture of prey, but to tear it apart while holding it down with powerful claws. Small birds, such as warblers, that pick insects and their larvae from the branches and leaves of trees, have rather short, fine-pointed beaks like forceps. Those like flycatchers and nighthawks that capture insects in flight have beaks with a wider gape than other birds of comparable size. They also have special feathers like bristles around their mouths. Hummingbirds have very slender beaks for probing into long-tubed flowers for nectar and for the insects that are feeding there. Seeds are a highly nutritious food, but most birds must either swallow them whole (jays) or forgo them as food. The finches and grosbeaks are specialized for husking and cracking

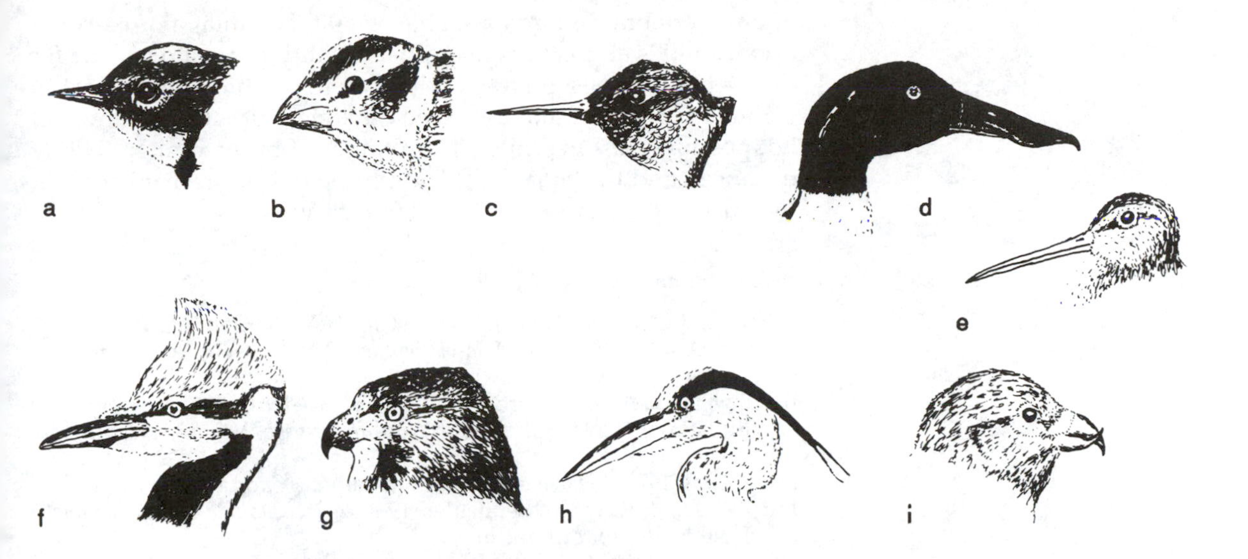

Figure 1.19. *Bird beaks adapted for different kinds of feeding: (a) yellow-rumped warbler (picking insects off foliage); (b) rose-breasted grosbeak (seed cracking); (c) ruby-throated hummingbird (nectar feeding); (d) shoveler (straining silt for small animals); (e) dowitcher (probing in mud); (f) pileated woodpecker (chopping into wood); (g) red-tailed hawk (tearing flesh); (h) great blue heron (spearing fish); (i) red crossbill (prying seeds from cones).*

seeds. Their beaks are short and powerful. Woodpeckers have long chisel-shaped beaks, adapted to chopping into wood for the boring insect larvae that are found there. Their tongues (except for sapsuckers) have backward-pointing spines for extracting the larvae from their burrows. The beaks of many birds that feed on fishes have been described, but ducks of the genera *Mergus* and *Lophodytes* (the mergansers) have beaks that are unlike those of other ducks, being slender and having serrated edges, reminiscent of the huge extinct pseudotoothed *Osteodontornis.* Woodcocks, many sandpipers, and especially the kiwi have long slender beaks that are used to probe deep into soft earth to capture worms or other invertebrates. The beaks of woodcocks are flexible at the tip and can be opened below ground. Many species of ducks use their beaks to filter small organisms out of the mud. Their flat beaks have matching serrations on upper and lower mandibles, between which the water is strained. Flamingos have the most elaborate filtering mechanism, which will remove tiny plants and animals from the water of their habitat. Oysters, mussels, and barnacles can be opened by the chisel-shaped beaks of oystercatchers.

Some birds use their beaks for display as well as for capturing and processing food. The beaks of toucans, hornbills, and puffins are bizarre in size, shape, and color. The plumage of a great many species of birds is colorful and boldly patterned. These conspicuous features have several different functions, and there is much discussion about these functions. Unfortunately, there are so many inconsistencies among different species that a study of individual cases is necessary for correct interpretations. All of the explanations involve one form of behavior or another, topics reserved for Part VI.

Because they have the greatest capacity for long-distance travel of any vertebrates, birds are able to take advantage of seasonal opportunities at great distances. This ability is most dramatic for the species that nest on the tundra, where productivity is great in summer because of the prolonged daylight, but where they could not possibly exist in winter. The abilities of birds to migrate long distances, and the behavioral and physiological mechanisms that they use, are discussed at length in Part V.

References and suggested further reading

Aymar, G. C. 1938. *Bird Flight*. Garden City, N.Y., Garden City Publishing.

Bellairs, A. d'A. 1970. *The Life of Reptiles*. 2 vols. New York, Universe Books.

Bigelow, H. B., & W. C. Schroeder. 1953. *Fishes of the Gulf of Maine*. Fishery Bulletin of the U. S. Fish and Wildlife Service, Vol. 53, Washington, D.C.

Bond, C. E. 1979. *Biology of Fishes*. Philadelphia, Saunders.

Burt, W. H., & R. P. Grossenheider. 1964. *A Field Guide to the Mammals*. Boston, Houghton Mifflin.

Cansdale, G. 1955. *Reptiles of West Africa*. London, Penguin Books.

Ditmars, R. L. 1937. *Snakes of the World*. New York, Macmillan.

Dorst, J. 1974. *The Life of Birds*. 2 vols. New York, Columbia University Press.

Duellman, W. E., & L. Trueb. 1986. *Biology of Amphibians*. New York, McGraw-Hill.

Eisenberg, J. F. 1981. *The Mammalian Radiations*. University of Chicago Press.

Ernst, C. H., & R. W. Barbour. 1989. *Turtles of the World*. Washington, D.C., Smithsonian Institution Press.

Feduccia, J. A. 1980. *The Age of Birds*. Cambridge Mass., Harvard University Press.

Forey, P., & P. Janvier. 1993. Agnathans and the origin of jawed vertebrates. *Nature* 361:129–34.

Gill, F. B. 1990. *Ornithology*. San Francisco, Freeman.

Goin, C. J., O. B. Goin, & G. R. Zug. 1978. *Introduction to Herpetology* (3rd ed.). San Francisco, Freeman.

Gorr, T., & T. Kleinschmidt. 1993. Evolutionary relationships of the coelacanth. *American Scientist* 81:72–82.

Gregory, W. K. 1951. *Evolution Emerging*. New York, Macmillan.

Hairston, N. G., Sr. 1987. *Community Ecology and Salamander Guilds*. New York, Cambridge University Press.

Jarvik, E. 1980. *Basic Structure and Evolution of Vertebrates*. New York, Academic Press.

King, F. W., & R. L. Burke (eds.). *Crocodilian, Tuatara, and Turtle Species of the World*. Washington, D.C., Association of Systematics Collections.

Lehninger, A. L. 1975. *Biochemistry*. New York, Worth Publishers.

Nowak, R. M., & J. L. Paradiso. 1983. *Walker's Mammals of the World*. Baltimore, Johns Hopkins University Press.

Orr, R. T. 1982. *Vertebrate Biology* (5th ed.). Philadelphia, Saunders.

Pough, F. H., J. B. Heiser, & W. N. McFarland. 1989. *Vertebrate Life* (3rd ed.). New York, Macmillan.

Romer, A. S. 1949. *The Vertebrate Body*. Philadelphia, Saunders.

Romer, A. S. 1966. *Vertebrate Paleontology*. University of Chicago Press.

Smith, P. W. 1979. *The Fishes of Illinois*. Urbana, University of Illinois Press.

Stebbins, R. C. 1966. *A Field Guide to Western Reptiles and Amphibians*. Boston, Houghton Mifflin.

Steene, R. C. 1977. *Butterfly and Angelfishes of the World. Vol. 1: Australia.* New York, Wiley.
Sterba, G. 1963. *Freshwater Fishes of the World*. New York, Viking Press.
Tomelleri, J. H., & M. E. Eberle. 1990. *Fishes of the Central United States.* Lawrence, University Press of Kansas.
Van Tyne, J., & A. J. Berger. 1959. *Fundamentals of Ornithology*. New York, Wiley.

2

Vertebrate sensing of the environment

Humans have the traditional five senses that make us aware of our environment and the changes in it: sight, hearing, smell, taste, and touch. Some other vertebrates are able to sense certain stimuli at levels that we cannot match, especially smell, taste, and hearing; others can sense kinds of stimuli that we are aware of only through instruments: weak electric fields and the magnetic field of the earth. A person able to sense either of these last two would surely have a sixth sense, or might even be thought to be capable of some form of "extrasensory perception." It is important to understand the uses to which different vertebrates put all of these senses, as much of field zoology consists in understanding those uses. This chapter begins with a consideration of the two senses that are most useful to humans, and for which we are superior to most vertebrates: sight and hearing. We can visualize our relative dependence on these senses by examining our degree of sympathy for people who have lost each. We tend to think of the loss of the sense of smell or of taste as almost comical, while recognizing the danger of not being able to smell smoke, for example. We sympathize with the deaf, invent hearing aids, and include sign-language inserts in some television programs. But for the blind, we go to great lengths to offer protection. The white cane elicits sympathy and voluntary help in a way that no other symbol of affliction does. After sight and hearing, we consider the means by which the remaining senses are exercised. That part of the chapter will take us into the (for us) unknown worlds that we can only imagine.

Sight

The eyes of all vertebrates are similar in basic structure and function. The eye consists of an enclosed, stiff-walled capsule with an inner surface (the retina) that is sensitive to light, a transparent front cover (the cornea) opposite the retina, devices for controlling the amount (intensity) of light admitted (the iris), and at least one device for focusing the light on the retina (the lens, and in terrestrial forms the cornea as well) (Figure 2.1). In image-forming eyes, such as all vertebrates have, it is necessary that the light entering be focused on the retina. This is accomplished through the property of light that in passing between two media with different refractive indexes, it is bent at the surface, as between air and water. The cornea has almost the same index of refraction as water, and the lens acts as the sole means of focusing in aquatic vertebrates. There are two ways in which that is accomplished. One is to position the lens properly by moving it toward or away from the retina, the method used by lampreys, fishes, and amphibians; the second is to change the shape of the lens by muscular contraction or relaxation at the edges, the method used by the amniotes. Terrestrial vertebrates must have rigid corneas that are parts of spheres, and the corneas perform an important role in focusing. The shape of the eyeball is maintained by the outer layer of tissue, the sclera, consisting of dense connective tissue in lampreys and mammals, but stiffened by either cartilage or bony plates in other vertebrates. The cornea is a continuation of the sclera, plus a layer of transparent skin. Common vision defects in humans arise from corneas that are not quite the correct shape, causing astigmatism, and the shape of the eye capsule itself. Farsightedness (hypermetropia) comes from a shortened eye, causing the focus to fall behind the retina; nearsightedness (myopia) comes from the opposite defect. As people age, the eyeball shrinks slightly, causing farsighted people to require stronger corrective eyeglasses for reading, whereas nearsighted people need less correction for driving. I am nearsighted and have had the strength of my eyeglasses reduced several times, as I am 76 years old.

The process of focusing on distant or close objects is accommodation. In humans, the muscle connecting the lens to the stiff sclerotic coat contracts to cause the lens to flatten for distant vision, or relaxes to allow the lens to resume its more rounded shape for viewing near objects. As one ages, the lens becomes less flexible, and accommodation is slower and more difficult, even if one's vision was perfect early in life.

Inside the sclera is a second layer of tissue, the choroid coat, that contains numerous blood vessels, essential for the maintenance of the rest of the eye, which lacks them. The choroid coat is pigmented and absorbs most of the light that reaches it through the light-sensitive retina. Some vertebrates, not normally including humans, have a highly reflective layer within the choroid coat, the tapetum lucidum. It is the source of "eye shine" in nocturnal mammals, and it serves to reflect the light back through the retina for greater sensitivity in low light. The amount of light entering the eye is controlled by the iris, a pigmented structure derived from both choroid and retinal layers. It contains both radial and

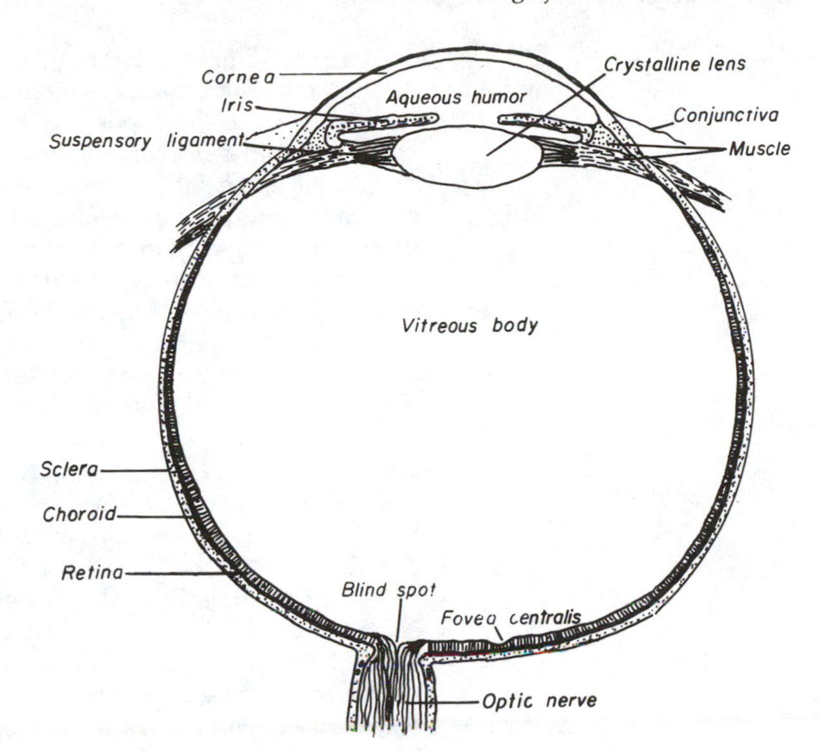

Figure 2.1. *Section through the mammalian eye, typical of many vertebrates.*

circular muscles that control the size of the pupil, the opening in the iris through which light enters. The shape of the pupil varies greatly in different vertebrates. In humans, which originally were diurnal, it is circular, as it is in some vertebrates (owls) that are active only at night. Mammals that have wide peripheral vision have pupils that are elongated horizontally; those that are active during both night and day tend to have vertically elongated pupils. Most remarkable are the nocturnal geckos, in which the pupil is a vertical, irregular slit that can be opened very wide at night (see Figure 1.8). It is thought that small "pinholes" in the slit permit the eye to operate on the principle of a pinhole camera during the day. If the hole of such a camera is small enough, the image is always in focus, the problem being the requirement of bright light.

All of the structures described so far have the function of delivering the proper amount of focused light to the sensitive inner layer, the retina. Curiously, the cells that are activated by light, the rods and cones, are on the side of the retina next to the choroid coat, or away from the direction from which the light comes. The nerve cells, making up the inner part of the retina, connect to the brain by passing through the outer layers of the eye to form the optic nerve. That peculiar arrangement is thought to have derived from the organization of the central nervous system in remote ancestral vertebrates. Both eyes contribute to both of the sides of the brain that interpret the stimuli as images. This is accomplished through the optic chiasma, where some of the nerve fibers from each eye cross to the opposite optic nerve.

The rods and cones contain pigments that dissociate in light and reassemble in the dark. The chemical process stimulates the cell to pass the stimulus on to the bipolar cell to which it connects, and thence on to a ganglion cell, which is the origin of a fiber of the optic nerve. Rods are approximately 100 times as sensitive as cones, but there is a trade-off, as cones have much greater resolving power, because their nerve connections average the stimulus over a smaller area. Hence, they provide much more acute vision in the presence of ample light. The distribution of relative numbers of rods and cones in the retinas of mammals is related to their activity patterns. Most mammals are nocturnal in activity, and many have retinas composed exclusively, or nearly so, of rods. Primates, including humans, are mostly diurnal and have retinas containing both kinds of cells, with cones concentrated near the center of the retina, and rods peripherally. A small area, the fovea, consists exclusively of cones, each connecting to a single nerve fiber, making for maximum resolution of the image. In humans, the fovea is centrally located because our eyes are directed forward, but a central location is not necessary, and some hawks have two foveas, as described later. Primates, like birds, most reptiles, and many fishes, have color vision, which is less common among other mammals. Color vision is a property of the cones, but the cellular basis of color vision is not well understood.

Visual fields of vertebrates

If we consider the positions of the eyes in the heads of different vertebrates, it becomes obvious that their visual views of the world must be quite varied and that many of them differ markedly from humans in that regard. With our eyes oriented in parallel forward, we have almost completely binocular vision; we see little with one eye that we do not see with the other eye. We share that trait with most primates and owls, and to some extent with cats. Early in the evolutionary history of primates, adaptation to arboreal life involved the ability to judge distances accurately, as in leaping from one branch to another. Not only was binocular vision necessary, but stereoscopic vision as well. Stereoscopic vision involves the perception of depth of field, or perspective. It is achieved as the brain interprets the angle between the lines of sight of the two eyes as they converge on the same object. The interpretation involves comparison of the movements of the separate eye muscles. The position of eyes on either side of the muzzle, as in many ground-living mammals, would have interfered with binocular vision, and as the sense of smell is much

Figure 2.2. *Two primates living in different habitats, illustrating muzzle lengths: (a) chacma baboon,* Papio ursinus, *a ground dweller, with an elongated snout; (b) dusky langurs,* Presbytis obscura, *forest dwellers, with flat faces. (Drawn from photographs in Nowak & Paradiso 1983.)*

less useful in arboreal life than on the ground, no great loss would have been involved in a shortening of the muzzle, with a reduction of the olfactory surface. The return from tree-living to the ground involved, in the case of our ancestors, an upright posture, and the nose was still removed from the source of most scents. It is noteworthy that baboons, monkeys that forage on the ground, have reversed the process and evolved enlarged muzzles, as compared with monkeys that live permanently in trees (Figure 2.2). Cats, owls, and, to a large extent, hawks, predators that hunt by sight, have also evolved binocular vision. The extent to which any of them have stereoscopic vision as well is unknown. The eyes of hawks are placed in such a way that they have both binocular vision and peripheral vision, and both are given high resolving power by the hawk's possession of two foveas.

Ground-dwelling mammalian hunters have limited horizons because of their low stature, and the Canidae (dogs and their relatives) have elongate muzzles with large olfactory membrane surfaces, as more information about their surroundings comes to them from the sense of smell than from sight. In contrast to predators, prey animals tend to have their eyes placed laterally and high on their heads, as is the case in rabbits. They have almost no binocular vision, but each eye covers a wide angle and sees above as well. They are especially sensitive to movement, but frequently cannot identify a potential predator if it is still. Perception is a highly subjective matter, and it is not known how the brain of such an animal interprets the messages from eyes that do not receive the same stimuli. It is possible that one or the other can be ignored. In one reptilian example, the interpretation is clearer. True chameleons (not the anoles of the southeastern United States that are frequently called chameleons) have the ability to move their eyes independently, and the strange feeling that it gives us to watch one of them reflects the way in which we perceive what our eyes "see."

Relative visual abilities among vertebrates

With stereoscopic vision in color, humans have the ability to see the world both in greater detail and at greater distances than almost any other vertebrate. The exceptions are the hawks and eagles, which can see small prey animals from high in the air, as they soar or hover. Their ability depends on a large number of small, tightly packed cone cells at the forward-directed fovea. They also have the most flexible necks of all birds, allowing them to scan a large area while retaining binocular vision.

The vertebrates that have high resolving power are all diurnal in activity, because the cones that provide the resolving power are much less sensitive to light than are rods. Thus, nocturnal animals can see in greatly reduced light, but the extra sensitivity involves a loss of visual acuity. Their retinas tend to be composed of rods, making them capable of seeing in very dim light, but also decreasing the resolution that they can achieve, and preventing them from having color vision. Domestic dogs have been bred to reverse some of the trends that have been described. Greyhounds have been bred to hunt by sight; pugs, pekingese, and bulldogs have greatly shortened muzzles and forward vision. Despite these artificially selected modifications, none of them has color vision, and most of them see relatively poorly.

Vertebrates that spend part of their time under water and part in air must have special features to allow them to do both, because the cornea's refractive index, being different from that of air, allows that structure to participate in focusing when the animal is in air. But because the refractive index of the cornea is practically the same as that of water, focusing depends on the lens, and in some cases on additional mechanisms when the animal is aquatic. All of these mechanisms exceed the capacities of humans, who must use face masks to see under water. Incidentally, loons, birds that chase fishes, use essentially the same mechanism as humans with face masks. Like other birds, they have an eyelid that is lacking in most mammals, the nictitating membrane, and in loons and diving ducks there is a transparent window in the membrane that functions as an additional focusing device. A second mechanism is the possession of a more flexible lens, permitting a range of focus that is adequate in both air and water. Such a lens is possessed by cormorants and by pond turtles as well.

Anableps, a surface-feeding fish of tropical America, has two pupils in each eye. The upper part of the lens, the focusing of

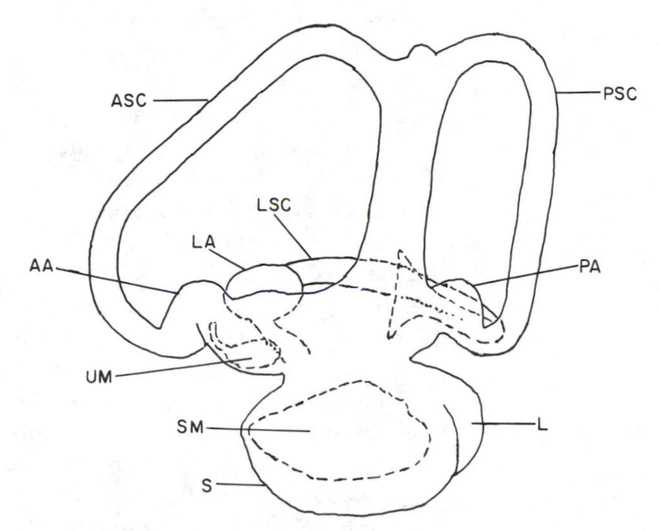

Figure 2.3. *The ear of a salmon,* Salmo salar. *AA, anterior ampulla; ASC, anterior semicircular canal; L, lagena; LA, lateral ampulla; LSC, lateral semicircular canal; PA, posterior ampulla; PSC, posterior semicircular canal; S, sacculus; SM, saccular macula; UM, utricular macula. (Redrawn from Platt & Popper 1981.)*

which is assisted by the cornea in air, is less rounded than the lower half, which must accomplish all focusing under water. The fish swims along exactly at the surface, "looking" up into the air and down into the water with both eyes simultaneously.

Hearing

Sounds are detected by homologous structures in all vertebrates, but unlike the situation for eyes, there is a clear increase in complexity of the ears that accompanies the progression from Agnatha to mammals. As might be expected from their possession of speech, humans are able to distinguish certain details of sound better than any other vertebrate. In other aspects of hearing, humans are surpassed by several other mammals, as well as by birds.

Structure

The organ by which sounds are detected is located in the posterior part of the skull and is part of the structure that detects changes in orientation, the organ of balance. In mammals, the part of this organ that detects sounds is larger than the organ of balance; in other vertebrates, the latter structure is larger. The simplest form of the organ of hearing is found in hagfishes. A somewhat more complex form is found in lampreys. In fishes, it consists of an enclosed two-part bag to the upper of which three semicircular tubes are attached at both ends (Figure 2.3). The whole is filled with fluid, the endolymph. The three semicircular canals constitute the organ of balance. Each lies in one of three vertically oriented spatial planes, and each is expanded into a bulb (ampulla) near one end. The bulb contains a group of special cells that have hairlike extensions covered by a gelatinous dome, the cupula, that extends into the cavity. Any change in orientation will cause the fluid in at least one of the semicircular canals to move, because they cover the three possible dimensions. Movement of the fluid deforms the cupula, stimulating the hair cells, which are innervated by the auditory nerve. Each hair cell has one long kinocilium and a bundle of adjacent stereovilli. The upper sac, to which the semicircular canals attach, is the utricle. The lower sac is the saccule. The utricle contains on its inner surface a sensitive area, or macula; the saccule also contains a macula and another sensitive area, the lagena. Each area is partly covered by an otolith, a flake of calcium carbonate. The otoliths have various shapes, which are species specific. Their function appears to be to respond to vibrations in the endolymph by producing a shearing action over the sensitive areas on the maculae. These sensitive areas have hair cells, and the arrangement of the cilia and stereovilli makes the cell sensitive to motion of the otolith from one particular direction. The vibrations are set up by sound waves in the water, either directly through the body of the fish or by way of the inflated swim bladders in one group of fishes, the Ostariophysi. These fishes have a series of small bones, the Weberian ossicles, that touch the anterior end of the swim bladder at one end, and the saccule at the other, thereby transferring sound waves in the gas-filled swim bladder into vibrations in the saccule.

Sound is the result of vibrations produced in either air or water, and its properties in the two media are quite different. It is easy to produce in air, where the waves travel at about 330 m/s. In water, much more energy is required, because of the density of the medium, but once produced, sound waves travel at 1,500 m/s. It is the waves that set up the motion in the endolymph. Sound has two properties, frequency, which determines the pitch, low or high, and amplitude, which determines the loudness. Frequency is recorded in cycles per second, or hertz (Hz), or thousands of cycles per second, kilohertz (kHz), a cycle being one complete period of the sound wave. All vertebrates are sensitive to both frequency ranges, and the amplitude necessary to produce a stimulus varies with the frequency. Many comparisons among vertebrates depend on that relationship, which is graphed as an audiogram (Figure 2.4).

The analysis of hearing in fishes encounters many practical difficulties because of the properties of sound in water. For example, sound is easily reflected from any surface, including the air–water surface, and the resulting background noise makes interpretation of the results of experimentation somewhat difficult. Nevertheless, it seems clear that fishes are sensitive to sounds as low as 100 Hz, and rarely above 1 kHz (the range for humans goes from 100 Hz up to 20 kHz). There appears to be much variation

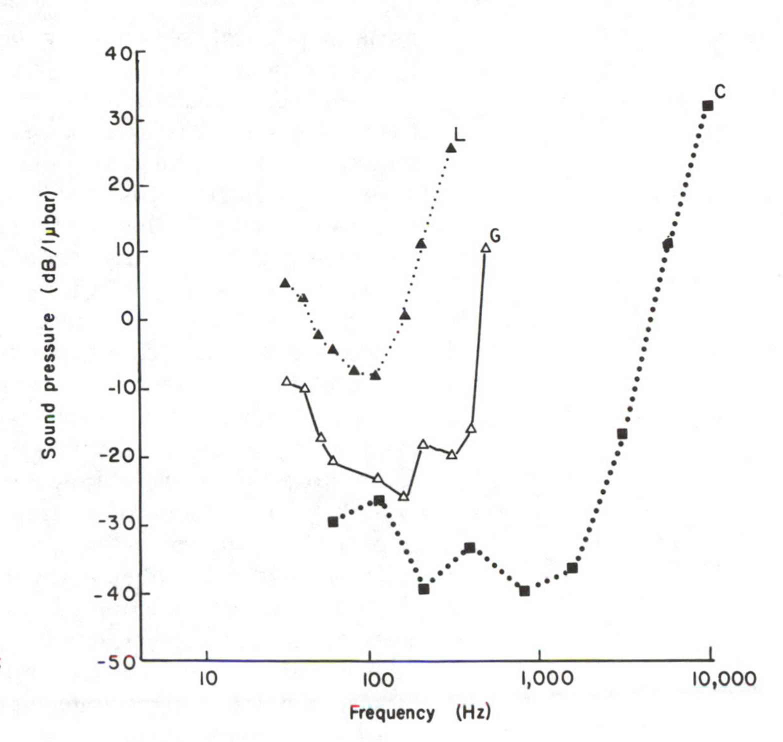

Figure 2.4. *Audiograms for three teleosts. L, dab,* Limanda limanda; *G, cod,* Gadus morhua; *C, catfish,* Amiurus nebulosus. *(Redrawn from Hawkins 1974.)*

between species in the amplitude necessary to produce a response at any frequency, but various experimenters have found major differences within a given species that apparently reflect differences in technique. Many fish species make sounds with specific timing, making it likely that the sounds are used in communication.

Frogs and salamanders have auditory organs that are more complex than those of fishes, but only frogs have both complex calls and a specific mechanism for transmitting airborne sounds to the sensitive part of the ear, and this description will concentrate on them. Airborne sounds have much less force than do sounds traveling in fluids. Therefore, to make the transfer, a large mechanical advantage is necessary. In frogs, this is accomplished by having the sound first cause vibrations in a large circular tympanic membrane (eardrum) and having that vibration transferred by a bone, the columella, to a small opening, the oval window, leading into the fluid-filled inner ear. Unlike fishes, frogs have two distinct layers of fluid, the endolymph, as in fishes, and a surrounding sac filled with perilymph. The endolymph is contained in a complex membrane much like that of fishes, with six comparable sensitive areas: three in the semicircular canals, one in the utricle, and two in the saccule. These last three, however, are covered by gelatinous membranes, rather than otoliths. These six areas are primarily sensitive to changes in orientation or acceleration, as is the case of the semicircular canals of fishes.

In small diverticula from the saccule there are two more sensitive surfaces: the basilar papilla, corresponding to those of amniotes, and the amphibian papilla, which is unique to Amphibia. The perilymphatic sac is closely applied to these two papillae, being separated from them by thin membranes. Each papilla consists of a patch of hair cells overlain by a tectonic membrane. The basilar papilla is simple in form, but the amphibian papilla is S-shaped, and there is a distinctive pattern of hair cells.

Recordings from electrodes placed in the respective branches of the auditory nerve reveal that the basilar papilla is stimulated by sounds of higher frequency than is the amphibian papilla, which responds to two separate ranges of frequencies. In different species of frogs, the respective ranges of frequencies correspond nicely to the particular calls of the frogs. For example, the amphibian papilla of the bullfrog, *Rana catesbiana*, which has a low-pitched call, is stimulated by sound frequencies of 250 Hz and 600 Hz, and the basilar papilla is stimulated by a sound frequency of 1,400 Hz. The papillae of the green tree frog, *Hyla cinerea*, respond to frequencies of 500 Hz, 1,200 Hz, and 3,100–3,800 Hz, respectively, corresponding to a higher-pitched call.

Salamanders and caecilians lack tympanic membranes, and in their cases sound is transmitted to the inner ear by vibrations through the forelegs or the body wall, respectively.

Reptiles and birds have middle and inner ears that are essentially similar, there being more differences among reptiles than there are between some reptiles and birds. The differences between various reptiles can be attributed to both historical and ecological factors. The historical factors come from the early separation of the reptiles into the anapsid turtles, the lepidosaurian lizards and snakes, and the archosaurian crocodilians. Ecological factors have affected hearing among lizards and among snakes.

The simplest reptilian ears are much like those of frogs, with a tympanum, a columella, and a very similar inner ear. The basilar papilla, contained in a small pocket of the saccule in frogs, is a somewhat longer structure contained in an enlarged pocket, the cochlear duct, in most reptiles. In crocodilians, the basilar papilla is part of a cochlear duct that is longer than the diameter of the rest of the inner ear (Figure 2.5). There is a rough correlation between the size (length) of the cochlear duct and the sensitivity of the ear.

For a long time it was part of the folklore of vertebrate zoology that snakes were deaf, an idea that apparently was based on an assertion in the early literature, rather than on any experimental testing. It is true that snakes have a low sensitivity to sounds, and it is also striking that their reactions to sounds depend not on the

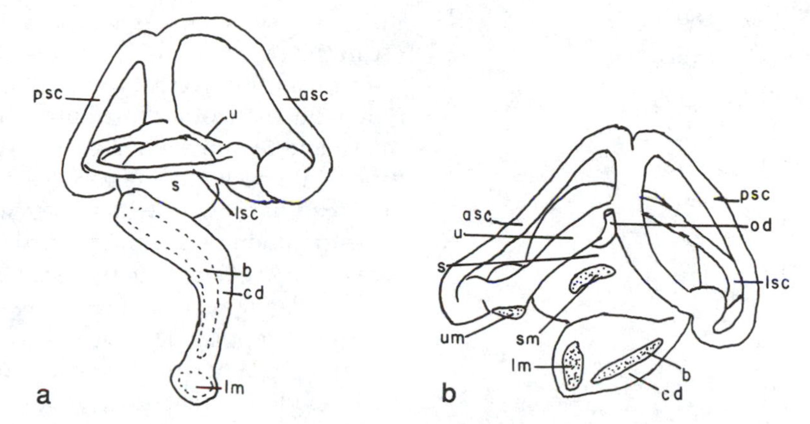

Figure 2.5. *Auditory labyrinth in two reptiles: (a) crocodilid; (b) scincid lizard. (Redrawn from Baird 1974.)*

proximity of the source of the noise to the snake's head but on the sound being transmitted through the elongate lung. In one snake that was tested, the maximum sensitivity was at 200 Hz, and sensitivity was very poor at frequencies above 500 Hz. Thus, the little that the animal may have heard was confined to low tones. It seems likely that vibrations coming through the ground are more important to snakes than are those coming through the air. Fossorial species (those living most of the time below ground) are more sensitive than are terrestrial species, and arboreal species apparently do not hear.

The hearing of turtles has not been investigated extensively, but one study produced an audiogram for *Chrysemys scripta*. It responded best to low sounds, in the range of 80 Hz to 1 kHz, but the sound pressure level required for a response was twice that for an unspecialized lizard, which in turn required a sound pressure level 1.5 times as great as did a gecko.

Lizards are more sensitive to sounds of higher pitch than are other reptiles, and they respond more or less equally to frequencies from around 500 Hz to 10 kHz. Above and below that range, higher sound pressure levels are required to cause a response. Geckos are nocturnal lizards, the name coming from the sound of the call of the species *Gekko gekko*. They are the only lizards that make sounds themselves, and they have been shown to respond to the calls of other animals. One species is attracted by the calls of crickets, and they have been observed remaining around the cavities from which male cricket calls attract females. As already stated, geckos are more sensitive to sounds than are other lizards, their audiograms showing maximum responses at two peaks of frequency: 700 Hz and 2 kHz.

Like geckos, crocodilians communicate by means of sounds, as in the roar of the male alligator and in the "peeping" noises made by hatching young, which bring the female parent to the nest mound, where she digs them out and carries them to water. Their maximum sensitivity is comparable to that of geckos and is greatest in the range from 300 Hz to 3 kHz.

The nearly universal use of vocalizations by birds, in conjunction with their other attributes, has made them the subjects of many studies on their hearing. Anatomically, the ears of birds differ little from those of reptiles, being especially similar to those of crocodilians, with which they share their most recent common ancestor among vertebrates. Birds have the tympanic membrane recessed more deeply in the skull than do reptiles, and its exterior opening is guarded by special feathers. The middle ear is much like that in reptiles. The cochlear duct is much longer than those of any reptiles other than crocodilians and a few geckos. There is good evidence that different sectors along the length of the cochlea are sensitive to different frequencies of sound, as is the case in mammals. With vibrations from the eardrum through the columella to the oval window, waves are set up in the perilymphatic fluid, contained in the scala vestibuli. These are communicated to the endolymph in the cochlear duct to cause appropriate parts of the tectorial membrane to stimulate hair cells below them on the basilar membrane. The pressure in the perilymph is relieved by passing around the end of the cochlea to the scala tympani and so to the round window, leading to the mouth or throat cavity.

Audiograms show that most birds are very sensitive to sounds, but to a quite narrow range of frequencies, from about 700 Hz to 7 kHz, the higher frequency being an abrupt cutoff (Figure 2.6). Owls are sensitive to sound intensities about 30 decibels lower than those detectable by most other birds, and also to a broader range of frequencies. In addition, the external openings of the ears of some owls are asymmetrically placed, apparently as an adaptation to accurate location of the source of noise. In other respects, the hearing capabilities of birds are close to those of mammals. They are slightly less able than humans to detect small differences in intensities of sounds at the frequency at which they are most sensitive. It was once thought that birds had a superior ability to resolve small gaps in timing of sounds. Recent work has shown that they are not different from humans in that regard.

The production of songs is so important in the lives of birds that one sould expect to find examples of matches between specific kinds of auditory ability and corresponding aspects of the songs. An example comes from song learning. Birds learn their songs by hearing adults, and their ability to reproduce a phrase accurately should match the accuracy of their auditory discrimination in frequency and intensity. Data are available for the canary and the parakeet, and the expectation is borne out (Table 2.1).

That does not mean that young birds exposed to various songs necessarily can learn only the song of their own species best. That is shown by a comparison of two closely related species: the

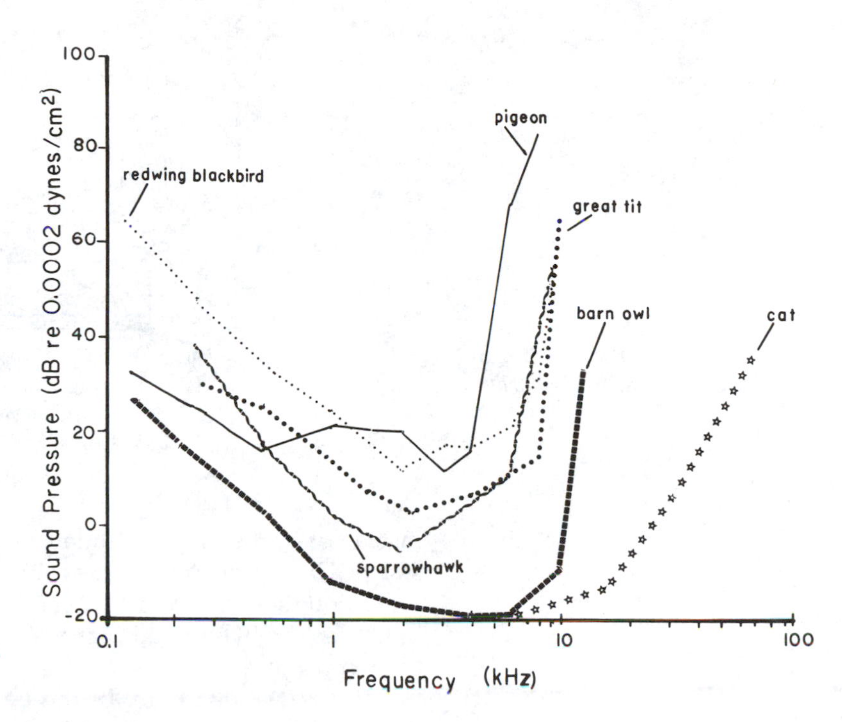

Figure 2.6. *Threshold curves for five bird species, compared with that for the cat. (Adapted from Fig. 13.26 in Manley 1990.)*

Table 2.1. *Comparison between accuracy of note production[a] and the ability to make auditory discriminations[b]*

Acoustic dimension	Species	Coefficient of variation (SE/$\overline{X}$)	Weber fraction
Frequency	Parakeet	0.008–0.020	0.007–0.009
Intensity	Canary	0.20 –0.80	0.35 –0.50
Duration	Parakeet	0.10	0.13 –0.20

[a]Expressed as the coefficient of variation.
[b]Expressed as "Weber fraction."
Source: Data from Dooling (1980), cited in Popper & Fay (1980).

swamp sparrow, *Melospiza georgiana,* and the song sparrow, *M. melodia.* Hatchlings of the two species raised in the laboratory under controlled conditions were exposed to the songs of both. Swamp sparrows learned only the syllables of their own species' song; song sparrows, on the contrary, learned the syllables of both species.

The ears of mammals differ from those of birds and crocodilians in two most important ways. The tympanum is connected to the oval window by a series of three bones, the auditory ossicles,

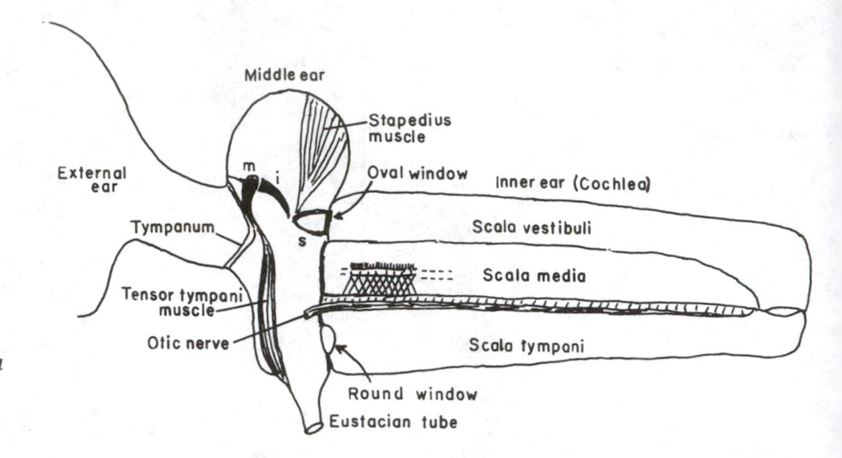

Figure 2.7. Simplified diagram, in two dimensions, of the mammalian ear. (Redrawn from Bench, Pye, & Pye 1975.)

instead of by the single columella, represented in mammals by the stapes. During the course of evolution, the hinging of the jaw has undergone several important changes from the original condition in fishes. The hyomandibular was replaced as the attachment for both upper and lower jaws to the skull by the posterior part of the upper jaw, the quadrate. In the evolution of the Amphibia, the hyomandibular became the columella, and that became known as the stapes in amniotes, because of its resemblance to a stirrup. The evolution from a grasping jaw to a chewing one in the transition from reptile to mammal resulted in new dermal bones (maxilla and premaxilla) replacing the quadrate in the upper jaw, and the mandible replacing what had been Meckel's cartilage in the lower jaw. The proximity of these original bones of the jaws to the opening of the ear allowed them to be incorporated into the hearing mechanism, the quadrate becoming the incus (anvil) and Meckel's cartilage becoming the malleus (hammer). The result is a three-part lever system transmitting the vibrations from the eardrum to the perilymphatic fluid via the oval window (Figure 2.7). The middle ear also contains two muscles, the stapedius and the tensor tympani. The function of these is to reduce the force of the transmitted vibrations when the sound is too intense. They thus constitute a safety mechanism protecting the inner ear.

The fundamental structure of the inner ear in mammals is much like that in birds, as shown diagrammatically in Figure 2.7. Actually, the cochlea is so long in mammals that it is coiled, an arrangement that makes it compact (Figure 2.8). The exception to this coiling is in the monotremes, which have cochleas that are curved, somewhat like those of birds, and probably reflect their reptilian heritage.

Studies using several different techniques have resulted in agreement that there is a progression in sensitivity to different

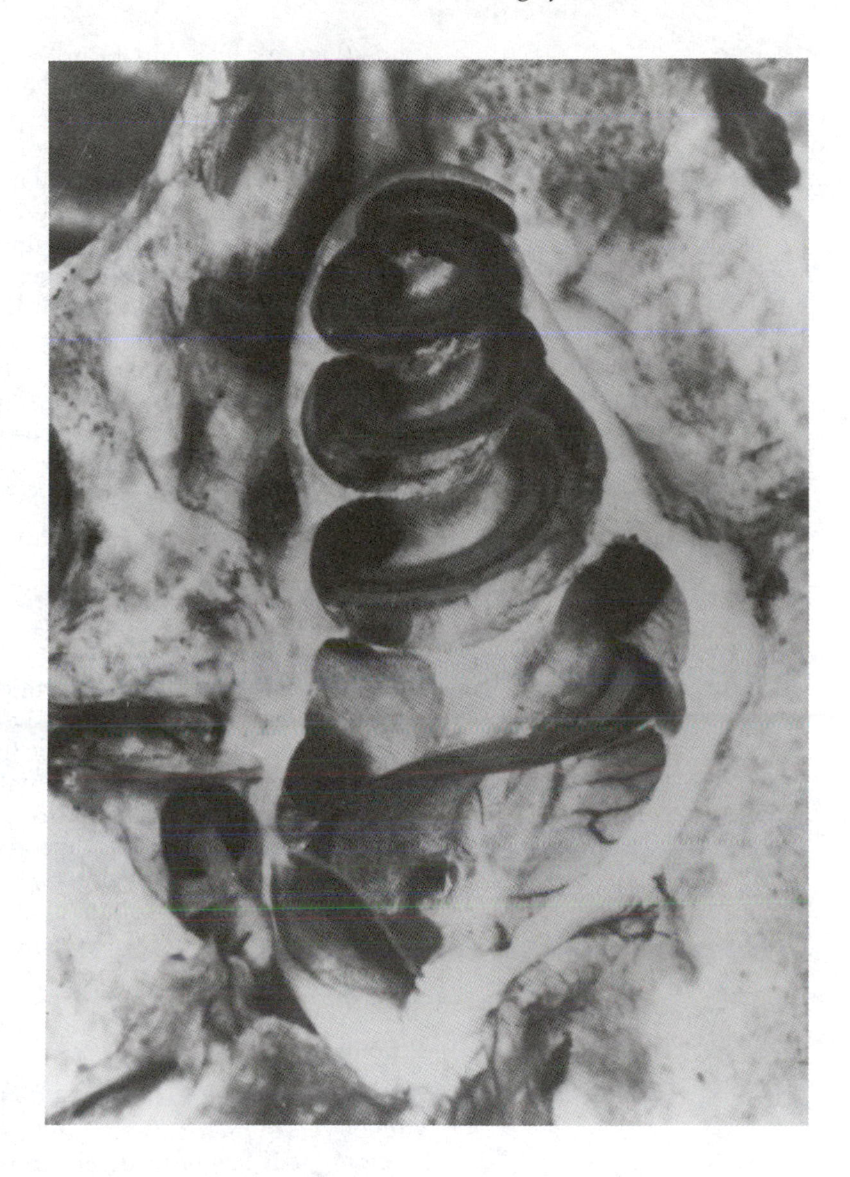

Figure 2.8. *Photograph of the cochlea of a guinea pig. (Redrawn from Bench, Pye, & Pye 1975.)*

frequencies of sound along the length of the cochlea from the oval window to the tip. The nearer to the oval window, the higher the frequency to which that part of the cochlea is most sensitive. In humans this is 20 kHz. Near the tip, the greatest sensitivity is to sound of 0.2 kHz. As people age, their ability to hear the highest tones declines, presumably because of deterioration of the hair cells nearest the oval window.

In general, mammals have excellent hearing, being as sensitive as or more sensitive than other vertebrates, and covering a greater range of frequencies than others, including the equally sensitive birds (Figure 2.9). Apart from their inability to hear frequencies higher than 20 kHz, humans have audiograms similar to those of most other mammals. In one respect, however, their ability far exceeds that of any other species of vertebrate. This is the ability to discriminate between very similar frequencies in the range between 200 Hz and 5 kHz (Figure 2.10). This ability seems clearly related to the use of speech, in which important nuances depend on very minor differences in tone.

The species of mammals with the most sensitive hearing are desert rodents, both in the ear East and in the American Southwest. Kangaroo rats, *Dipodomys* species, have enlarged auditory bullae (the round bony enclosures in which the auditory ossicles of the middle ear are housed). The net result is that whereas the combined levers of the human ossicles magnify vibrations from the eardrum to the inner ear by a factor of 18, the ossicles of *Dipodomys* magnify the vibrations by a factor of 97. This allows them to hear extremely soft sounds at low frequencies, down to 125 Hz, specifically the sound produced by the air flowing over the very soft feathers of approaching owls, or the scraping of the scales of rattlesnakes as they prepare to strike. The effectiveness of the mechanism was shown in experiments in which the bullae were partially filled with plasticine, which apparently reduced the magnification of the vibration of the ossicles. That reduced the animals' ability to hear experimentally produced sounds. Individuals thus impaired were unable to escape either predator, whereas normal individuals easily escaped both. It must be admitted that the experiment was not properly controlled, which would have involved carrying out a sham operation, opening the bulla without inserting plasticine. The operated animals did, however, behave normally in other regards.

Echolocation

The mystifying ability of bats to fly in darkness without colliding with objects in their path cannot be explained by unusually acute eyesight, as it has been demonstrated that their sight is poorer than that in other vertebrates. The fundamental problem was solved in the eighteenth century by Lazaro Spallanzani in Italy and Charles Jurine in Switzerland. Spallanzani conducted experiments by blinding bats and finding that they could fly around a room without difficulty. When he covered their heads, however, they were unable to fly without collisions. Jurine, having been told of Spallanzani's experiments, took them one step further and filled the ears of unblinded bats with wax. Those bats were unable to fly properly, and it was obvious that hearing was involved in their ability, but no one at that time could understand how the bat's hearing worked, when they themselves could hear nothing. It remained for an American, Donald Griffin, in the 1930s and

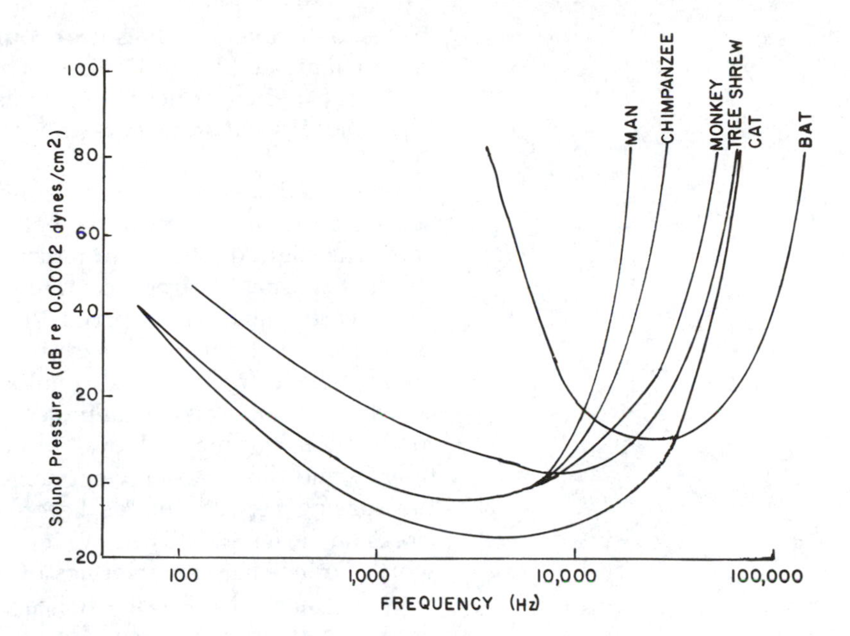

Figure 2.9. Audiograms for six mammals.

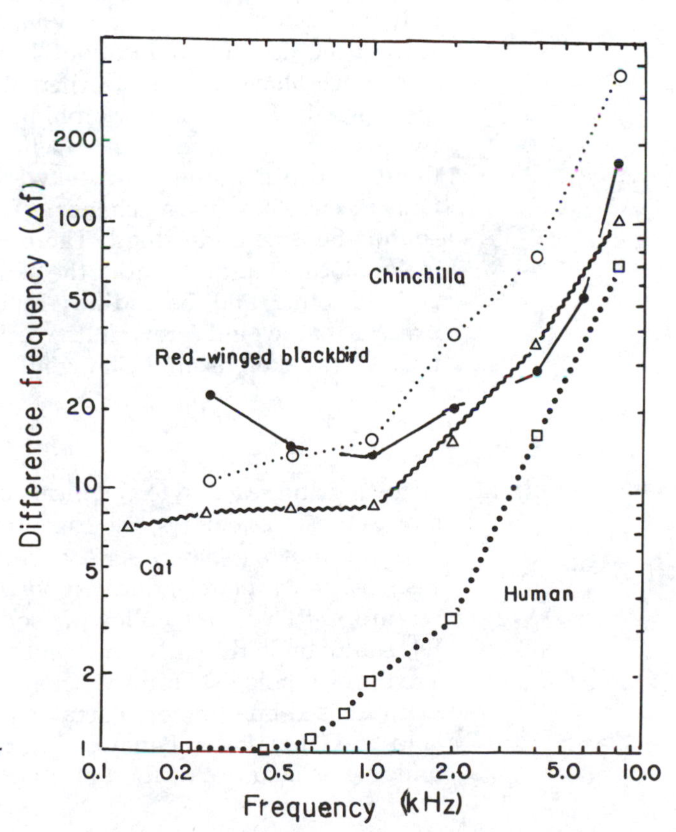

Figure 2.10. Frequency discrimination curve for humans compared with those for two other mammals and a bird. (Adapted from Dooling 1980.)

1940s, to discover that bats hear sounds at much higher frequencies than are audible to humans. That should not have been entirely unexpected. "Silent" dog whistles use the same principle. The small insectivorous bats project high-frequency sounds (30–70 kHz) at very high intensities (if their sounds were at frequencies we could hear, they would sound like piercing shrieks). These sounds strike any object in the bat's path and are reflected to its ears. The slight difference in return times to the two ears allows the bat to locate the object accurately. Griffin showed that bats can fly between piano wires spaced 1 ft (30 cm) apart. They not only can avoid objects but also can capture insects in the dark using this ability, which Griffin named echolocation, the ordinary word for which is "sonar." Bats do not project pure tones, but use frequency modulation, starting each brief burst of ultrasound at a high frequency, and sliding to a lower one. As a bat approaches an insect, the spacing between bursts becomes shorter and shorter as it closes on the insect. This allows it to locate its prey accurately. It is worth noting that some species of moths have produced counteradaptations. They are able to hear the calls of the bat, and sometimes can avoid capture by dropping suddenly. It is effective if done soon enough, but many bats can pursue accurately.

Members of other orders of mammals also use ultrasound, but only the Cetacea (dolphins and whales) approach the accuracy of bats in the use of echolocation. They use "clicks" that are not ultrasonic. Much has been learned about this from studying captive animals. Bottle-nosed dolphins can be trained to choose between two similar sizes of fishes while blindfolded. It appears that there is much communication between individuals by this means. Only recently have the mechanisms for producing and hearing the sounds become understood. There is a peculiar fatty body, the melon, located at the front of the skull that helps in focusing the emitted sound, and the auditory bulla is not fused to the skull, a feature that would prevent the echo from being transmitted through the skull itself, which would disrupt the process of echolocation.

Smell and taste: Chemical senses

Detecting and reacting to chemicals in the environment is important to most vertebrates, the exception being that relatively few birds are known to have a sense of smell, notably the kiwi, *Apteryx australis*, storm petrels, *Oceanodromus*, and especially the turkey vulture, *Cathartes aura*. Following some primitive experiments by J. J. Audubon in the early nineteenth century, more than 50 years ago the ornithologist Frank M. Chapman placed a carcass at night in an unused building on a small islet beside Barro Colorado Island in Gatun Lake, Panama. Turkey vultures recognized the building as a source of food and were seen perched on its roof,

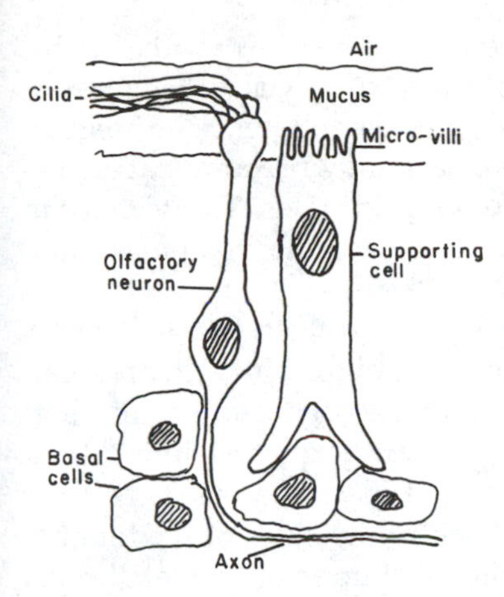

Figure 2.11. *Diagram of the cells of the mammalian olfactory membrane. (Redrawn from Dodd & Squirrell 1980.)*

where they had never been seen before. Extensive experiments by Stager (1964) confirmed the turkey vulture's use of odors in locating food and went far to show that other vultures rely on sight to find carcasses. The vertebrates most sensitive to chemicals are mammals and fishes, although both amphibians and reptiles secrete scents for use in communication.

Among mammals, the senses of smell and taste are in separate organs, but both are quite simple in structure. More information is available about olfaction than about taste. The olfactory organ, located in the nose, of course, consists of a highly folded tissue, the surface of which contains the sensitive cells (neurons), plus supporting cells in approximately equal numbers. Each sensitive cell (Figure 2.11) has a thin axon that joins other axons leading to the olfactory bulb of the brain. The dendrite is a narrow rod leading to the surface of the epithelium covering the organ. Its outer tip is expanded into a small bulb, from which a group of cilia arise. The cilia become narrower along their length, and lie in a covering of mucus. They support detectable tiny structures that are thought to be the actual sites that are stimulated by odors. A very large number of different odors can be distinguished by most mammals, and how that is accomplished is controversial. Dogs distinguish among individual people and can be trained to distinguish by scent the different emotional states, such as friendliness, fear, and anger, in a given person. They are able to do that even in the presence of a variety of other scents, as would be normal in any real situation. Such feats require extreme sensitivity, and it has been shown that humans require densities of volatile molecules in the air 2,500–1,100,000 times as high as dogs for detecting certain chemicals. Thus, we are effectively "blind" to much of the sensory world of our pets. Part of that greater sensitivity comes from the fact that most dogs have sensitive membranes 50–100 times as large in area as those in humans. The rest of the difference must come from extra sites on the cilia.

The uses of olfaction by mammals are many. Trailing prey by their scent is a feat at which canids generally excel. Many mammals use secretions, either from special glands or in urine or feces, to mark their territories. An unfamiliar member of the species is thus notified of the resident's territory, and enters at considerable risk in the case of wolves or male lions, for example. The herbivorous marsupial wombats place their scats (a euphemism for turds) on top of rocks at the borders of their territories. Presumably, that makes them more conspicuous. The musk used in making perfume is the product of a special gland in members of the weasel family. It is used in scent marking and, in the case of other members of the same family, skunks, as a defense mechanism. Many deer and antelope, as well as shrews, cats, and European rabbits, also mark their territories with scents.

Most mammals use scent to recognize the sex of another individual and to recognize the state of the sexual cycle of females. A female dog in "heat" (a period of ovulation and hence impregnation) followed by a pack of males is a familiar example. Female mice are extremely sensitive to the odor of males. The scent of a strange male will block the implantation of embryos in recently inseminated females.

It is commonly stated that our organs of taste ("taste buds") distinguish only four flavors (sweet, sour, bitter, and salt) and that all other flavors are actually sensed by the organs of smell. That opinion is not held universally, some authorities claiming that more complex flavors can be tasted. The connoisseurs of fine wines routinely let the wine flow back to the sides of the tongue for full appreciation. The taste buds are clumps of epithelial cells just beneath the surface of the tongue, each with a flagellum projecting into a pit on the surface. They are also found on other surfaces in the mouth and pharynx.

Among reptiles, snakes and lizards have well-developed Jacobson's organs for the sense of smell. These structures in the roof of the mouth are stimulated when the tongue is extruded into the air and returned to the mouth. Iguanid, agamid, and chameleontid lizards are nearly all diurnal, and most of their communication is by visual signals, but scincid, lacertid, teiid, and other related families depend more on tactile and olfactory stimuli, and gekkonids communicate vocally, as already described. The glands producing the relevant chemicals are found around the cloaca in most such species, but others use glands along the inside of the thighs. The cloaca is rubbed against the ground, and the attracted specimen of the opposite sex licks the area.

The scent glands of snakes that are most conspicuous to us are those associated with the cloaca. These are voided, along with the cloaca, when the snake is disturbed, resulting in an offensive odor along with the objectionable feces. The secretion does not appear to have an appreciable communicative function in most snakes, as far as other members of the species are concerned. Many species of the snake family Colubridae (the most abundant in numbers and species) have scent glands on the dorsal and lateral skin of the female, and the secretions of these glands are attractive to males during the mating season. They will follow a trail made by applying those parts of the skin of a female, but not a similar trail from a male.

Those families of turtles that are terrestrial or semiterrestrial have well-developed mental glands under the chin, and the secretions from these have been shown to elicit aggressive responses in males and a head-bobbing response in females similar to that seen in courtship. Male American alligators and Nile crocodiles secrete an oily musk from paired glands under the jaw, presumably to attract the females. The latter secrete musk from cloacal glands.

Several species of frogs have been shown to have a preference for the odor of water from their own breeding ponds over that from another breeding pond containing other individuals of the same species. That indicates not only great sensitivity but also a high level of discriminatory ability. Neither of these has been measured quantitatively. Otherwise, frogs communicate almost exclusively by their calls, and their communication through the chemical sense is rare or nonexistent.

Salamanders make extensive use of chemicals in communication, specifically in mating. That is primarily true of terrestrial or nearly terrestrial members of the family Plethodontidae, those that do not make a migration to water for breeding. The mating behavior of *Plethodon* was described in Chapter 1. The glands involved are most conspicuously the large mental gland under the chin, which becomes swollen in mating season, plus others on the dorsal surface, especially over the base of the tail. It appears that these glands function through direct contact with the skin, which in these lungless animals is richly provided with blood vessels for respiration. The small *Plethodon cinereus*, which has been shown convincingly to be territorial, marks its territory with feces and secretions from glands around the cloaca. An individual can distinguish its own scent from those of neighboring individuals and can even distinguish those in neighboring territories from total strangers. This is related to "dear-enemy" recognition, as described and discussed in Chapter 16. Plethodontids have nasolabial grooves connecting the nostrils with extensions of the lip. In detecting secretions, they repeatedly tap their noses on the substrate, and chemicals are carried into the nasal sacs. As they are lungless, that is the only function of the nostrils. Among aquatic salamanders, newts provide the best example of the use of chemical secretions in communication. As devotees of P. G. Wodehouse know, males of the European newts wag their tails at females in mating season. This appears to have the function of wafting secretions from cloacal glands in the direction of receptive females. Females of the red-bellied newt of California produce secretions that attract males, as was shown by an experiment in which sponges that had been soaked in water containing skin secretions from females attracted a long line of males from downstream. Sponges soaked in untreated water had no effect.

Tadpoles secrete chemicals that affect conspecifics. Placing an injured tadpole with a group in a pool will cause the others to move away, forming an unoccupied zone around the injured specimen. They are also able to sense food and will quickly form dense clumps where food has been added.

The chemical senses of fishes are located in various places, most prominently in passages in front of the eye. In most fishes, water is carried through this passage by swimming motions, plus, in some forms, muscular pumping. Inside, the water is carried over folds in a groove that originated in outpocketings of the brain of the developing fish. Other chemical detectors are located on fins or on barbels of catfishes and sturgeons. The most spectacular chemical sensing by fishes is that used by salmon. Fingerlings that have been marked in great numbers and released in the streams where they hatched have been recovered hundreds of miles at sea, and then recovered again several years later after migrating back to the same stream where they hatched. This was shown to be due to the chemical sense at a particular stage in early growth. The chemistry of that particular body of water was "imprinted" on the young fish and was not forgotten during its migrations. Fertilized eggs that were moved to an adjacent stream and released after the critical developmental period returned to the adopted stream, rather than to the native one. Returning individuals that had their nasal passages blocked with Vaseline went to streams at random.

Fishes respond to signals from each other, most notably in the case of alarm substances, as has been described for tadpoles. It has been demonstrated experimentally that an injured fish placed in the vicinity of normal specimens stimulates them to flee the area.

Touch and pressure

Our sense of touch is so diffused over our bodies that it is not usually thought of in the same way as other senses. Single cells of several kinds are responsible, but it is not known which kind responds to touch, which to pressure, and which to heat and cold. We do not even know specifically how pain is felt. We do lack the specialized long stiff hairs, known as vibrissae, that are possessed by many species of mammals. The vibrissae are the familiar "whiskers" on our household pets, and they provide information about a possible collision with an object before that object is reached. To understand the kind of information and the reaction, try stroking a cat in the normal fashion, and then touch the vibrissae in the same way. Most mammals have the vibrissae concentrated on the face, but the North American red squirrel *Tamiasciurus hudsonicus* also has them in various places over the body.

Pressure changes are much more important to fishes and aquatic amphibians than they are to terrestrial vertebrates. That is because the density of water transmits such changes much better than does air. Minute differences in current velocity (less than 0.1 mm/s) are detected by the lateral line organ, which is a tube just under the skin that begins on the head and continues in most forms along the side of the body to the tail. Openings in the tube are found at frequent intervals, and beneath these openings are structures very much like the units (neuromasts) that are found in the semicircular canals of the auditory system that has already been described. Neuromasts also occur as individual small organs of the same type, distributed primarily on the head. Each one consists of a series of pairs of hair cells covered by a jellylike cupula. The kinocilium of each hair cell arises on one side of the cell, the other side being occupied by a group of microvilli. The members of each pair of hair cells have the kinocilia placed in opposite positions. A pressure stimulus deforms the cupula in one direction and bends one of the two kinocilia toward the microvilli, and the other away from them. Pressure from the other direction has the opposite effect. The arrangement provides information as to the exact direction from which the pressure wave has come. Aquatic vertebrates can thus detect the presence of another nearby organism in still water and can also detect the presence of inert objects in running water, because they set up irregularities in the flow. Schooling fishes use the neuromasts on their heads to detect the locations of other members of the school and to maintain proper distances from them.

Heat sensing

All vertebrates are sensitive to heat and cold, but the pit vipers (Subfamily Crotalinae) are extremely sensitive to the differences in temperature between small mammals and their surroundings. They are able to detect the presence of a mouse that is more than 1 ft (30 cm) away from them. The sensing organ is a pit, one on each side of the head between the eye and the nostril. A sensitive membrane is located across the pit, and the nerves in the membrane are stimulated by temperature changes smaller than 0.005°C. Such snakes can strike accurately at moving prey even when all other senses are obliterated. The pit actually constitutes a primitive pinhole camera for the long infrared waves. All poisonous snakes of the Western Hemisphere, except coral snakes and some rear-fanged species, belong to that subfamily. Some nonpoisonous snakes also have heat-sensing organs that are placed on the upper or lower lip or both. Many of these are tree-living boas, specially adapted to eating birds.

Sensitivity to weak electric fields

The ability of various fishes to generate electric fields was described in Chapter 1, as was the functioning of that ability in informing them of their surroundings. The ampullae of Lorenzini compose the sensitive organ. These are similar to the neuromasts of the lateral line system, but they lack the cupula, and the two types lack either the microvilli or the kinocilium. They are sunk in the skin. With a few exceptions, all vertebrates below the amniotes have the ability to sense extremely weak electric fields, in the range of 0.2–2 microvolts per centimeter. These are the field strengths generated by the muscles of any living animal (not only those specifically producing stronger fields) and detectable in nearby water; they are also the field strengths generated by ions in ocean currents moving through the magnetic field of the earth, and such fields are detectable in the crust of the earth before earthquakes. The uses of such an ability in locating prey and avoiding predators are obvious. In an ingenious series of experiments it was shown that sharks could locate a hidden live fish by electrosensitivity alone, the stimulus being the electric currents generated by muscles moving the gill covers. The same mass of pieces of dead fish or a live fish shielded electrically elicited no reaction from the shark, but hidden electrodes generating the appropriate voltage were attacked. The potential to use this ability in navigation is discussed in Part V, as is the demonstrated ability of both pigeons and migratory songbirds to respond to magnetic fields.

References and suggested further reading

Baird, I. L. 1974. Anatomical features of the inner ear in submammalian vertebrates. Pp. 159–212 in *Handbook of Sensory Physiology*, vol. 5, W. D. Kneidel & W. D. Neff (eds.). Berlin, Springer-Verlag.

Bench, R. J., A. Pye, & J. D. Pye (eds.). 1975. *Sound Reception in Mammals*. New York, Academic Press.

Chapman, F. M. 1929. *My Tropical Air Castle*. New York, D. Appleton & Co.

Dodd, G. W., & D. J. Squirrell. 1980. Structure and mechanism in the mammalian olfactory system. Pp. 35–56 in *Olfaction in Mammals*, D. M. Stoddart (ed.). New York, Academic Press.

Dooling, R. J. 1980. Behavior and psychophysics of hearing in birds. Pp. 261–88 in *Comparative Studies of Hearing in Vertebrates*, A. N. Popper & R. R. Fay (eds.). Berlin, Springer-Verlag.

Fessard, A. (ed.). 1974. *Handbook of Sensory Physiology*. Berlin, Springer-Verlag.

Griffin, D. 1958. *Listening in the Dark*. New Haven, Conn., Yale University Press.

Hawkins, A. D. 1981. The hearing abilities of fish. Pp. 109–33 in *Hearing and Sound Communication in Fishes*, W. N. Tavolga, A. N. Popper, & R. R. Fay (eds.). Berlin, Springer-Verlag.

Kneidel, W. D., & W. D. Nuffield (eds.). *Handbook of Sensory Physiology. Vol. 5: Auditory Systems*. Berlin, Springer-Verlag.

Manley, G. H. 1990. *Peripheral Hearing Mechanisms in Reptiles and Birds*. Berlin, Springer-Verlag.

Müller-Schwartze, D., & M. M. Mozell (eds.). 1976. *Chemical Signals in Vertebrates*. New York, Plenum Press.

Nowak, R. M., & J. L. Paradiso. 1983. *Walker's Mammals of the World*, 2 vols. (4th ed.). Baltimore, Johns Hopkins University Press.

Platt, C., & A. N. Popper. 1981. Fine structure and function of the ear. Pp. 3–36 in *Hearing and Sound Communication in Fishes*, W. N. Tavolga, A. N. Popper, & R. R. Fay (eds.). Berlin, Springer-Verlag.

Popper, A. N., & R. R. Fay (eds.). 1980. *Comparative Studies of Hearing in Vertebrates*. Berlin, Springer-Verlag.

Romer, A. S. 1949. *The Vertebrate Body*. Philadelphia, Saunders.
Stager, K. E. 1964. *The Role of Olfaction in Food Location by the Turkey Vulture* (Cathartes aura). Contributions in Science, Los Angeles County Museum, no. 81.
Stoddart, D. M. (ed.). 1980. *Olfaction in Mammals*. New York, Academic Press.
Tavolga, W. N., A. N. Popper, & R. R. Fay (eds.). *Hearing and Sound Communication in Fishes*. Berlin, Springer-Verlag.
Walls, G. L. 1942. *The Vertebrate Eye and Its Adaptive Radiation*. Cranbrook Institute of Science, Bulletin 19, Bloomfield Hills, Mich.

PART

II

Classification

Placing objects into categories is essential to communication. It must be realized that the members of each category share certain characteristics that exclude all members of other categories. Starting with the usual opening of the game Twenty Questions ("Animal, vegetable, or mineral?"), we can narrow down the sizes of the categories and arrive at the proper category for any object, at least in principle. It is immediately obvious that the choice of characteristics is all-important. Thus, dividing vertebrates into two categories, those with scales and those without them, would produce a nonsense category including most (but not all) fishes, a few amphibians, all reptiles, and all birds, not to mention armadillos and pangolins, which have structures that resemble scales. The remaining category would include the rest of the fishes, all salamanders and frogs, and the rest of the mammals. Early Greek naturalists made a similar mistake by including whales and dolphins in the category of fishes because they are permanently aquatic and appeared to have fins.

The current practice of classification was worked out in the seventeenth and eighteenth centuries, culminating in the formal system devised by Carolus Linnaeus. He realized that animals and plants fell into hierarchically arranged groups, each group less inclusive than the one above it in the hierarchy. Putting organisms into the correct category requires knowledge of the structural details of each, and in recent years biochemical comparisons have become increasingly important. When the origin of structures or other traits is the same in two animals, they are said to be *homologous*. A common origin of structures in two or more animals frequently is determined by a study of developing embryos, as in the discovery that the bones in the ears of mammals and parts of the jaws of fishes have common origins, but comparisons with fossils have also been used. Examples of quite different but still homologous structures are the forelegs of lizards, the wings of birds, and the flippers of whales. In contrast, the wings of birds and those of insects have different developmental origins and are not homologous, although similar in function. Biologists since the time of Darwin have tried to represent true genealogical relationships in their classifications, rather than using classifications that may simply be convenient, but as discussed later, the task is not simple.

The smallest category in classification is a group of individuals that are at least capable of interbreeding and that therefore have no nonhomologous traits, except for those associated with distinguishing the sexes. Such a group is a species. Similar species share most, but not all, characteristics and are placed in the same genus, as described in the following chapter.

The naming of species and genera of animals follows rules laid down by the International Commission on Zoological Nomenclature, which updates rules at intervals of years, most recently in 1985. Without such rules there would be no uniform system of names, and communication among zoologists would be much more difficult because of uncertainty about the species being discussed.

3

Classification
Principles and problems

Introduction

There is intense controversy over the proper way to classify animals and plants. The controversy arises from the conflict between two desirable objectives of a classification: to show ancestor–descendant relationships and to have categories that can be used as an information-retrieval system, one in which a student can determine the proper category by examining the specimen.

We are not so fortunate as to have identity between ancestry and similarity of characteristics. The evolutionary pressures of aquatic life have given whales, sea turtles, and penguins quite similar flippers, although the ancestors of those groups had easily distinguishable limbs. These independently developed similarities are called *convergences.* Although it would be fatuous to claim that the similarities in flippers in this case have ever caused confusion, there have been cases in which convergences have led to mistakes in classification, and it should be noted that convergences are much more probable in biochemical traits than in morphological ones. Identical mutations of a gene in members of two distinct lines would constitute such a problem, and the only solution is to use enough different traits in our classification for such a convergence to be seen as an anomaly.

Since Darwin, it has been recognized that any category used in classification should include species that share a common ancestry, and exclude all species not sharing that ancestry. A more inclusive category could include the first one, plus categories containing species sharing a more remote ancestry with the first. In principle, the entire animal kingdom could have been treated in that way, but the principle became compromised by practical considerations. It was argued that sharing many common traits is more important in assigning organisms to a category than is strict estimation of ancestry. The classic example of the problem concerns the reptiles. Turtles, crocodilians, lizards, snakes, and sphenodonts have a great many characteristics in common, specifically the possession of scales and details of anatomy in the skeleton and circulatory system, as well as details of physiology. As the skeletal anatomy became applied to fossils and more became known of the fossil record, it became clear that the common ancestry of turtles and the rest of the reptiles was much more remote than was the common ancestry of crocodiles and birds. Moreover, the anatomy of fossils that were clearly ancestral to mammals agrees with that of reptiles. The inconsistency was argued away by claiming that reptiles represent a "structural grade" that is distinct from that of birds and mammals. That was a concession to convenience, as rearranging such major groups of vertebrates (the classes) would have caused considerable difficulty.

This chapter begins with a description of the classical arrangement of the categories that have been used in classification, followed by a discussion of the difficulties in applying the system. Most of the chapter is given to a discussion of the species in theory and practice.

Categories

Every high-school biology student learns that the kinds of animals and plants are called *species*, and that similar species are grouped together in a *genus*. (It should be noted that the word "kind" is used throughout this book in its original meaning: The antiscience creationists have attempted to coopt the word for a broader use, which is rejected.)

Each genus and its species are given scientific names, which frequently are based on Latin or Greek, partly because classical languages were much used by scholars when the system of naming them was devised by the Swedish botanist Linnaeus in the middle of the eighteenth century. There is no requirement that the genus and species have classical names, although when named for a person, the ending is customarily latinized; see, for example, the names in Figures 3.3 and 3.4. One reason for continuing the practice is the benefit that no taxonomists (people who classify animals and plants) will have the advantage of using their own language. By convention among biologists, both names are always italicized. These conventions are collectively known as the system of *binomial nomenclature*.

As similar species are grouped into a genus, similar genera are grouped into a *family;* similar families are grouped into an *order;* similar orders are grouped into a *class;* and similar classes are grouped into a *phylum*. The plurals of some of these words are unfamiliar, even to people who have a smattering of Latin. "Species" is both singular and plural; the plural of "genus" is "genera" (a fourth-declension word), and that of "phylum" is "phyla."

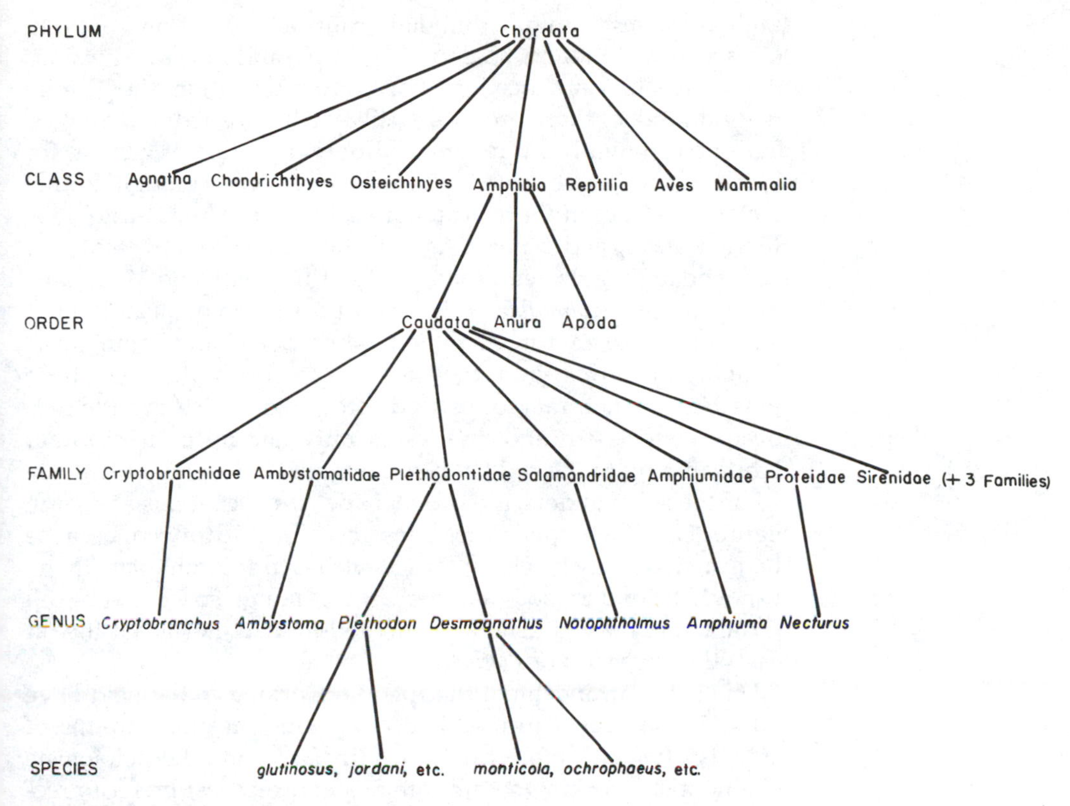

Figure 3.1. *A classification of some species of salamanders.*

Use of the system in practice

All biologists traditionally have used this system of classification because it is necessary to be sure that they are referring to the same kind in each instance, and because it is convenient to have increasingly inclusive groups of species. Figure 3.1 is a representation of the classification of some of the salamanders that will be discussed in later chapters on ecology. What this system looks like is a set of levels, and at each level each unit is equivalent to every other unit. Thus, each genus in a family will be equivalent to the others in its inclusiveness, as will each species within a genus. We can see immediately that something is missing: the vertebrates. The reason is that there are other groups of animals that are not vertebrates, but are more like vertebrates than they are like the members of other phyla. They and the vertebrates cannot be called classes, because that would force the inclusion of very dissimilar species in the same genus (e.g., *jordani* and *monticola*).

It is clear that nature is not organized exactly as taxonomists might like it to be. As far as the Chordata are concerned, the immediate problem can be solved by a subdivision between the phylum and the classes. Thus, the vertebrates are a *subphylum,* and the familiar classes belong within that group. The same kind of problem arises within most taxonomic groups, and new subdivisions of subdivisions are needed. For example, within the Family Plethodontidae, there are two easily distinguished subfamilies, and one of them has three quite different groups of genera, for which a new name, *tribe,* is used. Moreover, although formal names are not used for the categories, the four classes Amphibia, Reptilia, Aves, and Mammalia are known as the *tetrapods* (four feet), and among the tetrapods, the Reptilia, Aves, and Mammalia are known as the *amniotes,* from their possession of an amnion as embryos, as covered in Chapter 1. "Tetrapods" and "amniotes," though not fitting into the set of categories that has been described, are natural groups, and some taxonomists are attempting to have classification represent only natural groups, rather than adhering to the rigid set of categories in common use.

For the salamanders, a more realistic classification is shown in Figure 3.2. The formal categories have been dropped because there are too many levels for the categories to accommodate them. It should be understood that there is not enough space to show all of the twenty-seven genera of Plethodontidae, nor all of the (at least) thirty species of *Plethodon*.

This is the arrangement that people working in the field have found it convenient to use. Those working on other groups of animals often use more categories; in fact, some herpetologists (taxonomists working on amphibians and reptiles) prefer to recognize the Sirenidae as the Order Meantes, thus separating a group of gilled eellike animals that others include with the Caudata. This example raises the question of how anyone knows whether or not to divide one of the categories into two or more. That is not an easy decision to make, nor is any newly proposed arrangement uniformly acceptable. Clearly, such decisions are at least partly arbitrary, and they tend to be based on a choice between convenience and the intellectual satisfaction of having a more accurate representation of the ancestry of the different members of the group. The latter choice would make the science of classification like other parts of science, in that new information would require revisions of existing ideas, thus making all ideas subject to prediction, with the possibility of confirmation or rejection through the application of new discoveries. Such a seemingly desirable course would have the disadvantage of making classification less useful as a reference system. It would require learning the names of many more categories, rather than using a recognized arrangement that can be applied universally.

Arbitrary definition of all categories above species

The degree to which the categories in Figure 3.1 are arbitrary can be seen in the example of two genera: *Plethodon* and *Desmognathus*. As noted, there are at least thirty species of *Plethodon*. All are completely terrestrial, their eggs being laid in damp places on

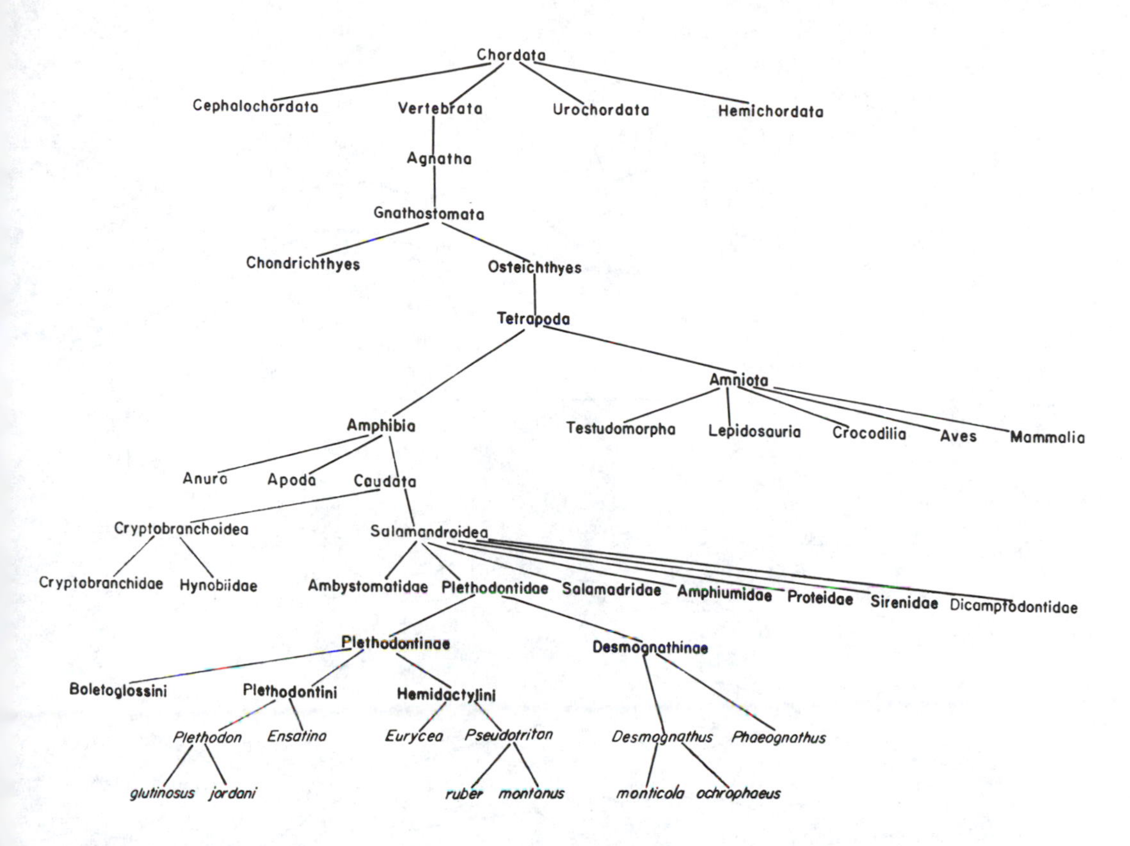

Figure 3.2. *A complete classification of some species of salamanders.*

land, without a larval stage. The largest species are only about 2.65 times as long as the smallest, and although there is a large array of color patterns, the only structural difference among them is that some of the smaller species are more slender, having more vertebrae than the larger species (Figure 3.3). In contrast, only twelve species of *Desmognathus* are recognized; most of them are semiaquatic as adults, spending part of their time in streams and part on land. Ten species are amphibious, with larval stages lasting from a few weeks to 3 years or more. The other two species, like *Plethodon*, are terrestrial, without larval stages. These habitat differences go with differences in both size and shape. The largest species is 4.12 times as long as the smallest; it is heavy-bodied, with a tail that is adapted for swimming, being flattened from side to side, and having a finlike keel that enhances the flat shape. Most of the more terrestrial forms are slender, with round tails, like *Plethodon*, but one of the smallest, *Desmognathus wrighti*, is stocky (Figure 3.4). Thus, with only a third as many species, *Desmognathus* encompasses much greater variety in size, shape, and habitat than does *Plethodon*. It is true that all species of *Desmognathus* have the same set of structural mechanisms for opening the mouth, a set that is unique to the subfamily, and no doubt that is the reason for keeping them in the same genus. Nevertheless, were it not for the species of intermediate habitat, shape,

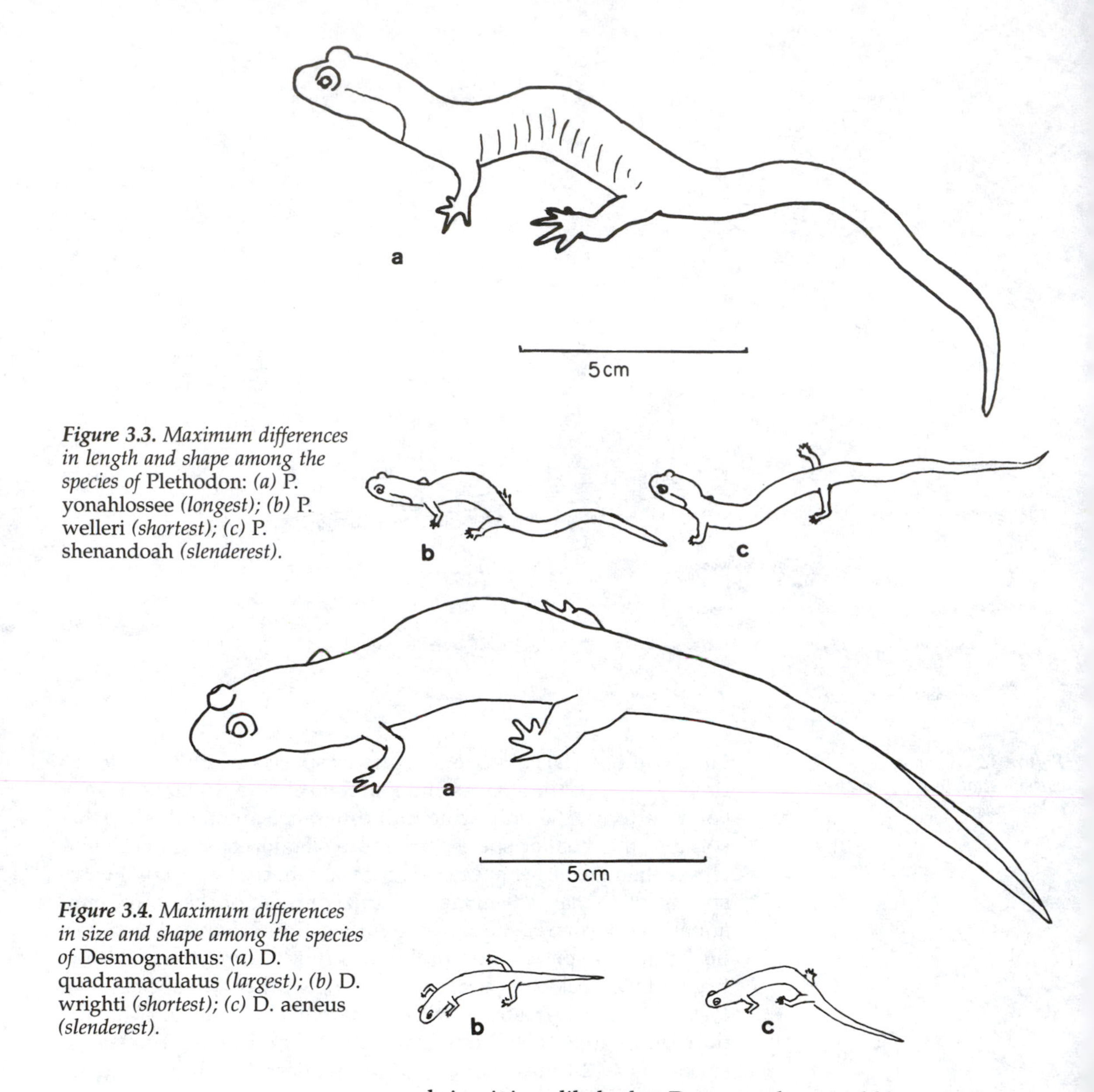

Figure 3.3. *Maximum differences in length and shape among the species of* Plethodon: *(a)* P. yonahlossee *(longest); (b)* P. welleri *(shortest); (c)* P. shenandoah *(slenderest).*

Figure 3.4. *Maximum differences in size and shape among the species of* Desmognathus: *(a)* D. quadramaculatus *(largest); (b)* D. wrighti *(shortest); (c)* D. aeneus *(slenderest).*

and size, it is unlikely that *Desmognathus* would be a single genus. To make the point even more emphatically, there is another species with the same mouth-opening mechanism, but with a somewhat different arrangement of nostrils, that is put in the genus *Leurognathus*. It is very similar in appearance to the largest *Desmognathus*. These examples show that the inclusiveness of genera and the boundaries between them are to a large extent arbitrary. It is for the reasons discussed that zoologists working with different groups of vertebrates find different kinds of arrangements convenient; for example, ornithologists (students of birds) are generally agreed that genera smaller than those used by herpetologists are convenient for their purposes. For example, in the Carolinas and Virginia, there are 84 species of Amphibia. They are distributed among 24 genera, for an average of 3.5 species per genus. In the same area, there are 189 species of nesting birds, representing 137 genera, or 1.38 species per genus – less than half as many as for the Amphibia. Also, 105 bird species are the only representatives of their genera, or 77% of the genera are represented in the area by a single species. In contrast, only 12 of the amphibian genera (50%) are represented by single species. There is, of course, no means of establishing which practice is preferable. An ornithologist could make a logical case for dividing *Plethodon* into at least three recognizable genera, and *Desmognathus* into three or four genera; a herpetologist could make an equally logical case for recognizing no more than four genera of wood warblers (Parulidae), rather than the thirteen currently listed from the area under discussion. Categories more inclusive than genera are subject to the same treatment. Only recently, the Dicamptodontidae have been made a family, separate from the Ambystomatidae. *It is important to note, however, that there is little or no disagreement about the number of species of either birds or amphibians present.* The reasons for the consistent treatment of species, in contrast to the inconsistent treatment of genera, are considered next.

An objective category: The species

Throughout the history of classification, the species has been regarded as an objectively definable category. The different kinds of animals and plants, many of which already had common names, were considered first as immutable acts of creation. That was an observably defendable concept so long as knowledge of nature was confined to a restricted area. Species seemed to be distinguishable on the basis of distinct gaps in appearance. As exploration of lands distant from western Europe brought in large numbers of unfamiliar specimens, the differences between some of the known species were found to be partly bridged by some of these new specimens, and other species were found to be more variable than had been thought. Nevertheless, these problems were regarded as minor nuisances. The accepted interpretation was still that species were separated by unbridgeable gaps. In support of that interpretation were the absence, in most cases, of organisms intermediate in appearance between recognized species and the sterility of hybrids when representatives of two species did mate. The classic example of a sterile hybrid is the mule, the result of mating between a male donkey (a jackass) and a female horse (a mare).

Darwin's work changed the way in which biologists thought about species. By amassing an overwhelming amount of evidence for the evolution of present-day species from forms that had been different in the remote past, and by proposing a mechanism (natural selection) whereby such changes came about, he gave the observable distinctions evolutionary significance and provided explanations for the nuisances that had interfered with the idea that species were "immutable acts of creation." Individual species that had been found to vary in appearance, and those that were not completely distinct in appearance, often were examples of geographic variation within species – specimens from distant parts of the distribution being different in appearance. Such examples came to be regarded as a stage in the evolution of two or more new species from a single ancestral one. The process is called *speciation*.

In this modern view, a species is a group of organisms that at least potentially can interbreed with each other and that collectively have an evolutionary future that is independent of the evolution of all other species. It is said to be reproductively isolated from other species. A spontaneous favorable heritable change (mutation) in one individual has the possibility of being transmitted to the ultimate descendants of any other individual within its species, but cannot be transmitted to any other species. Thus, our definition is radically different from the original, but it remains an objective one – very different from our definition of the genus or any other more inclusive category.

The two principal components of evolution are change through time and the division of lineages into independently evolving units, the species. Were it not for these divisions, the diversity of life that we observe would have been impossible. That is why the species and the origin of new species are so important.

Problems in applying the species concept

Despite the general agreement among zoologists that the species represents a conceptual entity that is easily accommodated in our current ideas about evolutionary theory, application of the concept to organisms in the field has met with five different kinds of problems. Three are practical, and two are conceptual. It should be emphasized, however, that these problem cases are exceptional. The vast majority of specimens, especially of vertebrates, are easily identified by experts, and most of them also by amateurs. Nevertheless, it is important to describe and explain the exceptional cases.

Practical problems

The first practical problem in identifying individual specimens comes from inadequate information. When enough facts are not available, classification is difficult. This difficulty can take at least three forms: unrecognized geographic variation, distinct species that look very much alike, and color phases within a species.

Unrecognized geographic variation

The history of the classification of the salamanders now known as *Plethodon jordani* provides a good example of the problems posed by geographic variation, a common phenomenon in which the different parts of a species' population have become physically isolated from one another on islands or mountaintops. The southern Appalachians were rather isolated from the rest of the country through the 1920s, roads being few and of poor quality. In 1901, one L. E. Daniels traveled across the Great Smoky Mountains between North Carolina and Tennessee, collecting snails. At the top, he also collected salamanders, and two of them belonged to a previously undiscovered species of *Plethodon*. They had red cheeks, but otherwise were uniformly dark gray or black. He took them back to Indiana, where a specialist described them and gave them the name *Plethodon jordani*. In 1906, Professor Frederick Sherman of Clemson College visited the Nantahala Mountains of North Carolina and collected salamanders that had red legs. He sent them to the U.S. National Museum, where they received the name *Plethodon shermani* in honor of their discoverer (Daniels had a salamander from a different genus named for him). Shortly thereafter, C. S. Brimley, director of the North Carolina State Museum, visited the mountains and collected salamanders that were similar, but with no red color, from the Balsam Mountains and from Grandfather Mountain. He described them as new and named them *Plethodon metcalfi* for a friend who was a prominent entomologist at North Carolina State College. Sherman continued his fieldwork, and in 1926 he found yet a different *Plethodon* from the foot of the Blue Ridge Mountains in South Carolina. These looked like *P. metcalfi*, but had conspicuous brassy-colored blotches on their backs. He sent them to Brimley, who named them *Plethodon clemsonae*. Thus, there were four different species, and anyone going to the places where they had been found originally could confirm their existence (they are not rare). In taxonomy, the describer of a new species is obligated to designate one specimen as the *type*, so that in the event that more than one species is mistakenly included in the collection, the correct name can be found by examining the type. The exact locality from which it came is the *type locality*, and collecting at such places is common practice, but can be misleading, as it was in this case.

During the late 1930s, the Civilian Conservation Corps constructed and improved large numbers of roads in the southern Appalachians, and by the end of World War II all parts of the mountains were accessible. By using these roads, I was able to go to previously inaccessible places, and in addition to finding some more different salamanders, I found that wherever two of the "species" came together, I found specimens intermediate between them. The existence of such specimens is taken as evidence that

Figure 3.5. Distribution of the described races of Plethodon jordani *in the region west of the French Broad River in western North Carolina, with adjacent parts of Tennessee, Georgia, and South Carolina:* 1, jordani *(red cheeks);* 2, shermani *(red legs);* 3, metcalfi *(gray dorsally);* 4, clemsonae *(brassy dorsal splotches);* 5, rabunensis *(white lateral spots);* 6, melaventris *(black throughout);* 7, teyahalee *(greenish yellow lateral spots). Areas in white are above 1,000 m.*

the two species are not reproductively isolated from each other, and hence cannot retain different specific names. By convention, the oldest name, *Plethodon jordani,* is now applied to all, regardless of any differences in appearence. The distribution of all of these forms is shown in Figure 3.5.

Species so similar as to be difficult to distinguish

The opposite problem from that posed by geographic variation within a species is encountered in examples of species that are so similar in appearance that identifying specimens is so difficult as to be almost impossible. A classic example among birds is the flycatchers of the genus *Empidonax.* Five species nest in eastern North America, with overlapping ranges. If specimens of all five species are examined, they do show some differences, mostly in shades of gray, greenish, and yellow, but none of the colors can be considered clear, and the colors of the different species grade into each other. The grayest of the group, the least flycatcher, is recognizably different from the yellowest, the yellow-bellied flycatcher, but the remaining species, the Acadian, willow, and alder flycatchers, provide a continuous gradation between the other two. Thus, if examination of specimens were the only method available to us, we might well conclude that they represented minor variations within a single widespread species. In fact, during migration, that is the only method available, and field identification is impossible for most individuals seen, which perforce are simply called *Empidonax*.

In nesting areas, matters are quite different. Most of the birds are easily identified by their calls. Although most passerines ("songbirds") use both visual cues and voices to identify conspecifics, these flycatchers apparently use only auditory cues, as follows:

Acadian flycatcher, *E. virescens:* pit! see
Least flycatcher, *E. minimus:* che bec!
Alder flycatcher, *E. alnorum:* fee bee! o
Willow flycatcher, *E. trailli:* fitz! bew
Yellow-bellied flycatcher, *E. flaviventris:* per wee!

This and similar examples emphasize our human dependence on vision, especially in classification, and our difficulty has little to do with the biology of the birds themselves. Despite their being called "sibling species" in works on evolution, there is no evidence that they are any more closely related than are the species of the wood warbler genus *Dendroica*, which are easily distinguished by their color patterns.

Color phases within species

A third problem arising from ignorance concerns color phases within some species. Snow geese, screech owls, and several species of *Plethodon* all come in two different colors. The genetic makeup of snow geese is better known than are those of the others.

For many years, two species of geese in the genus *Anser* were recognized, and listed in official checklists and guides to the birds of eastern North America. They were *Anser caerulescens*, the blue goose, which had been described and named by Linnaeus, and *Anser hyperborea*, the snow goose, described later by Pallas. Although the same size and shape, the two look quite different. The adult snow goose is pure white, with black wing tips; the adult blue goose has a dark gray body, with a white head and neck, and bluish gray wings with dark tips; some of them have white bellies. Lack of knowledge of either the nesting area or the winter range of the blue goose caused it to be considered a rare bird, whereas the snow goose was much better known. Actually, both are fairly common when looked for in the correct areas. The nesting area of the blue goose was discovered to be on Southampton Island, north of the Canadian mainland, and study of their nesting habits revealed that some of them mate with snow geese. Enough nests of such pairs were found to establish the fact that only two kinds of downy young progeny are possible: "dark" and "light." In adult plumage, reached at the end of their first winter, some of the "dark" birds have quite a lot of white on their breasts, bellies, and lower backs. Thus, three kinds should be recognized: dark-bellied blue, white-bellied blue, and white (Figure 3.6). The colors are controlled by a single pair of genes; the dark-bellied blues and the whites are homozygotes, and the white-bellied blues are heterozygotes. It had to be admitted that only one species is involved. As Linnaeus had founded the system of nomenclature, his name, *A. caerulescens*, had priority and remains the name of that one species.

The discovery of the genetics permits us the rare opportunity to determine the genotype of each individual in the field, and the number of each of the three genotypes can be counted easily. The point of doing so is to discover whether or not mating among them is random. It is not; there is always a marked deficiency of heterozygotes. The reason is that baby geese become attached at hatching to the first large object that they see; in nearly all cases that is the mother. Thus, they are most likely to pair with a similar goose. The phenomenon is known as imprinting.

As an interesting sidelight, many ornithologists actually prefer to use English names for birds, at least in North America, and the American Ornithologists Union publishes up-to-date checklists with both scientific and common names. Birds are so well known that their common names actually remain constant longer, on the average, than scientific names, which have to be changed when new observations show that they are no longer correct. No such rule applies to common names, and students of birds become attached to them. That happened to the name "snow goose" during the years when the blue goose appeared to be so rare that few people ever saw one. Then, when it became known that there is only one species, they kept the name snow goose, and they use that name in bird guides, even though the correct Latin name means "blue goose."

Speciation that has reversed itself

As we have seen, speciation (the process whereby one species splits into two or more) frequently involves geographic isolation of parts of the original species, followed by inability to interbreed should the isolates again come into contact. In the case of *Plethodon jordani* and *P. glutinosus*, that appears to have happened

Figure 3.6. *The three forms of the snow goose,* Anser caerulescens: *(a) "blue" homozygote; (b) heterozygote; (c) "snow" homozygote.*

over most of the southern Appalachians. In the Nantahala Mountains, however, they are not only able to mate; they behave like members of a single species. *P. glutinosus* occupies the valleys and lower slopes, and *P. jordani* the higher slopes to the tops of the mountains. On the intermediate slopes, a great variety of combinations of the characteristic colors is found, most having intermediate amounts of red on the legs and white on the sides, but some having much of both colors, and some having very little of either. A few individuals lack both colors entirely. Their existence and the existence of those with large amounts of both colors clearly represent recombinations of the genes after the mating of the original forms and demonstrate that the hybrids are fertile and produce fertile offspring.

This is obviously a case of complete hybridization, and an arbitrary decision has to be made about the names. One taxonomist did put the red-legged *jordani* into *glutinosus* (the older name), but most workers prefer to keep them separate, in recognition of the fact that over most of their joint distributions, hybrids are never found.

The open ring: One species that can seem to be two

In some cases, individual species are confined to habitats that are themselves distributed around areas unsuitable for inhabitation by the species. In such instances, these species often vary geographically in the manner that has been described, with different intergrading forms replacing each other around the inhospitable habitat. If they have spread from an origin in one part of the range, the geographic variants may be different when they reach the other end of the distribution. In a few cases, the now-overlapping forms do not intergrade, but behave as distinct species, with few or no hybrids. The classic example occurs among the salamanders

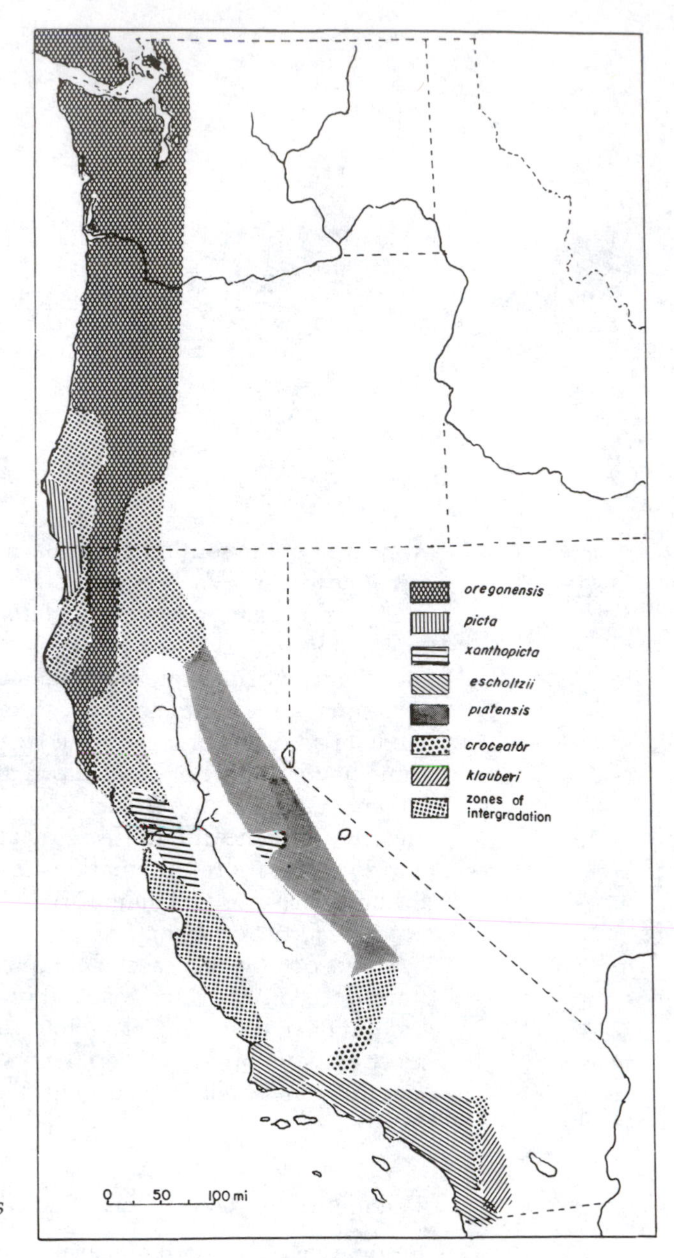

Figure 3.7. *Map of California, Oregon, and Washington showing the distribution of the "open ring" of subspecies of* Ensatina escholtzii. *(Redrawn from Stebbins 1949.)*

of the genus *Ensatina,* whose distribution encircles the Central Valley of California (Figure 3.7). These are forest animals, and the Central Valley does not provide acceptable habitat. Originally, like the geographic representatives of *Plethodon jordani,* several local forms were described as separate species. Later, when extensive intergradation was found, they were reduced to subspecies of a single species, *E. escholtzii,* although they differ from each other biochemically more than do the "species" of the *Plethodon glu-*

tinosus group, which is described later in this chapter. The interesting part of the story is in the fact that the two southernmost subspecies, *escholtzii* and *klauberi*, occur together with no hybrids in some places and a very low proportion in others. The logical explanation for the current state of affairs is that the genus originated in the forests of northern California or even farther north, and then spread more widely southward, the eastern part of the population spreading along the Sierra Nevada, and the western part along the coastal mountain ranges. When the two groups finally met, they were no longer able to exchange genes. An intermediate condition is to be found in central California, where representatives of the western subspecies *xanthopicta* managed to cross the Central Valley, apparently during a period of wetter climate that permitted some woodland. A population of *xanthopicta* is now found in the foothills of the Sierra Nevada, where it hybridizes with the Sierra subspecies *platensis* along a narrow belt that is only 1.4 km wide. Thus, the separation into full species is incomplete halfway down the state of California, but complete near the border with Mexico. Such situations appear to be easily understood, but may cause confusion when scientific names are applied.

Conceptual problems

Numerical taxonomy and "electrophoretic species"

The perceptive student will have concluded at this point that from a practical standpoint, to conclude that two kinds of organisms represent distinct, reproductively isolated species, they must be observed in the same area without the occurrence of individuals intermediate between them. If the two kinds do not overlap in distribution, the decision can be made in either of two ways: Mating experiments (with appropriate controls mating within the kinds) can establish their distinctness, or if such experiments are not feasible, a less satisfactory decision can be made, based on the amount of difference between related species that do coexist and are distinct. Even mating experiments have been known to give wrong answers. Two related species of wood mice, *Peromyscus leucopus* and *P. gossypinus*, have distributions that overlap in northeastern North Carolina and southeastern Virginia, and no intermediates have been observed there. Nevertheless, when caged together and given no choice, they mate and produce fertile offspring. The mating test can demonstrate that two forms belong to different species, but it cannot always demonstrate that they belong to the same species.

In the absence of definitive results from mating experiments or the existence of overlapping distributions, there is no certain criterion by which specific status can be determined. This situation has given rise to indirect methods involving elaborate statistical manipulations. The purpose of this mathematical approach is to estimate objectively the amount of similarity (or difference) between two or more populations. This use of mathematics is called *numerical taxonomy*. Originally, such analyses were used to group species into larger categories, but more recently, in conjunction with new biochemical techniques to be described later, they have been used to determine which groups of specimens belong to different species.

An essential element of the mathematics is the assumption that all characters used in the analysis are equally important. A "character," in this usage, is a single trait. The colors of the different populations of *Plethodon jordani* would be examples of characters. These populations differ in other ways – some being of different average sizes, others having different numbers of vomerine teeth. Many salamanders have groups of teeth in the roof of the mouth, these apparently having the function of helping the tongue hold prey in the mouth while swallowing is being undertaken. Some of these populations also differ in the relative sizes of males and females, a feature that is commonly associated with behavioral relationships between the sexes. Under the rules of numerical taxonomy, all of these traits (characters) would have to be given the same value, regardless of any importance in the biology of the animals. The argument for doing so is that if we have enough different characters to work with, any differences in "importance" will be overwhelmed. From that, it can be concluded that a large number of characters is needed for the method to succeed. It is also true that it is probably impossible to quantify "importance" in the first place. Giving all characters equal value greatly simplifies the mathematics. The frequency of occurrence of a particular character or set of characters is what is analyzed.

The result of the analysis is an assessment of the degree of overall similarity between individuals within populations and between different populations. The populations can then be connected in a branching diagram, with the lengths of the branches representing the amount of overall difference between the populations of individuals examined. Such a branching tree is called a phenogram. The one shown in Figure 3.8 represents the degrees of similarity between the many geographically adjacent populations of *Plethodon glutinosus* in the eastern United States. Each final branch in the phenogram represents a collection of ten to thirty specimens from a restricted locality (no more than 5 hectares in area).

The characters used in constructing Figure 3.8 are different from those discussed so far. They are identified by the biochemical technique known as electrophoresis. The amino acids of which all proteins are composed vary in the electric charges that they carry. When dissolved in water and subjected to a direct electric current of carefully regulated strength, some proteins will be propelled by

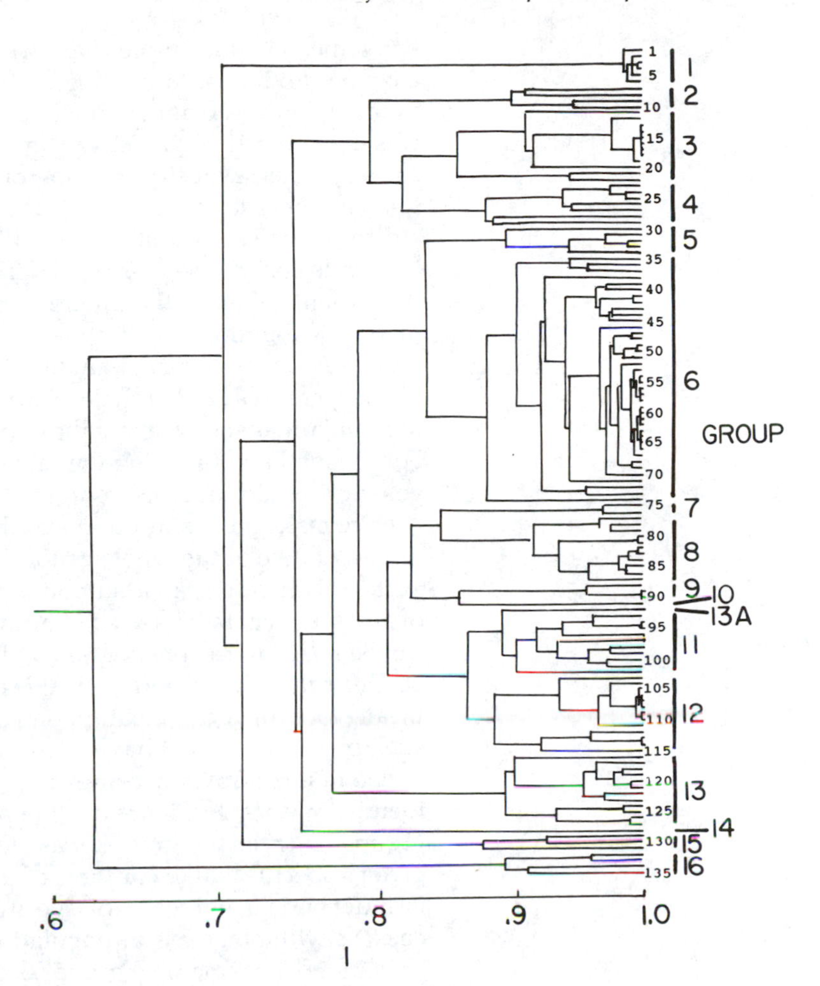

Figure 3.8. Phenogram showing calculated genetic similarities (abscissa) among 135 local populations of the Plethodon glutinosus group of species. Large numbers at right represent groups of local populations similar enough to be considered to belong to a single species. (Redrawn from Highton 1989.)

the current to distances related to the charge on their surfaces. If the original solvent has been spread thinly over a semisolid substance called a *gel*, and the protein solution is placed at one specific spot before the current is applied, the proteins will be at different distances from that origin when the current is turned off. As many proteins are enzymes, their positions can be revealed by allowing them to act chemically on the materials that they normally break down, followed by the use of stains that identify the specific breakdown product of each enzyme. Each specific kind of protein is synthesized under the ultimate direction of a single gene. The interest of all of this lies in the fact that a minor change in a gene can determine that a different amino acid will be placed at a specific place in the protein. If that amino acid has a charge different from that of the original, it can alter slightly where the enzyme will stop, and when that position is revealed by staining, it will be slightly different from the position of the original. If the gene locus is heterozygous, two bands will be seen, side by side.

For some gene loci, more than two variants can be found in the enzyme produced. Many different kinds of enzymes have been examined in this manner. The information from which Figure 3.8 was constructed involved twenty-two genetic loci, all of which showed heterozygosity in some of the thousands of individual salamanders examined.

The ecological meaning of the different variants of enzymes is known for only a very few cases. Therefore, the assumption that the variants of all of the different enzymes are equivalent in numerical taxonomy cannot be refuted at present. The frequencies with which they occurred among the salamanders from different places were analyzed with the use of a computer, which calculated the similarities shown along the horizontal axis of the phenogram. For the method of making the calculations, a book on population genetics should be consulted.

If we make the assumption that the twenty-two genetic loci are representative of the whole genomes (the entire genetic makeup) of the salamanders, an assumption that is implicit in the analysis of any set of characters used in numerical taxonomy, no objection can be raised to the phenogram. Other kinds of characters would be difficult to incorporate in the calculations, and there is no means by which the assumption can be tested. It is the uses of such phenograms that have caused at least two controversies.

The first controversy comes from the use of the phenogram to identify species. In Figure 3.8, the populations were classed into groups. Each of the groups was stated to be a species and was given a specific name. For three of the groups (1, 15, and 16) there is independent evidence of specific distinctness, because each coexists with other sets of populations without evidence of intergradation. The remaining groups occur in geographically separate areas within the overall range, covering most of the eastern United States, from New York State to central Florida, and from central Indiana to central Texas. No attempt was made to determine if intergrades occur where the geographic representatives meet. The basis for naming all of these species was the decision to use a similarity of 0.85 among the populations as evidence that they constituted a recognizable species. The decision was based on the observation that that level of similarity gave geographically congruent units. Note that this would place humans and chimpanzees in the same species, as their genetic identity is greater than 0.95.

The genetic basis of reproductive isolation is known in very few instances, and the reason for supposing that the amount of genetic difference is constant between reproductively isolated groups is based on the assumption that genetic differences arise at a constant rate. Even if they did, there is the additional assumption that the divergences of the groups being compared began at the same time. The general proposition is based on an experimental analysis of fruit flies in the *Drosophila willistoni* group of species and subspecies in South America, and the calculated similarities for different taxonomic levels in that group have been accepted for the *Plethodon* case. Given the radical differences in biological properties between fruit flies and salamanders, such as longevity and reproductive rates, that represents a major leap in justification. As chimpanzees and humans differ biochemically by much less than the salamander populations do from each other, it is easy to understand that the numerical taxonomy–electrophoretic "species" have met with less than universal acceptance.

To accept the arbitrary decision, we must also accept a redefinition of the species. We would no longer have a common evolutionary future that would be distinct from those of other species as a criterion, but would substitute some kind of average difference. Average differences abound among geographic representatives of single species, as described earlier for the forms of *Plethodon jordani* and *Ensatina escholtzii,* and they do blend into one another where their ranges come into contact. No evidence is presented that the remaining thirteen "species" in the *glutinosus* complex fail to blend into one another. Up until the 1960s, recognizable geographic races were called subspecies and were given latinized names. The practice followed in this case would raise the old subspecies to specific status and would greatly reduce the evolutionary significance of the species as a unit of classification. There would also be no category available to which to apply that significance.

A second controversy arises from the use of a phenogram to derive evolutionary relationships among the populations being compared. The originators of numerical taxonomy denied that it was permissible. Other taxonomists, however, have accepted as valid the approach known as cladistics. The arguments between the pheneticists and the cladists, and even among cladists, have been bitter. As the questions cannot be settled experimentally, the arguments involve assumptions and logic and are unlikely to be completely resolved. Cladistics depends on particular similarities among the populations, rather than on the overall similarity revealed in phenograms. Like many other new fields of science, cladistics has introduced a number of new terms that were intended to enhance precision of meaning without long explanations, but to the outsider this sometimes seems to produce unnecessary jargon. The principle requires a comparison with an "out-group," or a species related to those being analyzed, but clearly different from all of them. Characters not shared with that species but shared by the other populations are assumed to have arisen in a common ancestor that lived more recently than the common ancestor of the out-group and the populations being analyzed. Those characters are "shared derived characters," and identifying them at successive levels of inclusiveness eventually provides a

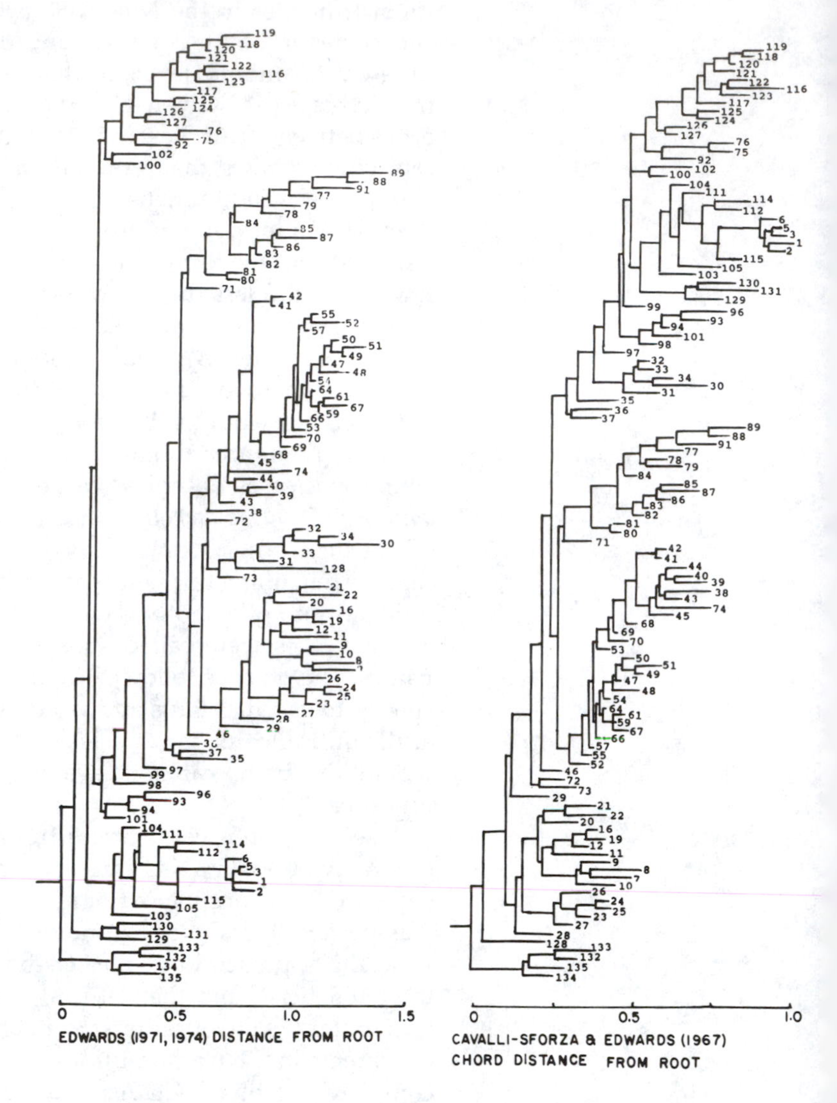

Figure 3.9. *Two different cladistic analyses of 115 of the 135 groups shown in Figure 3.8. These presumably show relationships among the species and thus reflect the amounts of time that they have been distinct. The out-group is group 16 in Figure 2.8, for which there is independent evidence of its specific distinctness.*
(Redrawn from Highton 1989.)

complete classification of the populations. It also provides a method of weighting characters in favor of those that are shared and derived. If the system worked perfectly, the final result would be a completely "nested" arrangement, with the same smaller groups always nested within the same common larger group, regardless of the computer program employed. Figure 3.9 shows that such is not the case, being the result of two slightly different "trees" resulting from two different cladistic analyses of 115 of the 135 local populations shown in the phenogram in Figure 3.8. The remaining 20 populations had to be omitted only because the computer could not accommodate so much information. Notice that the same groups of numbers tend to cluster together in the two cladograms. The differences come from choices that must be made in deciding which characters are the shared derived ones. Some of the choices are necessarily arbitrary, and preference is always given to the system of choices that involves the fewest steps in constucting the complete classification. The principle involved is that of *parsimony,* a procedure advocated by the philosopher William of Ockham in the fourteenth century. His point was that in choosing between explanations of natural phenomena, we should always choose the simplest until it is proved wrong. "Ockham's razor" is not a law of nature, but simply good advice on procedure. It has been successful when applied to questions of ancestor–descendant relationships, when the assumptions of the method are justified.

Cladistics provides an objective method for determining the branching points where new species arose, and that represents an important advance in methods of classification. However, it produces far more branching points than it is practical to handle in a system of classification. For example, within the genus *Plethodon* there are eight recognizable groups of species, which are related to each other in complex ways. To assign taxonomic categories to the various levels of relationship would require far more than the use of a category "subgenus" and would make the classification less than useful as an information-retrieval system, which is one of the purposes of having a classification.

More traditional classifiers recognize the value of having an objective method of finding relationships, but use cladistics after groups of similar organisms or groups of organisms have been identified by the consideration of all of their characters, not just those that can be shown to be shared derived ones. After these groups have been identified, cladistics can be used to test whether or not each group is derived from a single ancestral species, instead of being identified by characters that have arisen independently in separate lines of descent. Any valid group in a system of classification must include all of the descendants of a common ancestor, or else suffer the problem of arbitrary removal of subgroups. The problem is that subsequent statements about the evolution of such an incomplete group (called a *paraphyletic* group) are arbitrary to the extent that the group is arbitrary.

Parthenogenetic species

In three classes of vertebrates, Osteichthyes, Amphibia, and Reptilia, there exist "species" that contain no males. The females reproduce parthenogenetically, which means that they produce female offspring without fertilization. The offspring are identical genetically with their mothers, except in the rare case of mutations. It is easy to see that a favorable mutation arising in one of these females can spread no further than to the direct descendants of that female, and our concept of a species as a group of organisms with a common evolutionary future is not valid for an all-female group.

As a practical matter, these parthenogentic species are given scientific names in the customary manner, and that has caused little difficulty, because there is little variability in the appearances of the different female lines. Thus, different biologists working with them can be reasonably sure of the identity of the animals. That fulfills one of the purposes of binomial nomenclature, even though everyone recognizes the theoretical problem in species definition.

All of the parthenogenetic vertebrates have arisen through rare mating between otherwise distinct, sexually reproducing species. Instead of producing infertile hybrids that are incapable of reproducing, in these cases the genetic upset causes either the production of diploid ova that develop directly into female young or else some other cytological abnormality. Parthenogenetic lizards of the genus *Cnemidophorus* in the American Southwest feature direct production of young females. It has been possible to identify the two parent species by means of electrophoretic techniques and through skin grafts. The former will show the existence of heterozygotes at all loci where the parent species differ. The latter will result in rejection of the graft if the individuals differ genetically. They reveal the independent origins of different strains of the parthenogenetic species.

The parthenogenetic species in the fish genus *Poeciliopsis* are well known, and their origins have been investigated in detail. There are several normal, sexually reproducing species in this genus, which is restricted to the streams draining into the Gulf of California and the Pacific Ocean from western Mexico. It is easy to keep the origins straight because the whole system was worked out before they were given scientific names. The names of the parental species are incorporated into the names of the parthenogenetic ones. *Poeciliopsis monacha* and *P. lucida* are normal, bisexual species whose distributions overlap in parts of some of the streams in the area. Occasionally they hybridize, producing an all-female strain. These females mate with *P. lucida* and produce offspring, but again, they are all females. One might expect that as several generations pass, the females would become more and more like *P. lucida,* but that does not happen. This hybridization can be obtained in the laboratory, so the process is well known. What happens is that in the process of oogenesis (producing ova), all of the hybrid's *lucida* chromosomes are lost, and the ova contain only the haploid complement of *monacha* chromosomes. Thus, the *lucida* chromosomes are acquired anew each generation, while the *monacha* chromosomes are carried along as a clone. This all-female diploid species has been given the name *Poeciliopsis monacha-lucida* (Figure 3.10). It is a sexual parasite of *P. lucida,* as it cannot reproduce without fertilization by a male of that species.

There are two more all-female species in the area. They result from the fact that some of the gametes, presumably of *monacha-*

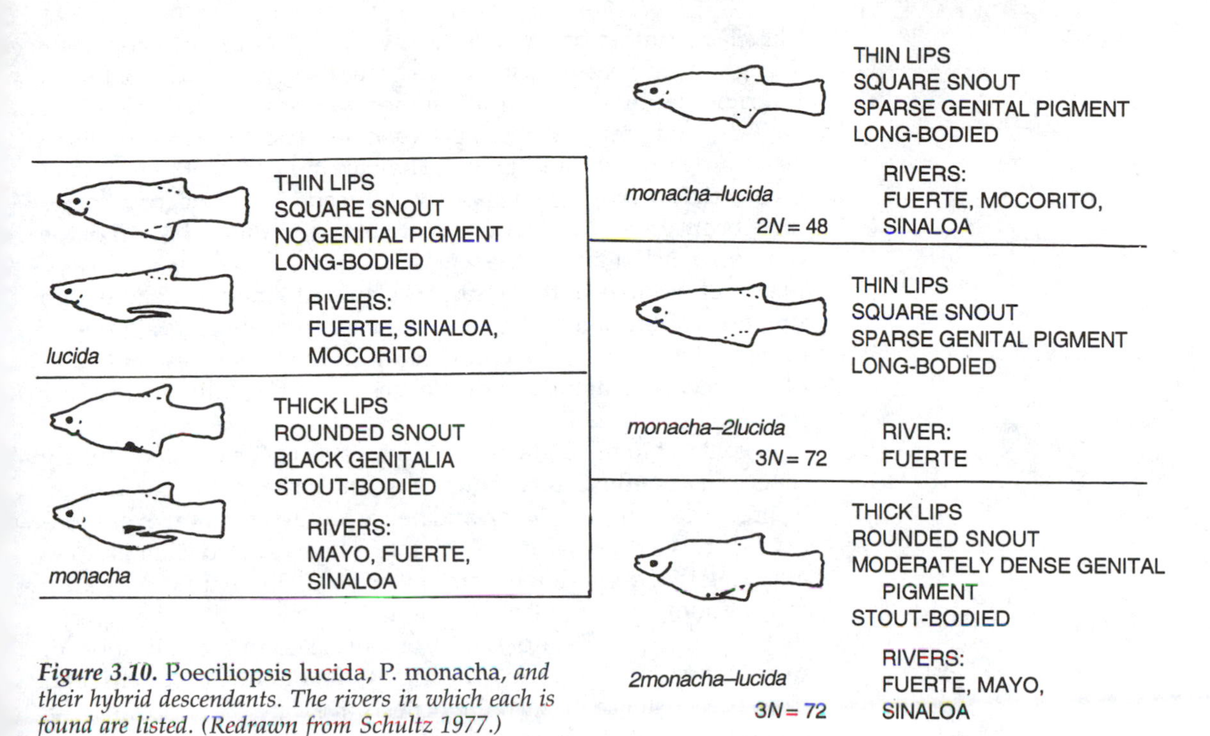

Figure 3.10. Poeciliopsis lucida, P. monacha, *and their hybrid descendants. The rivers in which each is found are listed. (Redrawn from Schultz 1977.)*

lucida, are occasionally diploid. The fertilization of one of these by a male of either parent species results in a triploid animal, which is always a female. Such females are named according to the dose of the parental chromosomes, either *2monacha-lucida* or *monacha-2lucida* (Figure 3.10). For production of young, these females must be stimulated by mating with a male of one of the parental species, but the male does not contribute any genetic material to the offspring. They are true parthenogenetic species, but like *monacha-lucida* are sexual parasites. Obviously, they must accompany one of the parental bisexual species.

Salamanders of the genus *Ambystoma* also have triploid hybrid forms that must mate with the appropriate male, and until recently the situation was thought to be like that of the triploid *Poeciliopsis,* the male contributing only the stimulus to lay eggs. It is now known that some fertilization does occur, and diploid as well as tetraploid individuals have been found. Thus, some transfer of genes into the parental species is known. How frequently fertilization of the hybrids occurs, relative to asexual reproduction, has not been established. Four bisexual diploid species are involved in various combinations: *A. laterale, A. jeffersonianum, A. texanum,* and *A. tigrinum.* The hybrids are found from Wisconsin to Québec.

Summary

The classification of animals follows a logical progression, even though the increasingly inclusive groups of species do not follow the rigid set of steps that are sometimes represented. There are well-defined groups that fall between the classical series: phylum, class, order, family, and genus. They are treated as subunits, and where that arrangement has not sufficed, new category names have been added in the appropriate places in the series. All of the categories listed are the results of decisions based on the convenience of workers in that part of the animal kingdom. There is no universal definition for "genus," for example, which is a group of species that resemble each other more than they resemble any other species – a definition that leaves much flexibility in deciding its boundaries.

In contrast to the higher categories, "species" has an objectively determined definition: a group of animals that are capable of interbreeding with each other, and hence have a common evolutionary future. Certain natural phenomena have caused difficulties in applying that definition to some examples. These difficulties arise from ignorance of the full situation, from reversal of the separation of one species into two, from the rigorous application of mathematics and molecular biology to taxonomic problems, and from the existence of parthenogenesis. Despite the problems, the species concept works remarkably well most of the time for most groups of vertebrates, and when it doesn't, we can nearly always find out why.

References and suggested further reading

Cooke, F., & F. G. Cooch. 1968. The genetics of polymorphism in the goose *Anser caerulescens. Evolution* 22: 289–300.

Dawley, R. M., & J. P. Bogart (eds.). 1989. *Evolution and Ecology of Unisexual Vertebrates.* Albany, N.Y., The University of the State of New York, State Education Department, New York State Museum.

Hairston, N. G. 1950. Intergradation in Appalachian salamanders of the genus *Plethodon. Copeia* 1950: 262–73.

Highton, R. 1989. *Biochemical Evolution in the Slimy Salamanders of the* Plethodon glutinosus *Complex in the Eastern United States.* Urbana, Illinois Biological Monographs, no. 57.

Mayr, E. 1963. *Animal Species and Evolution.* Cambridge, Mass., Harvard University Press.

Mayr, E., & P. D. Ashlock. 1991. *Principles of Systematic Zoology* (2nd ed.). New York, McGraw-Hill.

Schultz, R. J. 1977. Evolution and ecology of unisexual fishes. *Evolutionary Biology* 10:277–331.

Sneath, P. H. A., & R. R. Sokal. 1963. *Numerical Taxonomy.* San Francisco, Freeman.

Stebbins, R. C. 1949. Speciation in salamanders of the plethodontid genus *Ensatina. University of California Publications in Zoology* 48:377–526.

Wiley, E. O. 1981. *Phylogenetics: The Theory and Practice of Phylogenetic Systematics.* New York, Wiley.

PART

III

Vertebrate ecology

In our context, "ecology" will be considered to be the environmental factors influencing vertebrates and vertebrate populations. Some of these are obvious from the relationships between climate and the distributions of different species. This topic is discussed in Part IV. Here, we consider only the ways in which vertebrate species react to environmental changes and the ways in which they affect each other. "Ecology" is much broader than that, but the restriction at this point is justified.

It can be argued that any two species that occur in the same area must of necessity have some influence on each other. The argument goes as follows: Ultimately all animals depend on plants for their food. There is no escape from this, as all animals require energy, which they can obtain only from their food, and the only way in which energy can be converted into a chemical form (so as to be in food) is through photosynthesis. For our purposes, we ignore the energy obtained from hydrothermal vents on the ocean floor, and the invertebrate communities supported by the bacteria in such places. As far as we are concerned, green plants are the *producers* in the ecological context, and all other organisms are *consumers*. Not all consumers feed on green plants, of course. Carnivorous animals feed mostly on herbivorous animals. Moreover, there is a whole array of organisms that subsist on the dead bodies of other organisms. Most of these are bacteria and fungi, but many carnivorous vertebrates will eat carcasses, provided that decomposition has not gone too far, in which case the food is left, first to the vultures and certain insects, and finally to the bacteria.

In summary, we have producers and three classes of consumers (primary, secondary, and decomposers). The flow of energy through a terrestrial system is shown in Figure III.1. The losses are due to the fact that energy is required to keep the organisms alive, and this energy is lost through respiration. That is, the conversions are not 100% efficient. Unless chemically bound energy accumulates, or one of the categories within the circles declines, the losses must equal the gain in photosynthesis among the producers.

Some of the influences along the series are obvious. Producers are necessary for the primary consumers (herbivores), and these are necessary for secondary consumers (predators and parasites). The dependence works throughout the system, as is clear when we consider the use of matter instead of energy, as shown in Figure III.2. Other forms of matter have different cycles, but the principle is the same. There is a closed system, and producers are ultimately dependent on the decomposers. They are also affected in a less obvious way by the excreta of the consumer groups. That pathway indirectly allows the herbivores to be affected positively by their predators.

For the vertebrates, the coupling often is much tighter than the foregoing analysis indicates. This can be understood by a consideration of the numbers in Figure III.1. The numbers leading from the producers to the decomposers illustrate the reasoning. In terrestrial situations, except for a small percentage that is leached or decomposed by inorganic processes, all of the dead organic matter is consumed by the decomposers, because it doesn't accumulate from year to year. The implication of this for the decomposers is that these organisms must face a shortage. Therefore, if the members of one species succeed in obtaining dead organic matter, members of another species are deprived of it, at least to some extent. These decomposer species are in interspecific competition for a resource, and we can conclude that the decomposers are limited in abundance by the supply of that resource.

That being the case, consider the implications at the other end of the system. Only 1–10% of the producer biomass is consumed by the herbivores. The rest becomes dead organic matter. This surely means that interspecific competition among primary consumers (herbivores) is not universal, because so much producer biomass is left over. What determines the abundance of those species? Considering the numbers again, an average of 90% of the herbivore biomass is eaten by predators, a percentage that ought to have an effect on herbivore populations, and predation is seen as an important interspecific interaction. The predators and parasites are in the same situation as the decomposers, with nearly all of their food being eaten. It follows that interspecific competition occurs among the predators.

With minor exceptions, all vertebrates are either herbivores or carnivores, and we would expect that they would exhibit both competition and predator–prey relationships among them.

Before examples of these interactions can be described, it will be necessary both to discuss the ways in which different vertebrates are adapted to their physical environments and to illustrate the analysis of populations, because the interactions are really populational phenomena, and the responses of the individual species

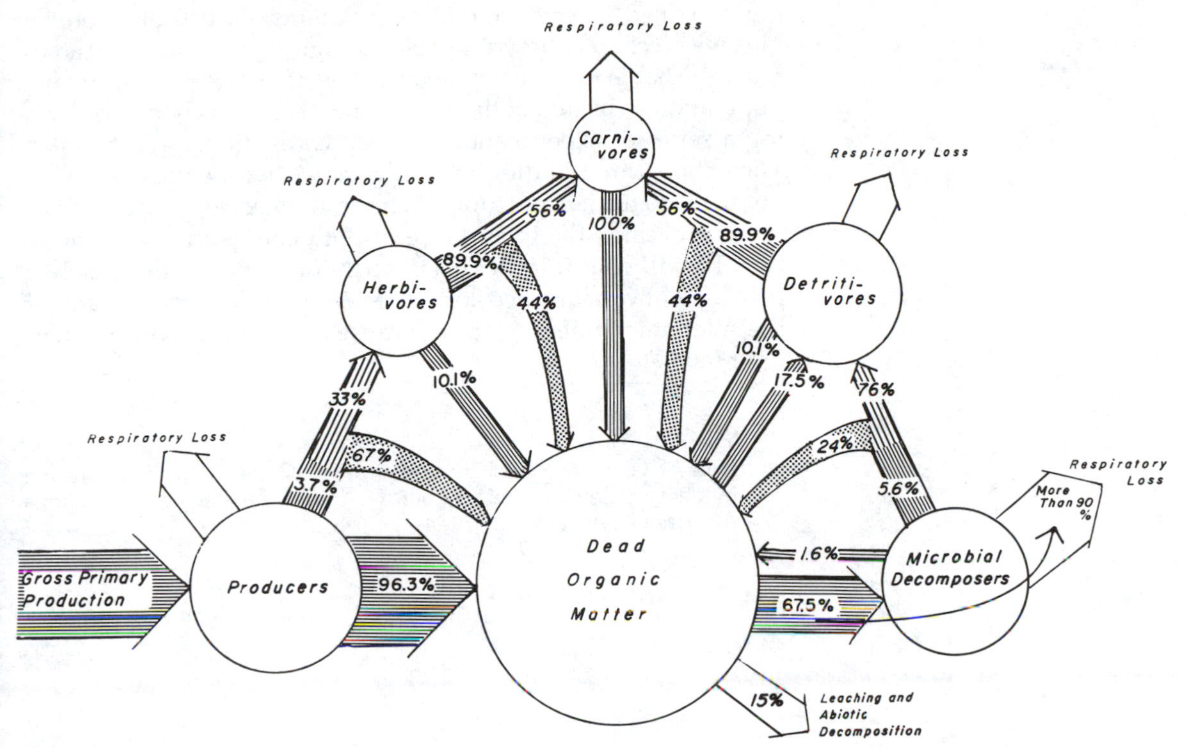

Figure III.1. Energy flow through a deciduous forest. (From Hairston & Hairston 1993.)

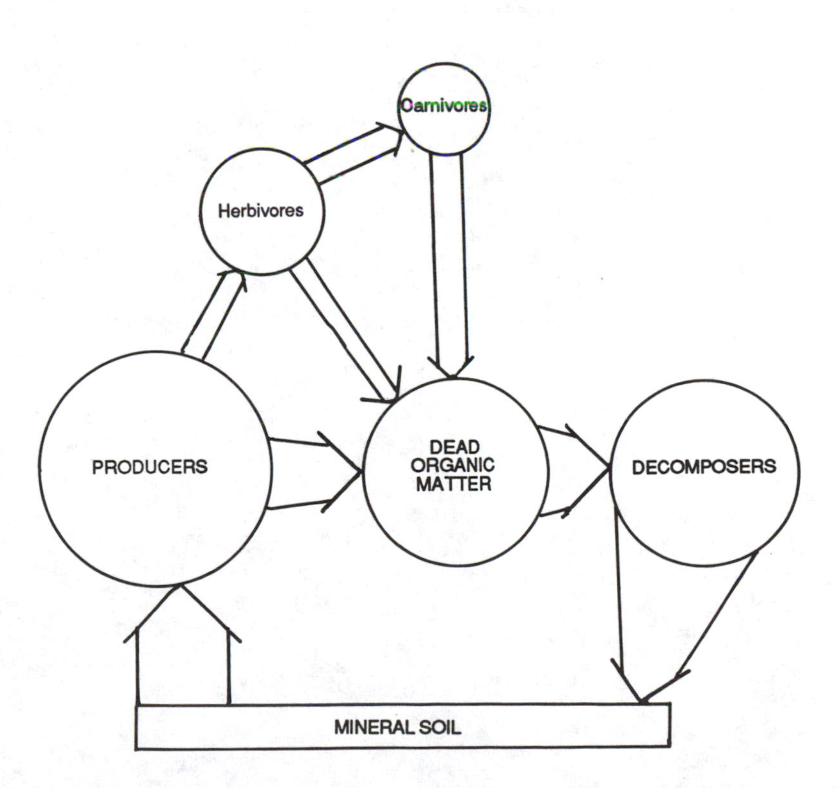

Figure III.2. A representative cycle of mineral nutrients in terrestrial ecosystems.

are described in terms of their populations. Part III therefore begins with a chapter describing physiological and behavioral adaptations. That is followed by an example of an excellent study of a population of lizards in the field. Next, there is a chapter describing a series of experiments in which competition was revealed. Then come three studies involving more than two species, as an approach to the general topic of community ecology. Those experiments revealed the importance of both competition and predation. Part III concludes with a description of one of the unsolved problems of vertebrate ecology: the regular cycles of abundance in the small microtine rodents, the cause or causes of which remain unknown.

Reference

Hairston, N. G., Jr., & N. G. Hairston, Sr. 1993. Cause–effect relationships in energy flow, trophic structure, and interspecific interactions. *American Naturalist* 142:379–411.

4

Adaptations to different environments

Vertebrates have evolved in many ways that allow them to cope with harsh environments and with environments that periodically become harsh. In this chapter, we first consider how different groups of vertebrates react to seasonal changes. Although different groups may face different problems during a given season, this part of the chapter takes up the seasons in turn, describing the activities that are characteristic for the various groups of vertebrates in each season. The second part of the chapter describes mechanisms for surviving environmental extremes.

Seasonal activity patterns

There is a common general problem facing all of the vertebrates in any location. They must adjust their activities to the changing conditions in successive seasons, and the solutions by different vertebrates tend to be similar for each season. Some of them, however, have capabilities that others do not, and their different methods are of interest. This part of the chapter takes up the vertebrates in a restricted area, the Middle Atlantic states, so that exceptions for other kinds of climates need not be considered until later in the chapter, although pertinent examples from other temperate areas are included.

Winter

In the temperate climates of midlatitudes, winter is the harshest season for most vertebrates. Only birds and bats are able to escape its rigors by migrating, and 25–30% of them do so. The remaining species of vertebrates must continue to find food, to have stored food when it was plentiful, or must reduce their demands on food or accumulated fat by hibernating or by becoming torpid for short periods, such as overnight.

"Hibernation" will be used here in the vernacular sense, referring to mammals that remain dormant ("asleep") for all or most of the winter months. This involves a drastic reduction in body temperature and slowing of metabolism, heart rate, and breathing. As described in Chapter 1, endothermy is expensive energetically, and the ability to reduce the body temperature to 2–3°C, as occurs in some hibernating rodents, represents a great reduction in energy consumption. The availability of suitable places in which to hibernate is important in energy conservation. Holes and cavities make good hibernaculae if wind and water are excluded. Hibernating mammals are found among several orders. The only true hibernator among Insectivora is the European hedgehog, *Erinaceus europaeus*. Among twenty-one species of North American Chiroptera, nine species hibernate in the general area of their summer activity, seven migrate south and hibernate there, and five live far enough south that they can find insects at frequent intervals, and they neither migrate nor hibernate. They may go through brief periods of unfavorable weather by partially reducing their body temperature, but only to 17°C. The Rodentia provide the majority of true hibernators, and this is especially true of the Sciuromorpha (squirrels). Among them, the marmots, including the woodchuck, *Marmota monax*, are the largest mammals that hibernate in the strict sense. Many species of ground squirrels, including chipmunks, hibernate. Among the Myomorpha, jumping mice (Zapodidae), some pocket mice and kangaroo mice (Heteromyidae), and some hamsters (Cricetidae) are known to hibernate. The annual changes in physiology, including changes in body mass and lowered body temperature, are not under the direct influence of the immediate conditions. It was found that those changes continued for 4 years in experimental golden-mantled ground squirrels *Spermophilus lateralis* under constant laboratory conditions (Figure 4.1). In that case, the physiological cycles continued with no outside stimulus, but in most such examples one annual environmental event serves to "reset" the cycle (see Chapter 15). The only group of rodents in which no members are known to hibernate is the Hystricomorpha, including the porcupine, *Erethizon dorsatum*. The only Carnivora that hibernate are the bears, and they are exceptional in that their body temperature only falls to 31–35°C. They nevertheless remain dormant for 3 months. It has been pointed out that mammals larger than a marmot would encounter severe problems in arousal, as that involves raising the body temperature back to normal, and would require an inordinate amount of time, as well as so much energy as to make it unprofitable to reduce the temperature as much as does a small rodent. Arousal of any hibernating mammal depends on two key physiological properties. The first is an increase in heart rate, which increases the flow of blood to the brown adipose tissue, the metabolism of which is the second key. Brown fat has the unique property that conversion of stored energy to heat is by

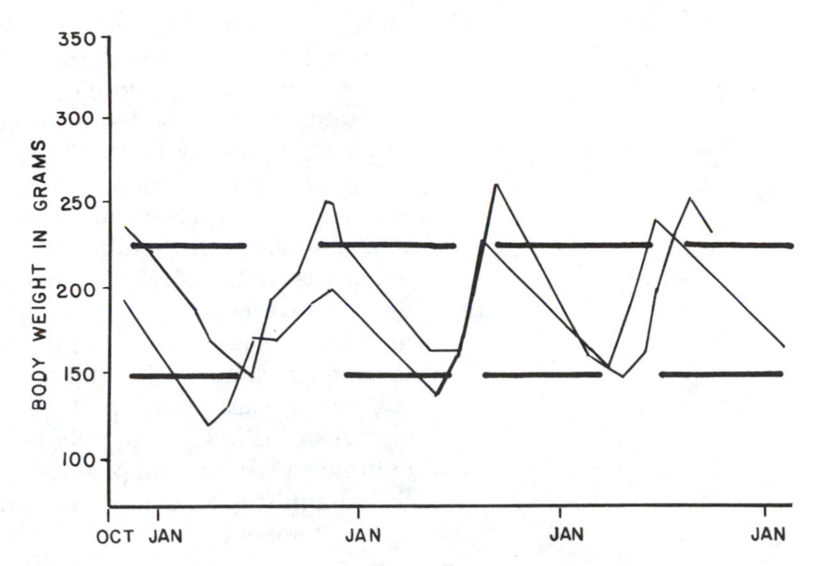

Figure 4.1. *Interrelationships of body weight (g) and whole hibernation periods (black bars) for two* Spermophilus lateralis *for nearly 4 years; upper individual at 12°C ambient, and lower at 3°C ambient, both with artificial photoperiod of 12 h. (Redrawn from Pengelly & Asmundsen 1972.)*

a much more direct metabolic pathway than is found in other tissues. The mitochondria contain a protein that short-circuits the usual production of adenosine triphosphate (ATP), allowing all of the energy accompanying proton translocation to be dissipated as heat. This heat then warms the blood returning to the heart, further increasing cardiac output and carrying warmed blood throughout the body.

It is clear that for hibernating mammals, in which the body temperature is 15–30°C below that of active animals, the temperature-regulating mechanism, as described later in this chapter, must be reset. This hypothalamic thermostat remains as sensitive as before, and for some hibernators the body temperature remains constant above freezing while the ambient temperature continues to drop. Not all mammals go into deep hibernation; many become torpid at higher temperatures than are found in the deep hibernators, and they are unable to arouse spontaneously from lower temperatures. For all of them, there is a substantial decrease in the resting metabolic rate. A small pocket mouse, *Perognathus californicus*, goes into torpor each night, its body temperature falling to 15°C. Its oxygen consumption in that state is less than one-third of that for its normal resting metabolic rate. For the bat *Eptesicus fuscus*, which hibernates with a body temperature of around 5°C, its oxygen consumption is only one-fortieth of that in its active state. The saving in energy is considerable, even for animals that remain in torpor only at night, and even counting the energetic cost of arousal, which is not negligible. For the 54 min during arousal of the pocket mouse, the oxygen consumed is nearly five times that consumed during more than 7 h of torpor. Among birds, only the poorwill, *Phalaenoptilus nuttallii*, of the western United States, enters true hibernation, but some small birds, such as chickadees and other tits, do engage in shallow hypothermia during cold nights.

Ectothermic vertebrates do not have the complex physiological mechanism for entering true hibernation, but there are physiological changes. Modification of the phospholipids in membranes for maintenance of greater fluidity at low temperatures, acclimatization of reaction rates, and altered lower lethal temperatures are all evidences of physiological compensation for cold. As the ambient temperature drops, animals' body temperatures fall as they go to areas below the frost line (for terrestrial vertebrates) or below the level to which ice reaches (for aquatic forms). The terrestrial salamanders *Plethodon jordani* and *P. glutinosus* disappear from the surface of the forest floor after the first air temperature of 0°C in autumn and do not reappear within the next 2 months, even with the stimulus of a warm rain. At the time of their emergence in spring, they appear to be more sensitive to immediate weather conditions, being active on the forest floor in warm weather and disappearing on cold nights. Fishes simply become inactive when the water is cold. The bluegill sunfish *Lepomis macrochirus* stops feeding (Figure 4.2). Herring have been found to follow rising temperatures in seawater, apparently in response to spring blooms of phytoplankton. The torpor of most reptiles appears to be directly under the influence of temperature, but there is evidence of an endogenous annual rhythm in the American alligator that is not present in the tropical caiman.

The birds and mammals that neither migrate nor hibernate must continue to find enough food to maintain their high body temperatures despite the cold. As small endotherms have a relatively larger surface area per body volume, winter is especially trying for chickadees, titmice, kinglets, creepers, and nuthatches. In addition to the large surface : volume ratio, they have thinner boundary layers than larger animals, making them prone to forced convective heat exchange, whereas larger species are affected mostly by radiative heat exchange. As they cannot forage at night, they may consume 70% of their fat during an especially cold night. If they maintained their normal daytime temperature of 40°C, they would need to consume more fat than they had. Their recourse is to become partly torpid by reducing their body temperature to 30°C and to begin foraging as early as possible. Small mammals, such as shrews, that do not hibernate are protected by burrowing under the snow cover, which insulates them from extreme low temperatures. It has been found that short-tailed shrews, *Blarina brevicauda*, in northern Michigan, where the ground is snow-covered nearly all winter, are less cold-stressed than are their conspecifics in southern Michigan, where snow does not remain on the ground throughout the winter and where

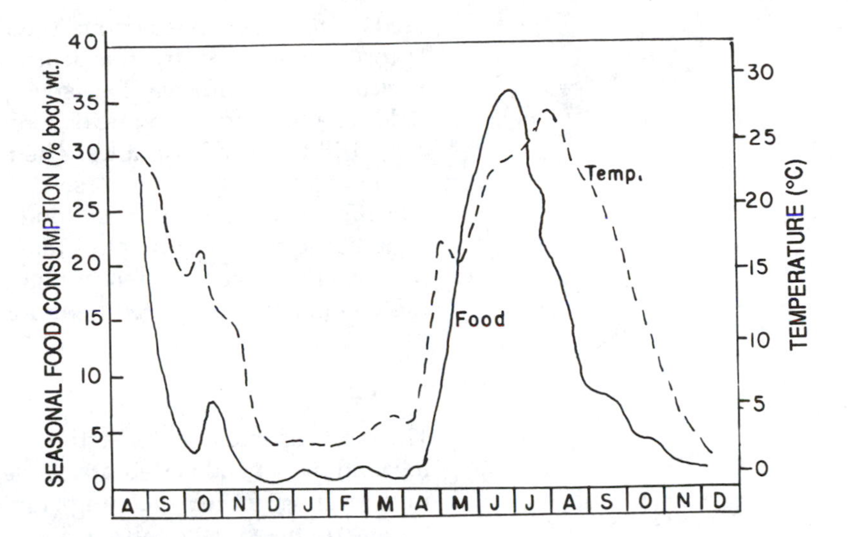

Figure 4.2. Dependence of food intake on temperature in the bluegill (Lepomis macrochirus). *(Redrawn from Lagler, Bardach, & Miller 1977.)*

the air temperature over bare ground is colder than that 500 km farther north under snow cover.

Some vertebrates store food for winter consumption, rodents being the most prominent. Like other sciurids, chipmunks, *Tamias striatus*, store large quantities of seeds and nuts, as those who operate bird feeders can attest. They arouse from hibernation at intervals during the winter to feed on the cache. Gray squirrels, *Sciurus carolinensis*, bury acorns and nuts, and in early spring can be seen searching for them. Apparently they can smell them, as it seems very unlikely that memory would serve for so many locations. Deer mice, *Peromyscus maniculatus*, store many small caches of seeds, to which they return during the winter. The most elaborate food storage practiced by any mammal is that of beavers, *Castor canadensis*, which feed on bark. Their pond, while serving as a refuge from predators, is also used to store branches and small logs under water. These are pushed into the mud and are retrieved through underwater entrances in the lodge during winter. Among birds, blue jays, *Cyanocitta cristata*, often store seeds obtained at bird feeders, and in one study they were observed to store 54% of the acorn crop in a part of Virginia. Clark's nutcrackers, *Nucifraga columbiana*, a member of the Family Corvidae of western North America, store the nuts of piñon pine in small ground caches in northern Arizona. They have been estimated to store two to three times as many of the nuts as are needed to supply their energy for the 6 months of winter. Black-capped chickadees, *Parus atricapillus*, store many seeds and have remarkable memories for where the stores are. In line with this spatial memory is an account of the species' ability to remember the way out of baited traps, and thus to return time and again for more seeds. The most conspicuous seed stores of any birds are the "acorn trees" used by the acorn woodpecker, *Melanerpes formicivorous,* of California. This species lives in groups, each group holding a territory with one or more trees into the bark of which they drill thousands of holes. The number of acorns stored (one per hole) is nearly always close to 320 per bird each winter. They consume a number of other foods, including insects, sap, and fruit, the acorns providing less than 10% of their energy demands between December 1, when storage is complete, and June 1, when the young have completed fledging.

Spring

The most conspicuous springtime activity of vertebrates is breeding, and the several vertebrate classes have characteristic patterns in the timing of their breeding (Table 4.1). Egg laying is taken as the start of breeding for all species except mammals, for which the start of breeding is taken as the average time at which they give birth. Some mammals breed continuously, especially small rodents. They are included in the total number of species, but not in the percentages. In all, 308 species of vertebrates, exclusive of fishes, in the Mid-Atlantic Region of the United States are evaluated in Table 4.1, which shows the percentage of species in each class that begin breeding in the indicated month.

The distributions of the start of breeding for mammals, birds, and reptiles are easily understood. Spring is the season when new growth is appearing, along with the insects, most of which feed on the new growth. These new resources are important for young animals, because it is vital for them to have enough food to grow rapidly. Young animals are vulnerable and suffer much higher death rates than do adults, and the faster they become grown, the better their general chances.

Mammals have an advantage over the other classes in not having to incubate eggs. They are thus further ahead of birds than it would appear from Table 4.1. From their pattern, it appears that April is the most advantageous month to start breeding for most vertebrates. More than twice as many species first give birth in April than in any other month. That coincidence is not achieved by synchrony of copulation. The first month of giving birth is not correlated with the length of the gestation period. That means that the timing of copulation is set independently in different mammal species that give birth at the same time. The shorttail shrew, *Blarina brevicauda,* has a gestation period of about 21 days, and the whitetail deer, *Odocoileus virginianus,* 201 days. The river otter, *Lutra canadensis,* has a period of active pregnancy of 63 days, but that is preceded by a period of delayed implantation of 7.5–8 months. All first give birth in April. There must be very different

Table 4.1. *Percentage of species of each class of vertebrates starting to breed in each month in the Middle Atlantic States*

Class	Total species	Month J	F	M	A	M	J	J	A	S	O	N	D
Mammals	45	2.2	0	15.5	44.4	8.9	17.7	2.2	0	0	0	0	0
Birds	157	0.6	0.6	1.3	30.6	47.1	17.8	1.3	0	0	0	0	0
Reptiles	60	0	0	0	6.7	21.6	45	13.3	11.6	1.6	0	0	0
Amphibians	46	4.3	10.9	17.4	15.2	10.9	13	2.2	2.2	4.3	2.2	2.2	2.2

stimuli or responses to the stimuli for copulation in the different species.

Birds cannot afford to lay their eggs earlier than they do, because the eggs must be incubated, but at the same time the adults must forage for food to maintain their own body temperature. However, they cannot leave the eggs for long, because the eggs can become chilled, especially in March, when they would have to be laid for the young to be hatched at the time that most mammals are born. Most birds must wait, because for each spring month there is an increase of 6°C in mean temperature over the previous month. That translates into a major energy gain, inasmuch as a small rise in temperature causes a significant increase in the basal metabolic rate, thereby greatly reducing the amount of food that would have to be obtained in colder weather, at an obvious danger to the eggs.

Reptiles are worse off than birds. Not being endothermic, they must allow their eggs to be incubated by environmental heat, which would greatly prolong the incubation period should they incubate in the spring. Eggs are even more vulnerable than young animals, being sought by many predators. Thus, reptiles must wait until the earth is warm. Even so, some hatchlings, such as those of the box turtle, *Terrapane carolina*, remain in the underground nest through the following winter.

Amphibia present an entirely different picture from the other three classes. There is no month when some species does not begin to breed, and the preponderance of reproductive activity is spread quite uniformly over 5 months: February to June. Some of that spread can be accounted for by the varying natures of the aquatic habitats used. Those breeding in temporary ponds are under pressure either to complete their development quickly or to start early. In the first category are some anurans, such as toads (*Bufo*) and spadefoots (*Scaphiopus*), that develop very rapidly and metamorphose at a size that is tiny compared with the adult. This forces them to lay eggs late in the season, to take advantage of the high temperature. In the second category are the mole salamanders. *Ambystoma maculatum* migrates to temporary woodland ponds as early as they are ice-free, usually February. If there is a dry spring, the larvae may not complete development before the pond dries, usually in late May or early June. *Ambystoma opacum* has pushed early breeding as far as possible. It mates and lays eggs in mid-September and October at sites where ponds will fill later, with the first rains after leaf fall in November. The leaves must fall first, because while they are still on the trees, evapotranspiration will remove water from the ground before it can accumulate in the temporary ponds. When the eggs of *A. opacum* become covered with water, they hatch immediately. The larvae thus begin growth much earlier than do those of *A. maculatum,* and in a normal winter they are large enough to be important predators on the larvae of *maculatum* when they appear. This normally gives *opacum* an advantage, and a higher percentage of its eggs reach metamorphosis than do those of *maculatum*. However, when the early winter is dry and cold, the eggs of *opacum* may be frozen before hatching.

Anurans breed in every month of the year, with some concentration in spring and early summer. Most species have extended breeding seasons.

A number of species of salamanders breed in or near streams, and temperature is relatively unimportant, as groundwater is constant in temperature, and that is the source of streams. Thus, stream species vary greatly in terms of the month in which they lay eggs, some doing so in each season.

Summer

As already described, summer is the time of egg laying for most reptiles. The majority of the species in the other classes have at least started breeding in the spring. For them, summer is variously a time for producing an extra brood, for "preparing" in some way for autumn, or for dispersal activity. In some species of mammals, such as chipmunks, the females born in the first brood in April may themselves reproduce in July or August.

The most obvious example of preparing for autumn is the molting of the feathers of birds. This is necessary because the feathers, being nonliving material, become abraded during normal activities and must be replaced, at a considerable cost in terms of increasing protein intake. The molt thus has several functions: to acquire new body feathers for winter protection, to renew the flight feathers of some species before the autumnal migration, and, for many species, to make males less conspicuous during a season when their bright colors would be a disadvantage in evading predators, those colors no longer being needed to advertise the possession of a territory as in the nesting season.

Postbreeding wandering is a conspicuous trait of herons, among which there is a northern movement. It has accounted for the northernmost distributional records of little blue herons, *Florida caerulea*, egrets, *Casmerodius albus* and *Egretta thula*, and other birds. Such dispersal occurs in many species, especially among the young, which are expelled from the territories or home ranges of their parents as soon as they are large enough to constitute a threat to the territory holders. Despite the danger that the dispersing young face when forced out of an area where the locations of resources and retreats from predators are known, unoccupied suitable habitat may be found and presumably must be found frequently enough for selection against the dispersal behavior not to operate. Some dispersal has been claimed to represent altruistic behavior – for "the good of the species." That is a problem that is discussed in Chapter 19.

Aestivation is the summer equivalent of hibernation. In areas with especially hot and especially dry summers, some species of mammals exhibit a torpor that is similar to hibernation. By secluding themselves, they reduce their water loss, but they cannot reduce their body temperatures below the ambient temperature, and thus their energy savings are much less than is the case in winter. An interesting example of aestivation is that of the common red-backed salamander, *Plethodon cinereus*, of eastern North America. Unlike its larger congeners, it is active above ground in early spring and late autumn, but is difficult to find during the summer months. This pattern of activity is unrelated to the available moisture, as summer rainfall is common. It has been found that, as expected, its energy consumption is positively related to temperature, but its digestive efficiency is inversely related to temperature in the range between 10°C and 20°C. Thus, it faces the difficulty that its foraging would have to be much more productive in the summer for it to acquire enough food to compensate for its loss in digestive efficiency. It is better for these individuals to remain inactive below ground during warm weather than to forage, even though prey items are most abundant then.

Autumn

The most conspicuous activity of vertebrates in autumn is migration, and birds are by far the best-known migrants. Part V of this book is devoted to migration, but it is most appropriate here to consider why some species migrate and some do not. In the Middle Atlantic states there are four distinct categories with respect to migration, all with similar numbers of species:

1. those nesting in the area, but leaving the area for winter (59 species)
2. those migrating to the area for winter (53 species)
3. those transient species neither nesting nor wintering in the area (58 species)
4. residents year round (57 species)

There are more species that have been recorded in the area irregularly, but these 227 species illustrate the points to be made. The observation that they do different things with respect to migration suggests that food may be available for some during the winter, but not for others. The kinds of foods can be divided into four general categories:

1. birds feeding mostly on insects or nectar (115 species)
2. birds feeding mostly on seeds or fruit (29 species)
3. birds feeding on flesh (21 species)
4. birds feeding mostly on aquatic or marsh life (62 species)

A species' placement in one of the categories is based on the preponderance of its food. It should be understood that for any one species, some of its food may well come from other categories. For example, the kestrel, *Falco sparverius*, is listed with the flesh eaters, although a significant fraction of its diet consists of grasshoppers, and the yellow-rumped warbler, *Dendroica coronata*, an insectivorous bird, feeds largely on myrtle and other berries in winter (hence the old name myrtle warbler). Despite such exceptions, there are some generalizations that can be made. In Table 4.2, each of the four categories of birds is apportioned according to the percentages of species taking each kind of food. In Table 4.3, each of the four categories of food is apportioned according to its use by the species in each migratory category. A study of the tables reveals in a general way the advantages of migrating. The transients and those leaving the area for the winter are largely insectivorous, with the second largest group among transients being those that feed on aquatic life. Among those species migrating into the area for the winter, the largest fraction is composed of those feeding on aquatic life. That food is not available in winter north of the Middle Atlantic states, and the large sounds, estuaries, and salt marshes in that area are open nearly all winter. Resident species appear to be able to find insects, or to substitute berries, during the winter.

Despite the observation that many residents are able to find insects during the winter, two-thirds of the insectivores as a group either are transients or leave the area for the winter. The area is most attractive to seed eaters, three-fourths of which either are residents or migrate to the area for the winter; the absence of prolonged snow cover would seem to be a major reason. The area's birds of prey are mostly residents; aquatic species migrate to the area for winter, whereas shorebirds, the other group feeding on aquatic life, are mostly transients. These last two groups nest in the far north, many of them on the tundra, where life in winter would be impossible for them. Other vertebrates such as polar bears, lemmings, foxes, musk-oxen, and wolves are unable to leave the tundra and must be able to survive in that climate. From

Table 4.2. *Percentage*[a] *of each migratory category of birds taking each kind of food*

Migratory category	No. of species	Type of food: Insects or nectar	Seeds and fruits	Flesh	Aquatic life
Residents	57	45	18	23	14
Migrate in	53	23	23	7	47
Transients	58	28	3	5	34
Migrate out	59	75	7	3	15

[a]Each row adds to 100%.

Table 4.3. *Percentage*[a] *of bird species taking each kind of food in each migratory category of birds*

Type of food	No. of species	Migratory category: Resident	Migrate in	Transient	Migrate out
Insects or nectar	115	23	10	29	38
Seeds and fruits	29	35	41	7	17
Flesh	21	62	19	14	5
Aquatic life	62	13	40	32	15

[a]Each row adds to 100%.

this analysis we can begin to understand the pressures for and against migration.

Coping with physical environments

Wherever there is food, some kind of vertebrate will be present to consume it. Only under the most extreme climatic conditions, such as those in central Antarctica and the absolute desert of northern Chile, are vertebrates absent, and in both cases it appears to be the total absence of liquid water that sets their limits. The gray gull, *Larus modestus,* is an exception. It does nest in the Chilean desert, but must forage at the nearby seacoast.

There are questions regarding how Arctic and Antarctic vertebrates meet the challenge of extreme cold (in which their bodies might be expected to freeze, with the disruption of cell structure by ice crystals) and regarding how desert vertebrates cope with extreme heat (in which they might suffer denaturation of their proteins or complete dehydration).

Cold

The degree of cold that an organism is able to survive depends on its normal environment. Tropical fish will die if the heater in their aquarium malfunctions during a cool night, but a fish of Antarctic waters, *Trematomus borchgrevinki,* lives at –1.8°C, the freezing point of seawater, despite the fact that the osmotic concentration of its body fluids would indicate conversion to ice at –0.8°C. This remarkable feat is due to the presence in its blood of a glycoprotein, which acts as an antifreeze. The antifreeze has the effect of lowering the freezing point of its plasma more than 1°C below its melting point, instead of being virtually the same, as it would be without the glycoprotein. Not all antifreeze compounds in fishes are glycoproteins. Two Arctic species have carbohydrate-free polypeptides.

A few vertebrates are able to tolerate the formation of ice in their body fluids. Most frogs and turtles spend the winter below the ice in aquatic habitats and thus are at temperatures above freezing, but two species of frogs, the gray tree frog, *Hyla versicolor* and the wood frog, *Rana sylvatica,* spend the winter in terrestrial habitats. They have ice in their body fluids, and apparently are able to tolerate it by having glycerol present (*Hyla*) or having elevated blood glucose levels (*Rana*). As already described, the newly hatched young of some turtles remain in their terrestrial nests until the following warm season. *Chrysemys picta,* the painted turtle of North America, may have 50% or more of its body fluids in the form of ice. Beyond these examples, retreat to aquatic or subterranean locations is the only recourse open to ectothermic vertebrates. The "denning up" of snakes in certain retreats used repeatedly year after year is well known.

Birds and mammals, being endotherms, have several mechanisms available that allow them to exist in cold environments. The first of these is insulation. Both feathers and hair are excellent insulators, and they have striking effects in reducing the radiation of body heat along a strong temperature gradient. In a series of experiments at Point Barrow, Alaska, fresh winter skins of a number of species of mammals were attached firmly over electric heaters, which were then regulated to a temperature of 37°C and placed in a room at a constant temperature of 0°C. The temperature gradient across each fur was measured after it had stabilized, and that was related to the thickness of the fur, which ranged, for different species, from 5 to 70 mm. The gradient was converted to a measure of insulation, the reciprocal of the energy transfer per square decimeter. For fur thicknesses of 5 and 50 mm (the thickness of the fur of a white fox, *Alopex lagopus*) there was a linear increase in insulating capacity. Fur thicker than that, found on Dall sheep, wolf, and grizzly bear, gave no further improvement. It was postulated that smaller animals would have their movements hampered if their fur were as thick as that of the fox. Those experiments were repeated in Panama on tropical mammals. There was a similar relationship between insulation and thickness of fur, but 35 mm was the thickest fur obtained, and at that thickness the insulation was only about half that for the snowshoe hare, *Lepus americanus,* with a fur of similar thickness. Much of the insulating capacity in these tropical species was attributed to the skin. Attempts to obtain values for Arctic birds, ptarmigan and snow buntings, were less successful, as measurements of the thickness of feathers could not take into account the fluffing normal to birds in the cold. The insulating capacity of the fur of aquatic mammals could not be evaluated in the same way, because in water the fur becomes filled with water and loses most of its insulating capacity. Harbor seals, *Phoca hispida,* and polar bears, *Thalarctos maritimus,* have thick layers of blubber under their skins. The insulating capacity of the blubber was measured with same apparatus immersed in ice water. Polar bear blubber was 55–60 mm thick, and seal blubber was 60–70 mm thick. Polar bear blubber had the lowest insulating capacity measured, and that of the seal blubber was about equal to that of terrestrial fur 20 mm thick. The aquatic species, however, had to meet a temperature gradient of only 40°C, in contrast to the terrestrial species' gradient of at least 80°C.

The maintenance of endothermy is helped by insulative fur or feathers, but in addition it requires a delicate mechanism to detect and respond to even very minor changes in body temperature. That is the function of the anterior hypothalamus, located at the base of the brain, immediately above the pituitary gland. The hypothalamus acts as a very sensitive thermostat; it contains both heat-sensitive and cold-sensitive cells, and it receives impulses from temperature sensors located in the skin, in other parts of the brain, and in the spinal cord. The hypothalamus controls body temperature by stimulating a number of responses. Radiation of heat from the skin can be reduced by vasoconstriction of the blood vessels, forcing most of the blood into internal organs; extra heat can be produced by shivering, which is muscular activity that is not coordinated and does not result in movement. Finally, temperature can be raised by increasing the rate of metabolism, or the oxidation of food or stored fat. It has been found convenient to compare different species by measuring their metabolic rates when there is no muscular activity. This is known as the normal resting metabolic rate. It is measured in terms of the amount of oxygen consumed or the amount of carbon dioxide produced by the resting mammal. As this varies with the size of the animal, it is converted to a value of 100 for each, and their responses to the ambient temperature can thus be compared directly as the value relative to 100 changes. For a well-insulated individual, the resting metabolic rate remains constant over a large range of ambient temperatures. For four different white foxes, the rates were measured repeatedly (Figure 4.3a), and no change was detected at any temperature between +30°C and –30°C. For two polar bear cubs, about twice as large as the foxes, but less well insulated, the rest-

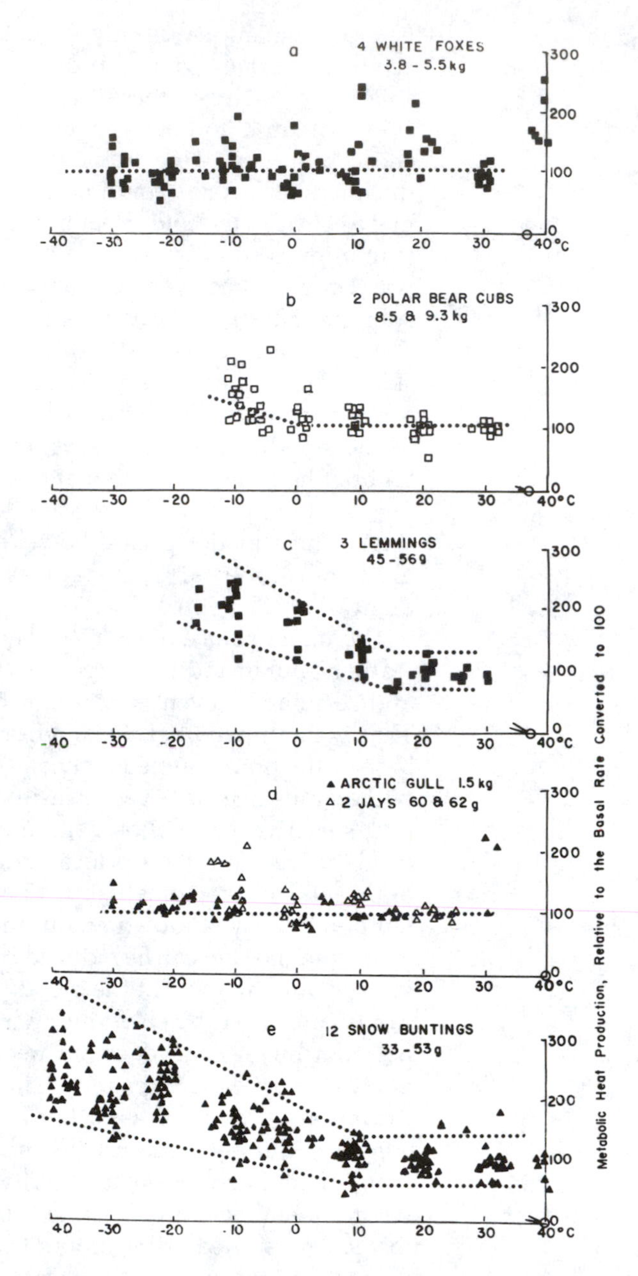

Figure 4.3. *Metabolic heat regulation by Arctic mammals and birds. Horizontal axis in degrees Celsius; vertical axis relative to resting metabolic rate, which has been normalized to 100: (a) 4 white foxes, (b) 2 polar bear cubs, (c) 3 lemmings, (d) Arctic gull and 2 jays, (e) 12 snow buntings. (Redrawn from Scholander et al. 1950.)*

ing metabolic rate remained constant between 30°C and 0°C; below that temperature the rate rose (Figure 4.3b). For three collared lemmings, *Dicrostonyx groenlandicus,* about 1% of the weight of the foxes, the rate rose at temperatures of 10°C and lower (Figure 4.3c). Three species of birds were tested. An Arctic gull, *Larus hyperboreus,* responded like the foxes; two gray jays, *Perisorius canadensis,* were similar to the polar bear cubs (Figure 4.3d); and twelve snow buntings, *Plectrophenax nivalis,* responded like the

Table 4.4. *Standard metabolic rates (SMRs) for house finches studied at different localities*

Population and season	SMR (mW/g)
Riverside, California	
Winter	16.47
Spring	16.02
Boulder, Colorado	
Winter	17.70
Spring	17.47
Michigan-Ohio	
Winter	17.92
Spring	16.58

Source: Root et al. (1996).

lemmings (Figure 4.3e). The temperatures below which the resting metabolic rates rise are the lower critical temperatures. The range above that is the thermoneutral zone. For the foxes and the gull, the width of the thermoneutral zone was not found in these experiments. This example shows that only in the case of the larger endotherms was insulation effective. The lemmings, snow buntings, and jays had to switch to increased metabolic heating at the moderately cool temperatures of 10–15°C. The term "resting metabolic rate," also called the standard metabolic rate (SMR), might be thought to imply some constant value, fixed for each species.

These rates, however, have been found to vary both seasonally and geographically within species. The small house finch, *Carpodacus mexicanus*, was originally distributed in tropical and subtropical areas of Mexico and the southwestern United States. It has spread naturally throughout the West Coast and as far east as Colorado. In 1940, some of these birds were introduced from southern California to Long Island, New York, and that population has increased and spread throughout the East from southeastern Canada and Michigan to North Carolina and Indiana. Those range extensions have exposed this species to more rigorous winter conditions than are found in its original habitat. In experimental tests of the extent to which birds from different locations could acclimatize to seasonal conditions, SMRs were determined for specimens from Riverside, California, from Boulder, Colorado, and from Michigan and Ohio, in spring and in winter (Dawson et al. 1983; Root, O'Connor, and Dawson 1991). Oxygen consumption in closed containers was measured. The results, expressed as milliwatts per gram, are shown in Table 4.4. There were no significant differences among the three groups in spring, but

Table 4.5. *Cold resistance in free-living house finches*

Population	Season	No.	Period over which homeothermy maintained at <60°C (min)
California	Spring	9	6.0
California	Winter	13	15.5
Colorado	Spring	6	8.8
Colorado	Winter	11	97.5

Source: Dawson et al. (1983).

Table 4.6. *Cold resistance of captive house finches during winter in southeastern Michigan*

Population	No.	Period over which homeothermy maintained at <60°C (min)
California	10	125.6
Colorado	6	138.7

Source: Dawson et al. (1983).

the SMR for California birds was significantly lower in winter than that for either Colorado or Michigan–Ohio birds. Thus, birds from populations that are exposed to rigorous winters respond by increasing their SMRs, showing that such rates are not fixed. In an earlier set of experiments it had been shown that they could also maintain constant body temperatures via increased metabolic rates longer than could those from a mild climate (Dawson et al. 1983). Birds in Riverside, California, and Boulder, Colorado, were captured in spring and winter, and all were exposed to extreme cold (–60°C) to determine how long they could maintain homeothermy (constant body temperature) (Table 4.5). Spring birds from Colorado were not significantly different from either spring or winter birds from California, but winter birds from Colorado maintained homeothermy up to sixteen times as long as did birds from California or birds from the same Colorado population in spring. Birds from both populations were transferred to an outdoor aviary in Ann Arbor, Michigan, and were subjected to the same test in January after 1–6 months in captivity. As can be seen in Table 4.6, they maintained homeothermy much longer than did the wild-caught birds, and there was no significant difference between the means for birds from the two localities. It was also found that there had been no increase in the insulation provided by the feathers. The implication of these results is that only direct responses to prolonged exposure to a different climate were involved, not any evolutionary change in genetic makeup.

Some small species improve their resistance to cold by behavioral means, such as huddling by flying squirrels, *Glaucomys volans*, and the construction of nests. The other nonmetabolic means of temperature regulation, the use of countercurrent heat exchangers between arterial blood and venous blood where feet, flippers, or the ears of jackrabbits are exposed to extremes of cold or heat, was described in Chapter 1.

Although there is no direct control of metabolism by the hypothalamus in ectotherms, there are indications of mediation of thermoregulatory behavior by thermally sensitive neurons in the brain stem. Thus, many reptiles maintain surprisingly uniform temperatures through the day. This is accomplished by changing positions with respect to sun and shade. Without the sun, there is little that they can do to maintain body temperature. However, one reptile, the python *Python molurus*, when incubating eggs, has been found to raise its body temperature by shivering, achieving as much as a 5°C increase above ambient.

Heat

Endothermic vertebrates have body temperatures in the general range of 35–40°C, or even higher for some birds. For most environments, that is high enough to be above the ambient temperature most of the time. A substantial advantage in maintaining a high temperature is that evaporative cooling, entailing loss of body water, is avoided in most environments. As described earlier, there are various mechanisms by which a normal body temperature can be maintained when ambient temperatures are significantly lower.

In some environments, specifically hot deserts, ambient temperatures rise above endothermic body temperatures, and that poses severe problems, as there is only one physiological mechanism by which the body temperature can be prevented from rising along with ambient temperature: evaporative cooling. To understand the problem, a simple equation for heat balance is needed:

$$H_s = \dot{H}_m + \dot{H}_c + \dot{H}_r - \dot{H}_e \qquad (4.1)$$

This says that the total heat stored is equal to the amount gained through metabolism, plus the amount gained or lost through convection and conduction, plus the amount gained or lost through radiation, minus the amount lost through evaporation. The amount of evaporation, as a percentage of body weight per hour, that is necessary to maintain a constant body temperature depends on the weight of the mammal (Figure 4.4). These data are for hot desert conditions. As mammals cannot tolerate water losses greater than 10–20% of their body weight, any mammal smaller than a jackrabbit will die if exposed to such conditions for more than an hour, at most. As mammals must keep their total heat approximately constant, and $\dot{H}_c$ and $\dot{H}_r$ are positive in hot deserts, and metabolic heat production cannot be lowered except

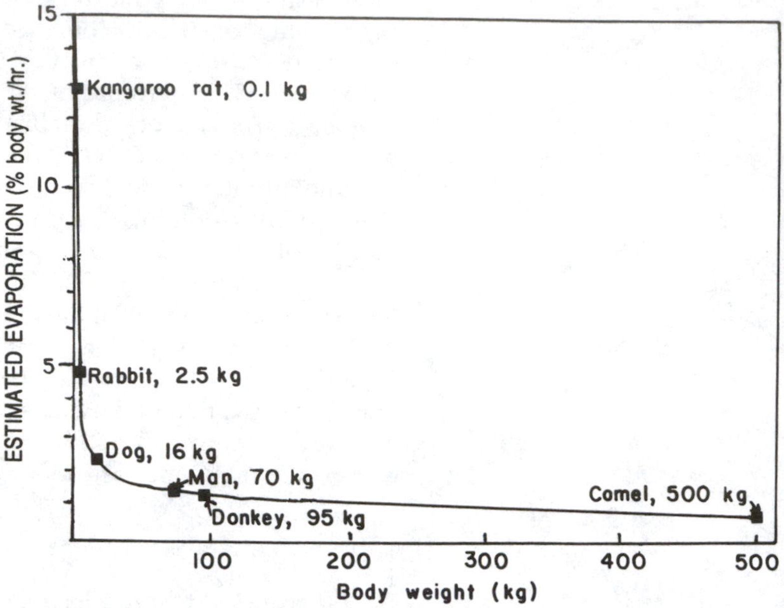

Figure 4.4. *Estimated evaporation necessary to maintain a constant body temperature in a hot desert environment for animals of various body sizes, calculated on the assumption that heat load is proportional to body surface. (Redrawn from Schmidt-Nielsen 1964.)*

through hibernation or torpor, $\dot{H}_e$ must increase as the ambient temperature rises toward body temperature (Figure 4.5). When the two are equal, heat loss through evaporation must equal metabolic heat production, and above that must exceed it. It is not surprising that jackrabbits (*Lepus alleni*) seek shade during the day. Smaller desert mammals must have burrows to which they can retreat during the day, and most of them are nocturnal, which of course allows them to avoid the problem. Ground squirrels, *Ammospermophilus leucurus,* however, are diurnal in activity, and they have been observed to remain outside their burrows for short times only, returning frequently to the burrows, where the temperature remains no higher than 27°C, both to lose body heat and to take advantage of the higher humidity there.

Desert mammals and birds are known to be lighter in color than conspecifics or related species in humid climates, and this has been attributed to the advantage of reflecting a larger proportion of the solar radiation, thus reducing heat gain in comparison with darker varieties. That the question is more complicated than that was shown in a study of representatives of two populations of golden-mantled ground squirrels (Walsberg 1990). Formerly classified as members of a single species, *Spermophilus lateralis* from northern Arizona and *S. saturatus* from the eastern Cascade Mountains of Washington differ greatly in color, *lateralis* being much the paler. As expected, the fractional reflectivity of the mid-dorsal region of *lateralis* was 0.30, 58% greater than that of *saturatus,* which was 0.19. When the solar heat loads transferred to

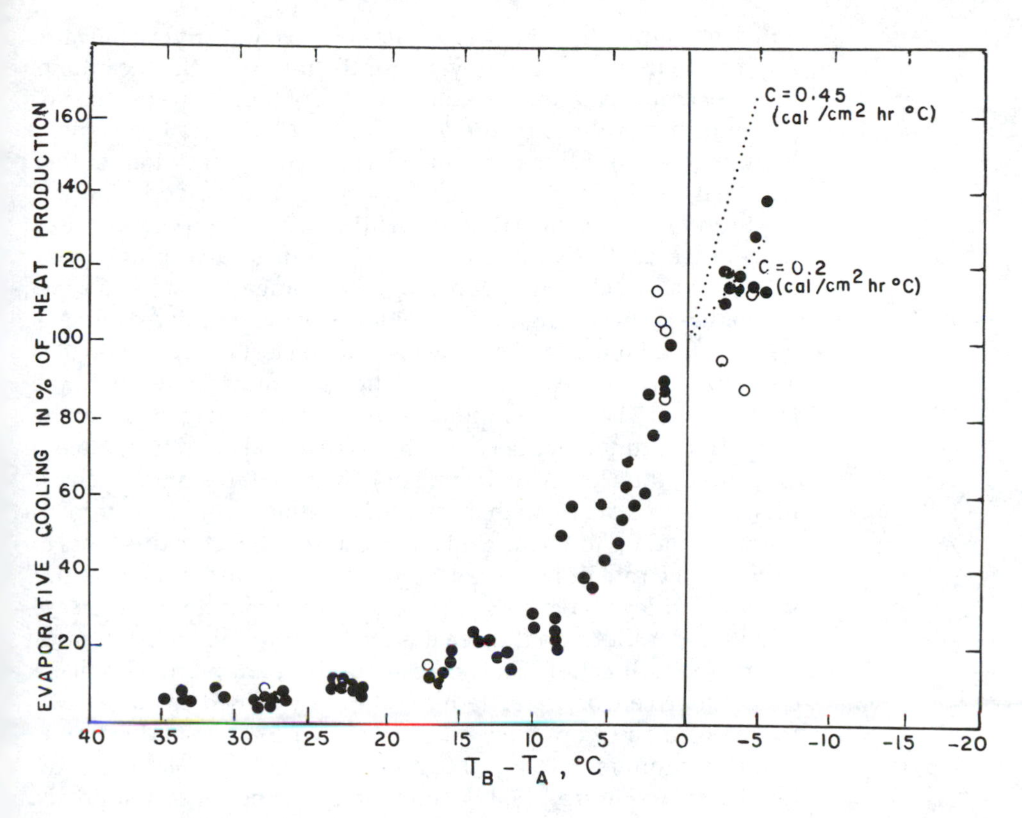

Figure 4.5. Evaporation as percentage of heat production by jackrabbits in relation to the difference between body temperature and ambient temperature. Open circles represent experiments in which body temperature changed more than 0.2°C during a 30-min run. (Redrawn from Dawson & Schmidt-Nielsen 1966.)

the skin were measured, however, they were not distinguishable. It was found that the difference in reflectivity was attributable only to the outer coat. The middle coat and inner coat are significantly thicker in *saturatus*, with the result that the mean radiative heat gain at the skin was measured as 54% for *lateralis* and 52% for *saturatus*, a negligible difference that is in the unexpected direction, given the reflectivity values.

Large mammals in the desert have problems somewhat different from those of small mammals. They cannot escape into burrows, but as described earlier, they do not have to lose such great percentages of their body water to maintain a constant temperature. Nevertheless, specific adaptations to desert conditions are necessary. Humans can keep their body temperature down by sweating, but within a day in the hot desert, a human would be prostrate from water loss unless that water could be replaced. They are not physiologically adapted to desert conditions, and must rely on cultural behavior patterns to survive. The best-studied large desert mammal is the camel, *Camelus dromedarius*, of the Sahara Desert. Contrary to popular legend, a camel does not store water, either in its hump or in a compartment of the stomach. It can survive for a week in the desert without drinking, but eventually it must have water. When water is available, a thirsty camel will drink as much as one-third of its body weight, giving the false impression that it is storing water for the future. In the meantime, the camel must balance Equation 4.1. One of the means for accomplishing that is through the term H_s, heat storage. The camel's body temperature fluctuates more widely than that of any other mammal not in torpor, especially when it is deprived of water (Figure 4.6). That means that during the day, its temperature rises to as much as 40.7°C, and during the night it falls to as low as 34°C. For a camel weighing 500 kg, the storage amounts to 2,900 kcal of heat. The saving in the avoidance of evaporative cooling is equal to five liters of water. The heat can be lost at night, which is always cool in the desert, by conduction and radiation without the loss of water. The high temperature during the day means that the temperature difference between the camel's body and the ambient is reduced, thus lowering its rate of gain from the environment. It may come as a surprise that camels have thick fur, 50–65 mm on the back, and up to 110 mm on the hump. The effect of this fur is to decrease the rate of heat flow from the environment. The ventral surface and legs have short hair, only 15–20 mm long. These are folded under the animal when it is resting, minimizing heat transfer from that direction. The net result of these adaptations is that a camel deprived of water reduces the heat gained from its own metabolism and from the environment (to be stored and later lost to the environment at night) to about one-half the total for a watered animal (Figure 4.7). Like humans and other large mammals, camels sweat as a means of evaporative cooling. A few mammals cool themselves by salivating and licking their fur, but that is an inefficient method. More species, including many carnivores and some small ungulates, use panting. They mostly lack sweat glands, although dogs have such glands on their feet. Panting would be almost self-defeating if the process did not involve shallow breathing, because the consumption of oxygen would raise the metabolic rate and hence the temperature, and the excess amount of carbon dioxide lost would raise the alkalinity of the blood. Panting by a dog involves a shift from 30–40 breaths per minute to 300–400, without an intermediate stage. It has one advantage over sweating in that it does not involve loss of salts, because the saliva is retained.

Desert birds lose heat by evaporation through the skin, despite their lack of sweat glands. This can amount to 45–75% of their total evaporation, the remainder coming from the trachea, the bronchial passages, the buccal cavity, and even the corneas. The major exception to cutaneous evaporation occurs in the ostrich, *Struthio camelus*, in which it amounts to less than 2% of the total evaporation. Some birds have been observed to lift the wings away from the body under hot conditions, and this is attributed to the near absence of feathers under the wings. Most of their heat must be lost through evaporation, either by increased respiration (oxygen consumption has been observed almost to triple as the

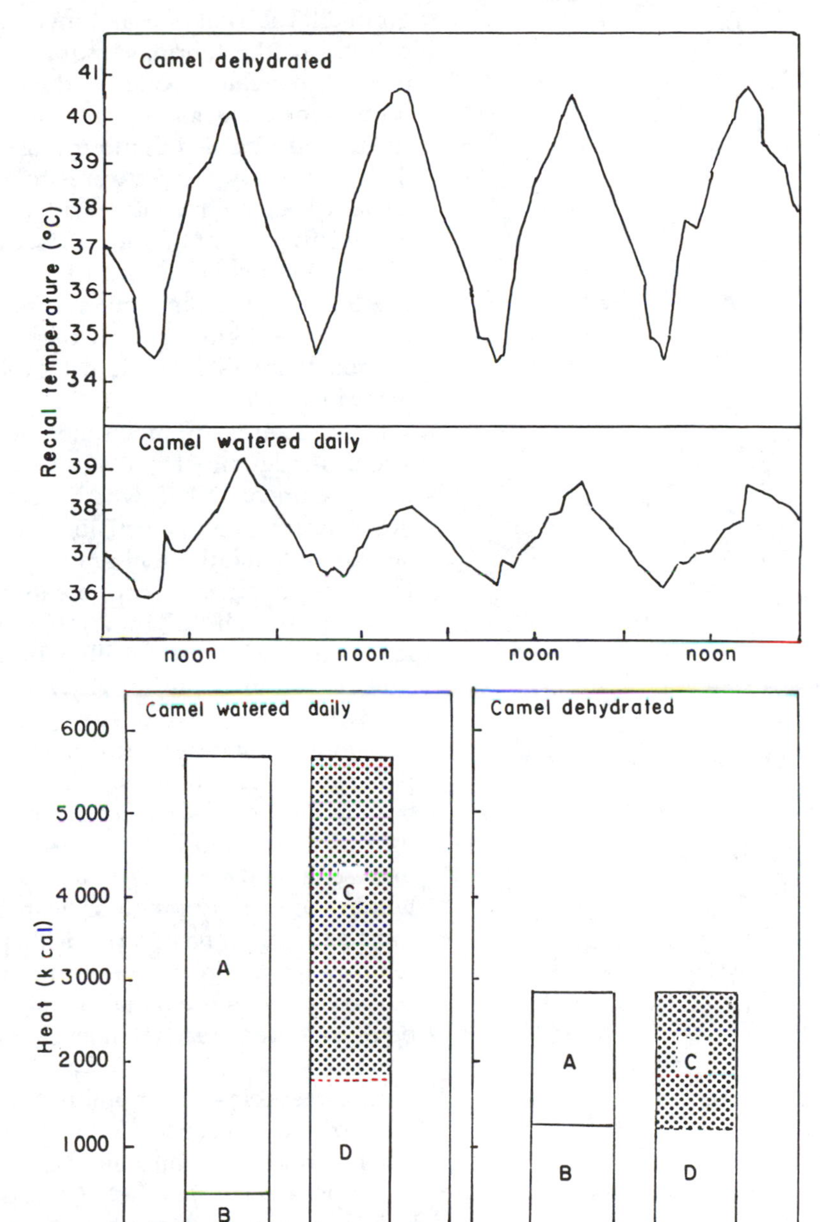

Figure 4.6. Changes in rectal temperature of a camel over 4 days when it was watered daily and 4 days when it had been deprived of water. (Redrawn from Schmidt-Nielsen 1964.)

Figure 4.7. Environmental heat gain by a camel, calculated as the difference between total heat load and metabolic heat production. The total heat load is determined as the sum of heat dissipated (A) and stored heat (B). The metabolic heat (D) can be subtracted from the total, leaving an estimate of the environmental gain (C). In a dehydrated animal, the environmental heat gain was much smaller than that in the same camel watered daily. (Redrawn from Schmidt-Nielsen 1964.)

temperature has been raised from 27°C to 42°C) or by fluttering the loose skin under the throat. As birds' normal body temperatures are higher (40–42°C) than those of mammals, they have a greater tolerance for desert temperatures. Nevertheless, small and medium-sized birds restrict their diurnal activity to early morning and late afternoon, and remain in any shade during the rest of the day. Large desert birds spend much of the day soaring at heights of 700–1,000 m, where the temperature is 7–10°C below that at ground level. That does not involve a large expenditure of energy because of the thermal updrafts characteristic of that environment. The greatest potential danger from high ambient temperatures is for the eggs, which cannot tolerate temperatures greater than about 40°C. Adult mourning doves, *Zenaida macroura*, reduce their own body temperatures to a mean of 38.3°C, and by continually keeping in contact with their eggs during the hottest part of the day, prevent the egg surface temperature from rising above 38.6°C. As the only means of keeping their body temperature so much below the air temperature of 44–45°C is by evaporative cooling, these birds must experience severe dehydration before evening temperatures fall to a level that will permit the bird to go for water.

Desert lizards are much more tolerant of heat than are snakes. The desert iguana, *Dipsosaurus dorsalis*, remains active at a body temperature of 42.1°C, which would be harmful or lethal to other desert lizards, and is well above the 38.5°C that they maintain in an experimental thermal gradient. Preferred temperatures for ectotherms in other environments range from 15°C for sockeye salmon, through 22°C for rainbow trout and 27°C for western toads, to 32–35°C for the alligator (Hainsworth 1981). Most lizard species are diurnal in activity, and desert forms must adjust their behavior so as to avoid excessive temperatures. They do this by remaining in shade or returning to burrows or other retreats. They pant at increasing rates with increases in temperature, but this has been found to be relatively ineffective for keeping body temperatures at tolerable levels. Snakes are nocturnal animals, especially in deserts, and they are unable to survive at temperatures well within the activity range of lizards. The sidewinder rattlesnake, *Crotalus cerastes*, has a critical thermal maximum of 41.6°C, the temperature at which it loses muscular coordination. For practical purposes that is the lethal temperature, because if its body temperature were to reach that level, it would be unable to escape to a retreat.

A few species of amphibians exist under desert conditions. Most of those recorded as living in deserts live in streams and other permanent water surrounded by deserts, but spadefoots (*Scaphiopus hammondii*) live where the only water comes from sporadic and brief rains during a limited part of the year. The adults are able to dig deep burrows, where they spend most of their lives. During the period when rain can be expected, they move to the top of the burrow, and they come out immediately with any rain. They feed as rapidly as possible, and if enough rain has fallen to make temporary pools, they breed and return to their burrows. Development is rapid, and metamorphosis occurs within 3 weeks.

In the foregoing pages, the effect of evaporative cooling has been stressed. As that depends on a supply of water, and water is scarce or totally lacking in deserts, vertebrates there have evolved a number of adaptations to that situation. Ectotherms have a large advantage over endotherms because they do not generate large amounts of body heat that must be eliminated. In dry air at 23°C, two desert reptiles, the chuckwalla, *Sauromalus obesus*, and the desert tortoise, *Gopherus agassizii*, lose only 0.3% and 0.2%, respectively, of their body mass, daily, as water through the combination of evaporation from the skin and respiration. For the chuckwalla, 66% of its water loss is through the skin, and for the tortoise, 76%. The fraction lost is minuscule compared with those for the mammals, shown in Figure 4.4. Note that the figures for the reptiles are per day, and those for the mammals are per hour. An average desert tortoise weighs 3.5–4.0 kg. Both reptiles are herbivorous and are able to replace their water losses by feeding on succulent vegetation. Cactus plants contain as much as 80% water.

Unlike many desert reptiles, most birds must drink water to survive, and therefore they are restricted to locations sufficiently close to surface water that they can reach it at intervals of no more than 1–2 days. Those that apparently do not require drinking water are the raptors, which obtain water from the flesh of their prey, and some galliform species that can obtain water from the vegetation on which they feed. Birds share with reptiles the ability to excrete uric acid, instead of the urea excreted by mammals. Uric acid is excreted as a paste, with very little loss of water. Urea must be dissolved, and although some mammals excrete unusually high concentrations of urea or excrete it as low-solubility urate salts, water is lost from the body in the process. Desert rodents excrete urine whose urea concentrations are 3.5–6 times that for humans. Removing as much water as possible before feces are deposited is a second water-saving device. Kangaroo rats, *Dipodomys spectabilis*, lose 2.53 g of water for each 100 g of grain eaten; white rats lose 13.6 g. Even more remarkable is the ability of kangaroo rats to meet their entire water requirements from dry seeds or grain. There is a small amount of free water in barley at a relative humidity of 20%, but the rodent gets 90% of its water from that obtained in metabolizing the grain. They can live without either drinking water or eating vegetation, and they do not lose weight in the process.

References and suggested further reading

Dawson, T., & K. Schmidt-Nielsen. 1966. Effect of thermal conductance on water economy in the antelope jack rabbit, *Lepus alleni*. *Journal of Cellular Physiology* 67:463–72.

Dawson, W. R. 1984. Physiological studies of desert birds: Present and future considerations. *Journal of Arid Environments* 7:133–55.

Dawson, W. R., R. L. Marsh, W. A. Buttemer, & C. Carey. 1983. Seasonal and geographic variation of cold resistance in house finches *Carpodacus mexicanus*. *Physiological Zoology* 56:353–69.

Gans, C., & W. R. Dawson (eds.). 1976. *Biology of the Reptilia*, vol. 5. New York, Academic Press.

Hainsworth, F. R. 1981. *Animal Physiology. Adaptations in Function*. Reading, Mass., Addison-Wesley.
Hill, R. W., & G. N. Wyse. 1989. *Animal Physiology* (2nd ed.). New York, Harper & Row.
Hochachka, P. W., & G. N. Somero. 1984. *Biochemical Adaptation*. Princeton University Press.
Lagler, K. F., J. Bardach, & R. R. Miller. 1977. *Ichthyology* (2nd ed.). New York, Wiley.
Lyman, C. P., J. S. Willis, A. Malan, & L. C. H. Wang. 1982. *Hibernation and Torpor in Mammals and Birds*. New York, Academic Press.
Pengelley, E. T., & S. J. Asmundsen. 1972. An analysis of the mechanisms by which mammalian hibernators synchronize their behavioral physiology with the environment. Pp. 637–61 in *Hibernation and Hypothermia: Perspectives and Challenges*, F. E. South et al. (eds.). Amsterdam, Elsevier.
Prosser, C. L. 1991. (ed.). *Comparative Animal Physiology* (4th ed.). New York, Wiley-Liss.
Root, T. L., T. P. O'Connor, & W. R. Dawson. 1991. Standard metabolic level and insulative characteristics of eastern house finches, *Carpodacus mexicanus* (Muller). *Physiological Zoology* 64:1279–95.
Schmidt-Nielsen, K. 1964. *Desert Animals: Physiological Problems of Heat and Water*. Oxford University Press.
Schmidt-Nielsen, K. 1990. *Animal Physiology: Adaptation and Environment* (4th ed.). Cambridge University Press.
Scholander, P. F., R. Hock, V. Walters, F. Johnson, & L. Irving. 1950. Heat regulation in some arctic and tropical mammals and birds. *Biological Bulletin* 99:237–58.
Scholander, P. F., V. Walters, R. Hock, & L. Irving. 1950. Body insulation of some arctic and tropical mammals and birds. *Biological Bulletin* 99:225–36.
South, F. E., J. P. Hannon, J. R. Willis, E. T. Pengelly, & N. R. Alpert. 1972. *Hibernation and Hypothermia, Perspectives and Challenges*. New York, Elsevier.
Walsberg, G. E. 1990. Convergence of solar heat gain in two squirrel species with contrasting coat colors. *Physiological Zoology* 63:1025–42.
Walsberg, G. E., & K. A. Voss-Roberts. 1983. Incubation in desert-inhabiting doves: Mechanisms for egg-cooling. *Physiological Zoology* 56:88–93.
Yousef, M. K., S. M. Horvath, & R. W. Bullard. 1972. *Physiological Adaptations in Desert and Mountain Animals*. New York, Academic Press.

5

A field study in population ecology
The rusty lizard

Understanding the population dynamics of any species requires knowledge of the proportion of individuals surviving to each age and the average number of young produced per female during each time interval after the female became separated from her mother, whether as an egg or as a newborn individual. The requirements are thus easily stated, but the information can be difficult to obtain. With that information it is possible to calculate the rate of increase or decrease of the population, the size of the population at some future time, and the expected proportion of individuals in each age class. If we have independent observations of the population size and the proportional distribution of age classes, the method permits a check on the estimates of birth rates and survival rates at each age used in the initial calculations. As an example of the use of the method, I cite the excellent fieldwork by the late W. F. Blair on a population of the rusty lizard, *Sceloporus olivaceous*, on his property near Austin, Texas, from 1952 to 1956, and my calculations based on his published data. One of the strengths of Blair's study was that he was able to capture alive virtually every lizard on his 4-hectare (10-acre) property each year.

Egg laying took place from late April to late August. Thus, the lizard population consisted of distinct age classes. A few of them lived as long as 5 years, but Blair was able to identify only four classes by size: hatchlings of the summer in which they were laid as eggs, yearlings, 2-year-olds, and those older than 2 years.

For yearlings and older individuals, he obtained survival data by clipping the toes of each in a unique combination. Survival of that age class over the next year was determined by the number of marked lizards that were recaptured the next summer. These adult lizards had an average survival of 21.8% per year, there being no detectable differences in survival of the different age classes.

The eggs are laid in nests dug into the ground by the female.

Nest predation, mostly by snakes, was heavy, only 23.4% of the nests having escaped, but 93.6% of the eggs that were not eaten hatched. Survival from egg laying to yearling had to be estimated indirectly. Blair knew from his annual census the number of females in each age class, and he dissected specimens captured away from his property to determine the number of eggs per clutch for females of each age. That was found by counting the number of eggs in the oviducts, as these would be the ones laid. He also determined the number of clutches laid per female during each season. From that information he calculated the total number of eggs laid on his property each year. The number of yearlings, divided by the number of eggs laid the previous year, gave the survival from egg to 1 year of age. Blair's calculations gave survival rates for the 4 years of the study as 2.57%, 2.85%, 2.82%, and 5.57%, for an average of 3.45%.

Yearling females laid approximately one clutch, and older females laid an average of four clutches in each season. Blair obtained enough data on clutch size, in addition to the stated frequencies, to determine the fecundity during each reproductive season. For the analysis of such data, the findings are always stated in terms of the number of female eggs laid per female of each age class, because males, of course, produce no eggs. Therefore, clutch sizes are divided by 2, on the assumption that the numbers of female eggs and male eggs are equal.

Two methods of analyzing the data are available, depending on whether or not reproduction is continuous. In the case of the rusty lizard, reproduction continues for 4 months, or one-third of each year. Thus, either method can be used. The continuous reproduction model is somewhat easier to follow and is the method used here. It requires that the data on survival and reproduction be assembled in a life table, which consists of four columns: age, designated as x; survival to each age x, designated as l_x; fecundity during the period of which x is the midpoint, designated as m_x; and the product of the successive entries in the l_x and m_x columns. The crude life table, from the data described, is shown in Table 5.1.

The contribution of each age class to the next generation is the entry for that age class in the last column. The sum of these products is the net reproductive rate, which is conventionally symbolized by R_0. This is the ratio of increase (or decrease) per generation, provided that each entry in the life table remains constant over the years. If it is equal to 1.0, the population is calculated to be in equilibrium, with no significant change from generation to generation. If R_0 is greater than 1.0, the population is increasing, and if it is less than 1.0, the population is decreasing.

The censuses of yearling and older lizards showed 250, 216, 201, 178, and 197 individuals for the successive years 1952–6. Thus, qualitatively, the prediction of a declining population, based on observations of survival and reproduction, is correct. Population

Table 5.1. *Life table for* Sceloporus oliveaceous, *based on data in Blair's study*

Age in years (x)	l_x	m_x	$l_x m_x$
0	1.0000	0	0
1	0.0345	5.65	0.1949
2	0.0075	36.60	0.2752
3	0.0016	49.00	0.0784
4	0.0004	49.00	0.0196
Sum			0.5681

Source: Blair (1960).

analysis should be better than qualitative, however. Was the decline in numbers predicted correctly? Here there is a problem. The predicted decline is in terms of a generation, but the observed censuses were year by year. If we knew the number of years in a generation, we could solve the problem.

A potential solution comes from this equation:

$$R_0 = \lambda^T \tag{5.1}$$

where λ is the ratio of increase per year, and T is the number of years in a generation. Thus, if we have either λ or T, we can find the other. Fortunately, there is a famous equation in population ecology that allows us to use l_x and m_x to calculate λ:

$$\Sigma \frac{l_x m_x}{\lambda^x} = 1 \quad \text{or} \quad \Sigma l_x m_x e^{-rx} = 1 \tag{5.2}$$

which is more familiar to ecologists, and the two are the same, because $\lambda = e^r$. The equations have no direct solution for λ or r, but can be solved by trial and error. Use of the data in Table 5.1 gives a solution of 0.754 for λ, and approximately 2.0 for T. Thus, over the 4 years, the population should have declined to 0.754^4, or 32.27% of its original size of 250 individuals, or 80.67.

Inasmuch as the observed census of 1956 showed that there were 197 individuals present, the predicted value of about 81 is unacceptably low. Something is wrong with the values in Table 5.1. It should be noted that Blair did not make use of either method of analyzing the data for a quantitative prediction, and he was unaware of the discrepancy. The numbers in Table 5.1 were taken directly from statements in his text. A careful search of the detailed data in his tables and figures revealed two sources of differences between his text and the details. In summarizing the data, he had given rounded figures, which did not seem unreasonable when the text was read.

The first discrepancy came from the fact (ignored by Blair) that there was some movement of lizards onto the property from outside. These immigrants made up an appreciable number, and he included their potential reproduction in estimating the number of eggs laid on the property each year. That meant that the marked lizards were only part of the survivors, as there was no reason to suppose that emigrants were not as numerous as immigrants. Thus, if the immigrants are included as representing survivors that were not counted because they were off the property, corrections can be made in the estimated survival. The correction raises the estimated survival from egg to maturity to 0.0528, and the survival of adults to 0.251 per year.

A second revision that is necessary involves estimated reproduction by yearlings. Not all breed, but others lay eggs more frequently than the once per year stated in the text. The fraction breeding (0.956) can be found in one of Blair's tables. In another table, 66 egg-laying yearlings were cited, which represented a total of 69 in the age class. They produced at least 79 broods. Hence, the average number of clutches can be raised to 1.145, and the number of female eggs per yearling can be raised to $(11.3 \times 1.145)/2 = 6.36$. These can be apportioned to different parts of the 4-month breeding season by the proportion reaching breeding size.

As egg production declines through the season, the m_x values for older age groups must also be adjusted from data given in one of Blair's figures. By assuming constant death rates during and between the breeding seasons, l_x values can also be adjusted. Instead of even years of age, the median age at egg laying for each season is used. The corrected values are given in a new and more detailed life table (Table 5.2). Using the same equations as before, the new constants are $\lambda = 0.9554$, $r = -0.0456$, and $T = 2.000$ years.

Four years were involved, and we expect λ^4, or 0.833, as many lizards in 1956 as there were in 1952. This is 208.25, which is to be compared with the observed 197 – a much more satisfactory estimate than was the first attempt.

The expected proportional distribution of age classes, called the "stable age distribution," can be calculated in the following manner: With the proportion of the zero age category set at 1.000, the representation of any other age group x is $l_x e^{-rx}$. This provides another opportunity to cross-check Blair's detailed observations on survival and reproduction, because he gave the observed age distribution. Taking the time immediately before egg laying, as the youngest age group is approaching 1 year of age, the results are as given in Table 5.3. The agreement with the observations is outstandingly good.

This study and the mathematical analysis of the data illustrate both the importance of careful fieldwork and the value of the life table approach to population analysis. The analysis shows that what appear to be minor discrepancies between the text and the detailed data have important implications for the predicted fate of

Table 5.2. *Life table for* Sceloporus oliveaceous, *corrected according to detailed data in Blair's tables and figures*

Age in years (x)	l_x	m_x	$l_x m_x$
0	1.000	0	0
0.167	0.215	0	0
1.167	0.0415	6.36	0.264
2.167	0.0133	36.60	0.487
3.167	0.0028	49.00	0.137
4.167	0.0005	49.00	0.025
Sum			0.913

Source: Blair (1960).

Table 5.3. *Calculation of the stable age distribution and the proportion of adults in each age class compared with the observed proportions*

			Fraction in each age category	
x (yr)	l_x	$l_x e^{-rx}$	Expected	Observed
1.000	0.0528	0.0553	0.0553/0.0761 = 0.727	0.731
2.000	0.0152	0.0167	0.0167/0.0761 = 0.219	0.211
3.000	0.0030	0.0034	0.0034/0.0761 = 0.045	0.058 (3.000 and 4.000 combined)
4.000	0.0006	0.0007	0.0007/0.0761 = 0.009	
Sum		0.0761		

the population. Blair's observations were even better than he thought.

The constant r, called the "intrinsic rate of natural increase," is used in many mathematical models of population phenomena, including population growth, interspecific competition, predation, and models of community structure. As the best measure of how well a population is adapted to its physical environment, it can be used to compare growth rates for different populations of a given species under different physical conditions, and populations of different species under the same physical conditions.

Reference

Blair, W. F. 1960. *The Rusty Lizard. A Population Study.* Austin, University of Texas Press.

6

Field experiments on competing vertebrates

Nearly all experiments on interspecific competition are undertaken because someone thinks, on the basis of field data, that certain species are competing; otherwise, the probability of negative results would be so great that the effort involved could not be justified by the likely value of the study. As has been suggested, this reason for conducting such experiments may bias our impression of the frequency with which competition occurs in nature, but the cases are so numerous that it is worthwhile to present some examples.

Terrestrial salamanders

Many experiments testing for expected competition have arisen from the observation that two or more species shared a common resource, but some have come from detailed examination of the ecological distributions of two species. Such was the case for the altitudinal distributions of two species of salamanders that I studied in the Black Mountains of North Carolina. I made a series of transects up the sides of the mountains, stopping every 30.5 m (100 ft) of elevation to search for salamanders. The results for the genus *Plethodon* showed that two of the species had almost non-overlapping distributions, although for the mountain range as a whole the overlap in elevation was more than 518 m (1,700 ft) (Figure 6.1). The vagaries of taxonomy have been such that the two species involved have each suffered a name change. As discussed in Chapter 3, *Plethodon metcalfi* has been shown to be conspecific with *P. jordani*, the older name, and *P. glutinosus* has been divided into sixteen species, the one in the Black Mountains now having the name *cylindraceus.* My interpretation of the distributions was that the two species were in intense competition, because there was no vegetational or soil boundary corresponding to the separation. It should be explained that these species are completely terrestrial, with no larval stage. There is no

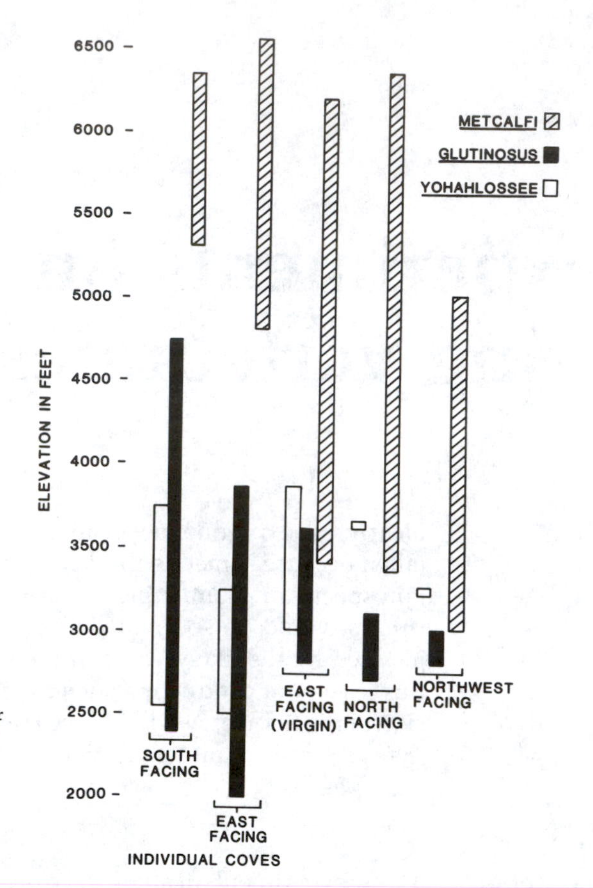

Figure 6.1. *Vertical distributions of three species of* Plethodon *in the Black Mountains of North Carolina, based on observations in individual transects. (From Hairston 1949.)*

migration to a pond or stream for breeding as is the case with many other genera.

Notice that competition is a sufficient explanation for the observed distribution, but it is not a necessary explanation, because there is at least one alternative. If *cylindraceus* is intolerant of cold and *jordani* is intolerant of dry conditions, these two factors could explain what we see, because the higher elevations are cooler and wetter, and the lower ones are warmer and have less rainfall.

Similar studies in other parts of the southern Appalachians revealed the interesting situation shown in Figure 6.2a,b. In the Great Smoky Mountains, the situation is exactly like that in the Black Mountains, with very little overlap on any transect, but in the Balsam Mountains the overlap is almost complete – more than 1,400 m. Clearly, if interspecific competition is sufficient to account for the distributions in the Blacks and Smokies, we have to conclude that competition is too weak in the Balsams to affect the distributions of the two species. The situation in the Balsams also eliminates the alternative explanation suggested for the situation in the Blacks.

All of this is plausible and interesting, but the discovery of two different situations meant that all explanations were in question.

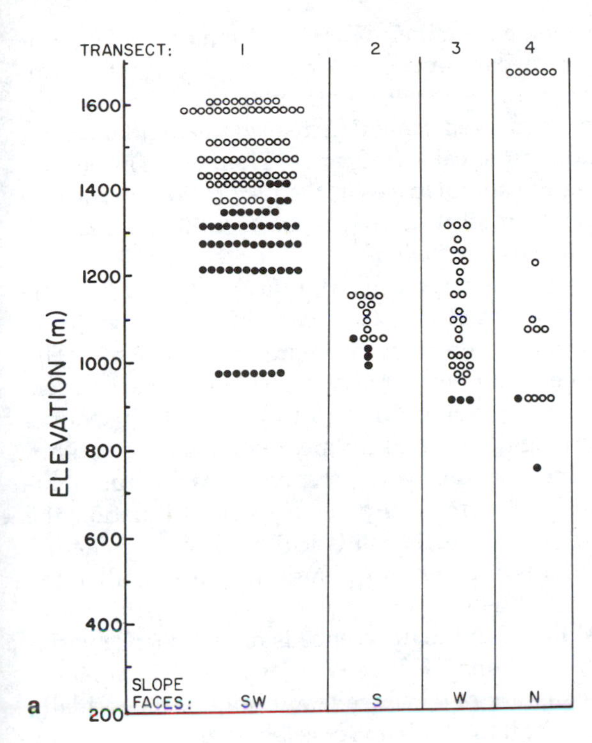

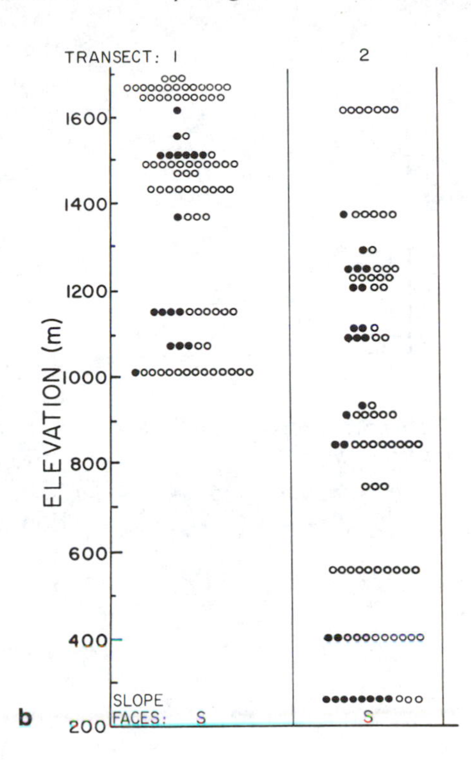

Figure 6.2. *Vertical distributions of two species of* Plethodon *in (a) the Great Smoky Mountains and (b) the Balsam Mountains of North Carolina. Symbols are specimens, except for transect 1, for which the symbols at each elevation represent means.*

The problem called for an experiment, and eventually I conducted the obvious one. It consisted in selecting areas in both the Smokies and the Balsams that were at the same elevation and otherwise were as much alike as possible (same direction of slope, same kinds of trees) and setting up permanent plots in each area. From two plots at each location *jordani* was removed, and from two plots *oconaluftee* (formerly *glutinosus*) was removed. Other plots were left as controls, where nothing was done, except for counting the salamanders present. Counts on all plots were made at night, because the animals are active above ground at night and it is not necessary to disturb the habitat by turning logs and stones. If the hypothesis of competition was correct, the species left on the experimental plots should respond to the removal of its competitor by increasing in abundance in the Smokies much more than in the Balsams. The minimum life cycle of *jordani* is 4 years, and that of *oconaluftee* is 5 years, and therefore the experiments were continued long enough for each species to respond by changes in all parts of the population: five seasons of salamander activity, from May 1974 until September 1978.

It should be noted that *jordani* is six times as abundant as *oconaluftee* in the Smokies, and eight times as abundant in the Balsams. Thus, one might expect a quicker response by *oconaluftee* to the experimental manipulation. The results were as predicted by the original hypothesis. As suggested, results were not obtained immediately, but where *jordani* was removed from plots in

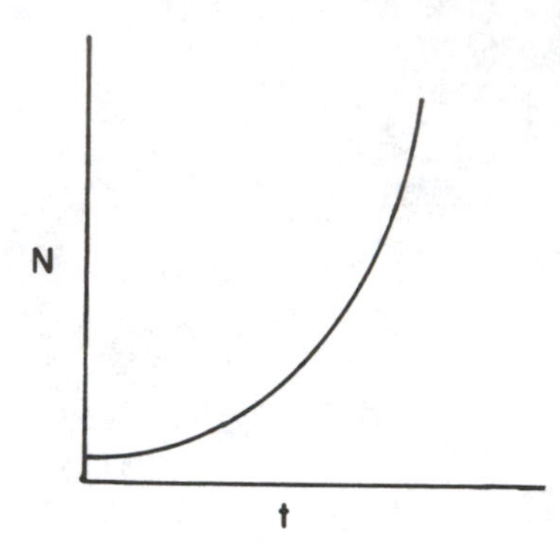

Figure 6.3. *Exponential growth of a population.*

the Smokies, the average number of *oconaluftee* increased significantly above the average number on the control plots for the third, fourth, and fifth years of the experiment. On the plots from which *oconaluftee* had been removed, *jordani* increased in abundance, but not enough to claim statistical significance. There were important and statistically significant increases in the proportions of the two youngest year classes, indicating that reproduction had been increased by the removal of *oconaluftee*.

What do we mean by "statistically significant"? In each of the six observation periods each year (two each in May–June, July–August, and September) there were two searches of each plot. The average number of each species for each treatment for each observation period was thus made up from four counts. If the counts differed greatly among themselves for any average, we could not have much confidence that the average represented the true situation. When two averages being compared are in that situation, the difference does not mean much. Statistical analysis is a mathematical method of finding out how sure we can be that the difference is real. The answer is always expressed in this form: What is the probability that the difference is due to chance variation among the observations? If there is any variation at all, there is always *some* probability of a chance result. Biologists normally use the convention that if the probability is less than one in twenty (<0.05), the difference is accepted as "real" or "significant." They accept the fact that one time in twenty they will be mistaken.

In the Balsams, to my surprise, similar results were obtained. In the case of *oconaluftee* where *jordani* had been removed, however, the increase did not occur until the last year of the experiment, and the increase in the proportion of young *jordani* was less than that in the Smokies. Thus, although some competition was detected, it was less intense than in the Smokies, and the result was observed much later in the experiment. The results were as predicted in the original hypothesis: Competition is much more intense in the mountain range where the altitudinal overlap is narrow than in the range where the overlap is wide.

More can be made of the results of these experiments than the test of the initial hypothesis. We can use the numerical data to understand the theory of interspecific competition. To understand the theory, we start with population dynamics. If we imagine a population growing freely, we can see that a graph of numbers against time will become steeper indefinitely. Thus, if the population doubles each month, then after 1 month there will be twice as many as at the start, after 2 months four times as many, and so forth (Figure 6.3). Mathematically, this process is described by the equation for exponential growth:

$$\frac{dN}{dt} = rN \tag{6.1}$$

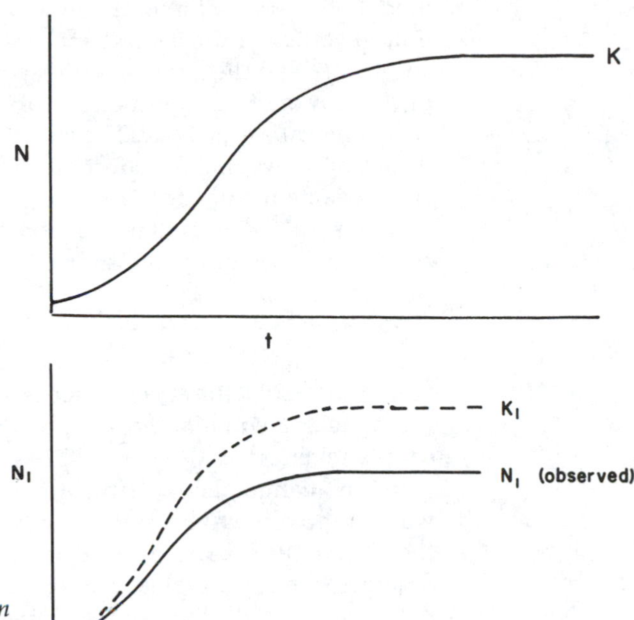

Figure 6.4. Density-dependent growth of a population, according to the logistic equation.

Figure 6.5. Growth of a population in interspecific competition, according to the Lotka–Volterra equations for competing species.

where N is the number of individuals present at time t, dN/dt is the instantaneous growth rate of the population at time t, and r is the "intrinsic rate of natural increase" – a constant for a species under specified conditions, as explained in Chapter 5. Quite obviously, no population grows like that for more than a short time. When it is tried experimentally, we find that the rate of population growth per individual present, $dN/N\,dt$, does not remain constant as it does in the exponential equation, but declines as the population increases. That results in a graph of the population history that is S-shaped, so that eventually the population growth stops, and the population remains constant (Figure 6.4). To describe such a growth curve, we need a mathematical term that allows each newly added animal to decrease the growth rate per individual:

$$\frac{dN}{dt} = rN\left(\frac{K-N}{K}\right) \qquad (6.2)$$

where K is the maximum number of individuals that the environment can sustain. Thus, as N increases towards K, $(K - N)$ approaches zero, as does the whole right side of the equation, making the growth rate zero also. This is the *logistic equation*, which expresses intraspecific competition. Now consider interspecific competition, which is our concern with the salamanders. Let us imagine a second species, the effect of which is to decrease the growth rate of the first and to prevent it from reaching its K (Figure 6.5). This can be expressed mathematically in the simplest model of interspecific competition by adding to the logistic equation the effect of the second species. Call the number of the first species N_1, and the number of the second species N_2. Similarly, the intrinsic rate of natural increase of the first species is r_1, and its "carrying capacity" is K_1. The equation should also contain some means of comparing the effect of a single individual of N_2 with the effect of one member of N_1. Thus, we multiply N_2 by a constant that makes the conversion. This constant is the competition coefficient, which we designate as α:

$$\frac{dN_1}{dt} = r_1 N_1 \left(\frac{K_1 - N_1 - \alpha_{1,2} N_2}{K_1} \right) \tag{6.3}$$

where α is read: "the effect on species 1 of species 2." Clearly, there is a matching equation for the growth of species 2, with the subscripts reversed.

In the salamander experiment, the plots were not fenced, and it was not possible to keep the removed species completely out of the plot. In the case of the effect of *P. jordani* on *P. oconaluftee* in the Smokies, the control plots contained an average of 32 *jordani*, but the experimental plots still contained an average of 14. Keeping the density down by 18 per plot resulted in an increase of an average of 3.5 *oconaluftee* above the controls. Thus, the effect of each *jordani* was 3.5/18 of the effect of one *oconaluftee*, or 0.194. That number is $\alpha_{o,j}$, or the effect on *oconaluftee* of *jordani*. For the Balsams, the comparable figure was 0.14. It was also possible to calculate the effect on *jordani* of *oconaluftee* by projecting the added young individuals into the future to give increased densities of *jordani* where *oconaluftee* had been removed. Thus, all of the α values could be calculated.

It should be clear that we did not observe the populations while they were growing, which must have taken place long ago. The counts on the control plots showed that there was no consistent trend. Therefore, we can ignore the parts of the equations that describe the situation before the populations stopped growing, and that allows us to set dN_1/dt and dN_2/dt equal to zero. When that is done, and we multiply both sides of the equations by K_1/r_1N_1 or K_2/r_2N_2, we obtain

$$0 = K_1 - N_1 - \alpha_{1,2}N_2 \quad \text{and} \quad 0 = K_2 - N_2 - \alpha_{2,1}N_1 \tag{6.4}$$

In these equations, we know N_1 and N_2 from the average counts on the control plots, and we can also calculate the α values. That allows us to calculate K_1 and K_2. The complete set of counts and the experimentally determined values of the constants in the equations are given in Table 6.1. It should be pointed out that the values of K_j and K_o are correct only if there are no other species that reduce the rates of population increase. That apparently was the case for the other eight species of salamanders that were present on the plots, because neither the removal of *jordani* nor that of

Table 6.1. *Average densities of* Plethodon jordani *and* P. oconaluftee *on control plots, and calculated values for the constants in the competition equations for the experiments in the Great Smoky Mountains and the Balsam Mountains*

	Smokies	Balsams
N_j	32.03	44.58
N_o	5.42	4.53
$\alpha_{o,j}$	0.194	0.14
$\alpha_{j,o}$	2.255	0.63
K_j	45.27	47.43
K_o	11.69	10.93

oconaluftee had any effect on their populations, and it is likely that the reciprocal relationship also holds. It is noteworthy that the values of K in the two areas are very similar, indicating that the locations chosen were equally favorable for both species.

The experiments demonstrated that competition was the explanation for the altitudinal distributions, but revealed no evidence as to the resource for which the salamanders compete. An obvious possibility was food, but that was shown not to be the limiting resource by comparing the stomach contents of the two species in the two mountain ranges. Had food been the object of competition, the food items taken should have been less alike where competition was weaker. The reverse was found: Where the two species were in more intense competition, the food of the two species of salamanders (different species of arthropods and other invertebrates) was less alike than where competition was less intense. A second experiment was performed, with *jordani* being transplanted between mountain ranges to test for its direct effect on *oconaluftee* in the unfamiliar area. Specimens transferred from the area of intense competition (Smokies) to the area of less intense competition (Balsams) had a strong negative effect on *oconaluftee*. Had the two species in the Balsams evolved to be less competitive, there should have been little effect, because the *oconaluftee* should have been tolerant to any form of *jordani*. The reciprocal transfer of *jordani* from Balsams to Smokies gave the reverse result: The Smokies *oconaluftee* increased in abundance, just as they did when the local *jordani* were removed. That suggested that the Smokies *oconaluftee* were superior competitors compared with the *jordani* from the Balsams. Laboratory experiments revealed that both species from the Smokies were as aggressive toward each other as they were toward their own kind, whereas in the Balsams intraspecific aggression was much stronger than was aggression toward the other species. The combination of results demonstrated that in the Smokies, evolution had resulted in increased competitive ability. The aggression evidently is advantageous in protecting a place. Thus, the limiting resource is space, but the use made of the space is not known. They do inhabit burrows and are remarkably faithful to the immediate vicinity. In a study lasting an entire season of activity, sixty-four individuals (forty *jordani* and twenty-four *oconaluftee*) were marked with fluorescent dye and followed by repeated night searches. For all observations, the areas traversed averaged less than 2 m^2 per individual.

Desert rodents

Small rodents occur in great variety in the southwestern United States. The family showing the greatest diversity is the Heteromyidae. They are specialized for arid climates and feed exclusively on seeds. They never need to drink water, having the physiological ability to subsist on water that is derived metabolically from their food. All heteromyids have fur-lined cheek pouches. Species of the genus *Dipodomys* (kangaroo rats) have long hind legs and tails, and that makes it efficient for them to progress by hopping rather than by running. Some twenty to twenty-five species of Heteromyidae occur in the four-state area of Nevada, Utah, Arizona, and California. A second family of rodents, the Cricetidae, is also common in the area. These feed on insects and snails, in addition to seeds, and thus are less specialized in diet than are the Heteromyidae.

Despite the great diversity in the area, no more than six species of these two families occur in any single habitat. That kind of observation is common, and the conventional explanation is that competition prevents additional species from coexisting with the residents. In the case of the rodents, that suggestion is strengthened by the fact that in no case are any two species of the same size (Figure 6.6). A famous proposition in ecology holds that consistent size differences among coexisting related species represent an evolutionary response to interspecific competition. It is thought to be one form of the general phenomenon of niche partitioning. The proposition seems plausible, but mathematical "proofs" are unconvincing (see Part VI). The theory is that species must be different in their utilization of a resource that is present in a limited supply, or one of them will displace the other by interspecific competition, it being almost impossible that they would be exactly equally efficient in using the identical resource. Thus, natural selection favors slight differences between them in their use of the resource. In the case of the rodents, it was thought that species of different sizes would each use seeds of a particular narrow range of sizes and would be more efficient in using that size of seeds than would the other species in the habitat. Removal of seeds from the cheek pouches of a number of species confirmed the sugges-

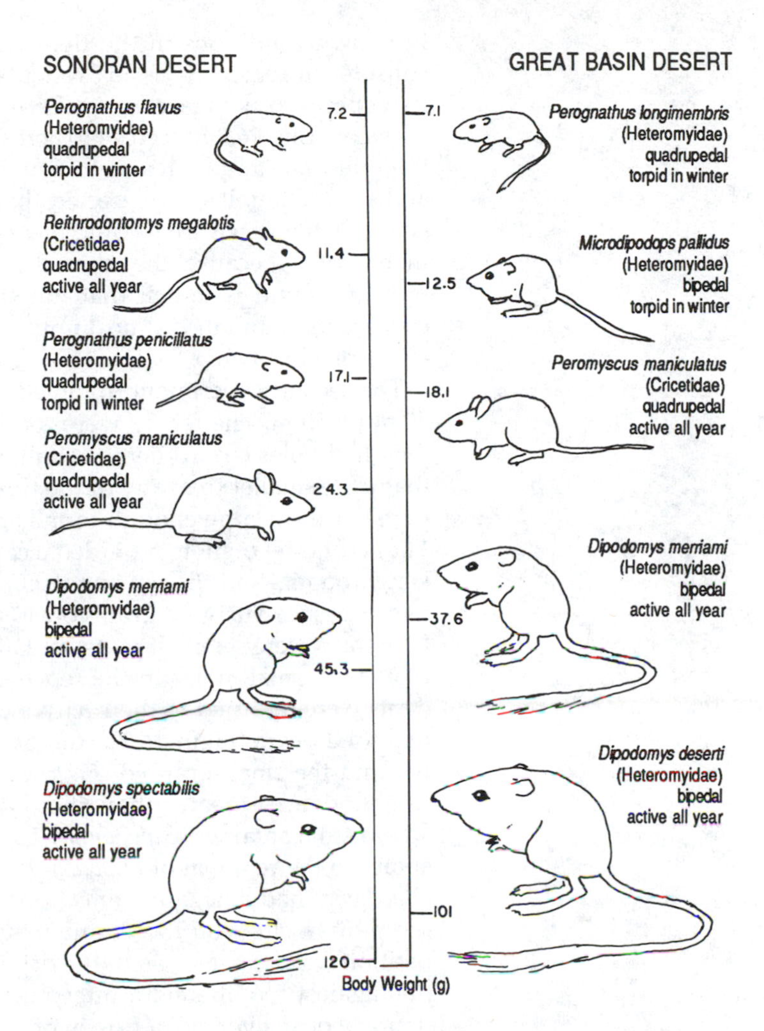

Figure 6.6. *Sizes of small granivorous rodents in two deserts in the southwestern United States. (Redrawn from Brown 1975.)*

tion, as there were close correlations between the size of the seeds in each pouch and the size of the animal, and the same correlation was found among the rodents of two distinct desert types.

It should be explained that seeds are more abundant, especially in the desert, than is generally supposed. When one hears of the desert blooming after a rain, that is because of the presence of the seeds of annual plants that have been stimulated to germinate, grow, and flower. The seeds of these annuals are adapted to remain dormant until the next rain, which may come several years later. It is this large supply of seeds that is exploited by the rodents.

Thus far, we have only the hypothesis that the supply of seeds limits the abundance of the rodents and that competition for seeds is important to the community. That hypothesis has led to several experiments. Unfortunately, not all experiments yield significant findings, especially those conducted in the variable and only partly known conditions in the field. The first experiment we shall consider in regard to desert rodents involved removing each of two species of kangaroo rats (*Dipodomys*) from replicated areas, each covering 16.2 hectares (40 acres). Those unfenced plots were thought to be ample in size to avoid the problem of immigration and emigration. It was expected that if they were competing, the species left on each area would increase in abundance. The experiment failed because the animals, despite inhabiting burrows, moved around so much that the species that was removed recovered its abundance almost immediately, leaving no time for the experiment to work.

The second experiment involved eight fenced areas measuring 50 m by 50 m. The fences were too high for rodents to scale, but they had holes cut in them that allowed passage only to rodents that were smaller than any of the three species of *Dipodomys* (four plots) or were large enough for all species of rodents (four plots). The second set of plots provided a control on both the exclusion of kangaroo rats and the fencing itself. The smaller rodents were of two types: exclusively granivores (seed eaters), and omnivores. Live traps were set in all plots, and all kangaroo rats caught in the plots with small holes in the fences were removed. All other rodents were returned to their native plots. The experiment continued for 4 years (Figure 6.7). Removal of *Dipodomys* was successful, and the small granivorous rodents responded by increasing to abundances greater than they attained on the control plots, which still contained *Dipodomys*. The small omnivores did not respond to the removal of *Dipodomys*. The experiment showed that *Dipodomys* had a negative effect on the small granivores. The absence of an effect on the omnivores was an indication that competition was for seeds, as in the original hypothesis. As a precaution against too liberal an interpretation, it should be noted that removal of an average of fifteen of the large *Dipodomys* resulted in the addition of only an average of around three or four of the small granivores.

A further experiment was conducted to test the hypotheses about the importance of the abundance and sizes of seeds. Eight of the plots were used, and the experiment involved the addition of 96 kg of seeds to each plot each year. There were two sizes of seeds, and two schedules for addition of seeds. The experimental design proved unnecessarily complex, as the results were the same for all plots: a significant increase in the abundance of *Dipodomys spectabilis*, the largest species, significant decreases in the abundances of the two smaller species of *Dipodomys* (*ordii* and *merriami*), and no significant effects on the remaining species of rodents. One might have expected a general increase in response to the addition of so much food. The findings indicate that there was some kind of interference by *D. spectabilis* on the other species. In any event, the consumption of the additional food did not

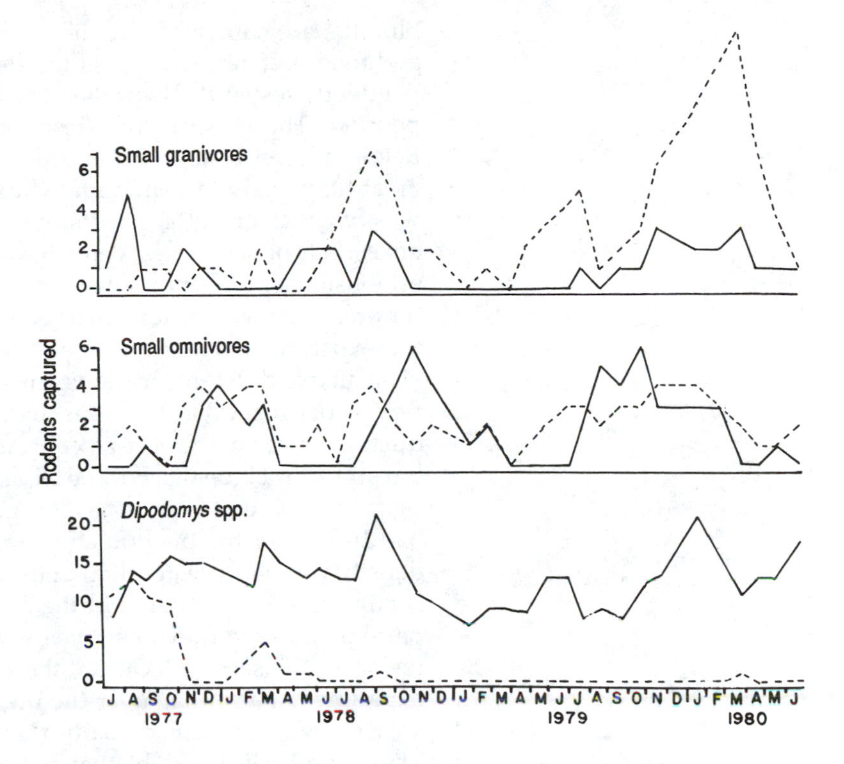

Figure 6.8. Population histories of small rodents during the first 4 years of an experiment involving removal of Dipodomys, *which is larger than the other species. (Redrawn from Munger & Brown 1981.)*

increase in proportion to the calculated benefit from an increase in the supply of the supposed limiting resource. Direct suppression of the closest relatives was more easily detected than was a response to the addition of food.

Forest birds

Students of birds contributed much ecological theory between 1945 and 1975, but the theory was based on the interpretation of observations, rather than on any experimental tests of hypotheses. Birds are more difficult subjects for experimentation than are most animals, at least partly because they cannot be confined to experimental areas. It should not be surprising that some of the experiments that have been attempted have yielded inconclusive or negative results. The two experiments to be described here were both conducted in Europe, and they involved real ingenuity.

Great tits and blue tits

Tits (closely related to American chickadees) have been studied intensively both in Britain and on the continent of Europe. They are abundant and relatively tame and are convenient for some experimental purposes because they nest in holes, thus being amenable to manipulation through their use of nest boxes provided by the experimenter. Great tits are the largest of the group;

blue tits are somewhat smaller, but have much the same habits and food preferences as great tits. Two individuals working independently tested the hypothesis that the two species are in competition. The experiments described here were carried out in Belgium. Nest boxes with entrance holes large enough to admit great tits can also accommodate blue tits, and the question asked was whether or not great tits were excluding blue tits from the boxes. Sets of nest boxes were observed over a number of years to establish the number of blue tits using the standard boxes, both in the areas that were to remain as controls and in those to be used in the experiment. The experiments were conducted in successive years in two different locations, thus providing a rare example of a field experiment that was repeated. The experiment involved reducing the size of the nest-box entrances so as to exclude great tits, but still admit blue tits. For the experiment in the first location, the density of blue tit pairs was 72% greater in the first year than it had been over the previous 5 years, and in the second year the density was 66% greater than in the 5 years before the experiment began. In the control area in the first year there were 10% fewer pairs than in the previous 5 years, and for the second year, 23% fewer. In the second location, there were 77% more pairs in the experimental area than over the previous 5 years, and the control area showed 24% fewer pairs than over the previous 5 years. There can be little doubt that the great tits had been excluding blue tits from the nest boxes before the entrance holes were reduced in size.

Magpies and jackdaws

For many years there has been a common implied assumption that related species are likely to be in competition, and that assumption has been verified experimentally frequently enough to justify its continued acceptance as a provisional hypothesis. As earlier parts of this chapter have shown, most competition experiments have dealt with related species. The salamanders, kangaroo rats, and tits are each congeneric, and many of the other rodents are in the same family as the kangaroo rats. The bird family Corvidae, to which our crows and jays belong, is well represented in Europe. Two common species of about the same size that are not shared with eastern North America are the magpie and the jackdaw. Their nesting habits differ, the former building nests in trees, and the latter nesting in holes. In most competition experiments, one species is removed and the other is monitored for possible increases in density, reproduction, growth rate, or other indication of improved conditions. In an unusually ingenious experiment, jackdaws were added to the territories of magpies by the method of setting up nest boxes for them. A similar number of magpie territories without additional nest boxes served as controls. The method was successful, all of the boxes being occupied. The reproductive performance of the magpies, on territories with and without jackdaws, was assessed in terms of clutch size, number of fledged young, number of broods with at least one fledged young, and mean nestling weight. The significant effects all occurred after hatching, as survival and growth of the nestlings.

More food is required during the feeding of nestlings than at any other time. Therefore, the effect of the jackdaws most probably was through competition for food. In the control territories, there were more broods with two or more fledged young, more fledged young per successful brood, and more fledged young per breeding attempt, and the mean nestling weight was greater than that in the presence of jackdaws. The other traits showed no significant effects. Nest sites obviously were not involved, and no interspecific aggression was observed. In the second year, the experimental and control areas were reversed, so that the performances of many of the same females could be compared under both conditions. For those females, the number of fledged young was significantly greater in the territories without jackdaws, but the findings regarding the start of laying, clutch size, and clutch weight were alike, showing significant and nonsignificant effects similar to the within-year experiment. This was an ingenious repetition, with the added benefit of comparing the performances of the same females under both conditions.

Lizards

Many studies have been conducted on lizards with the special purpose of looking for evidence of interspecific competition, either at present or as having provided the selective pressure to evolve niche partitioning. It is safe to say that the majority of such studies concluded that competition is and has been important. In marked contrast to the purely observational work, experiments testing for interspecific competition have yielded either negative results or equivocal findings. The experiment that yielded the clearest result was conducted by the late D. W. Tinkle in a streamside habitat in the Sonoran Desert in Arizona. He studied the ecological distributions of three species over four summer seasons and compared his observations with those from another study area in Utah, where two of the species were lacking. *Sceloporus undulatus* was more arboreal in the Arizona location, which he attributed to the presence of *S. clarki* and *Urosaurus ornatus*. He divided that location into an experimental area, from which he removed *S. clarki* and *U. ornatus*, and a control area, where he repeated the same observations on ecological distribution, abundance, size, and survival of *S. undulatus* that he had made in the area from which the supposedly competing species had been removed. The experiment was carried out for 3 years. There was no evidence that any of the observations differed for the two areas.

Whatever caused *S. undulatus* to have an unusual ecological distribution in the Arizona location, it was not competition from the two related species of lizards.

Two other experiments in the southwestern United States involved the removal of each of two species to observe the effect on the other, anticipating that each would respond favorably to the removal of its supposed competitor. In the Big Bend National Park in Texas, the two species were *Sceloporus merriami* and *U. ornatus*, both small lizards living exclusively on rocky habitats surrounded by unsuitable terrain. Two of the habitats were designated for removal of *S. merriami*, and two for the removal of *U. ornatus*, with two serving as controls, where the same observations were made as on the removal habitats, the control sites being otherwise undisturbed. The experiment continued for 4 years. The results as reported differed among years, and between the two species. For *U. ornatus*, better survival through the first year of life was reported on the *merriami*-removal habitats from 1974 to 1975, but on the control habitats from 1976 to 1977. For first-year *S. merriami*, survival was reported to be better on the *ornatus*-removal habitats from 1975 to 1976. Adult *ornatus* survived better on the control sites from 1974 to 1975 and from 1976 to 1977, but the reciprocal result was found for *merriami*. There are problems with the statistical interpretations, and there is some question as to whether or not any attention should be paid to the claimed statistical significance of the survival data.

Similar interpretational difficulties arise in the analysis of growth rates. Two of the years, 1975 and 1977, were unusually dry; 1974 and 1976 received above-normal rainfall. The growth rate of *U. ornatus* was reported to be significantly higher on experimental areas for the two dry years pooled, but not for the two wet years. No experimental effect on the growth rate of *S. merriami* was observed. A complete explanation cannot be given here, but the statistical analysis was again flawed, through the confounding of differences among replicate areas and among years.

The amount of fat stored prior to hibernation was studied by collecting animals in both experimental and control habitats at the end of the experiment in 1977, and by collecting "control" animals in 1976 from habitats away from the experimental area. There was thus information for the lizards on "control" habitats in wet and dry years, and for those on experimental habitats in a dry year. For *U. ornatus*, more fat was stored by those in the experimental habitats in the dry years than on the controls; the wet year for controls was like the experimentals in the dry year. For *S. merriami*, the only difference was between wet and dry years, with no difference between experimental habitats and controls. The overall claim was for an effect on *U. ornatus* due to experimental removal of *S. merriami* in dry years. Considering the statistical problems and the absence of a reciprocal effect on *S. merriami*, that claim should be viewed skeptically, as should the result reported in a similar experiment in an Arizona woodland – some effect on part of one population due to experimental removal of the "competitor" in a dry year.

References and suggested further reading

Brown, J. H. 1975. Geographical ecology of desert rodents. Pp. 315–341 in *Ecology and Evolution of Communities*, M. L. Cody & J. M. Diamond (eds.). Harvard University Press.

Brown, J. H., D. W. Davidson, J. C. Munger, & R. S. Inouye. 1986. Experimental community ecology: The desert granivore system. Pp. 41–61 in *Community Ecology*, J. Diamond & T. J. Case (eds.). New York, Harper & Row.

Ehrlich, P. R., & J. Roughgarden. 1987. *The Science of Ecology*. New York, Macmillan.

Hairston, N. G. 1949. The local distribution and ecology of the plethodontid salamanders of the southern Appalachians. *Ecological Monographs* 19:47–73.

Hairston, N. G., Sr. 1987. *Community Ecology and Salamander Guilds*. Cambridge University Press.

Hairston, N. G., Sr. 1989. *Ecological Experiments. Purpose, Design, and Execution*. Cambridge University Press.

Munger, J. C., & J. H. Brown. 1981. Competition in desert rodents: An experiment with semipermeable exclosures. *Science* 211:510–12.

7

Groups of vertebrate species (communities)

Thus far, the ecological situations described have involved only one or two species. Vertebrates live together in much larger combinations called natural communities. As discussed in the introduction to Part III, the interactions among the species in a community can take the form of either competition or predation, or both. In this chapter, experiments are described that involve the interactions of more than two species. The numbers of species involved still are not large, but the experiments do represent the manipulation of groups that are large enough to approach the composition of communities.

Niche partitioning among sunfishes

Niche partitioning and its correlates have occupied the attention of many ecologists during the past three decades. In most discussions and presentations of observations it has been an implicit assumption that differences in the uses of a common resource among coexisting species have arisen as evolutionary responses to the pressure of competition through natural selection. There has been little discussion of whether these differences are fixed genetically, as would surely be the case for morphological differences (e.g., sizes of beaks), or whether the species can respond in a facultative way to the presence of competitors. In a series of experiments on three species of sunfishes, such facultative responses were observed. These species, *Lepomis macrochirus,* the bluegill, *L. cyanellus,* the green sunfish, and *L. gibbosus,* the pumpkinseed, exist together in ponds and small lakes in eastern North America. The experiment was conducted in southern Michigan, where a number of circular experimental ponds, 29 m in diameter and a maximum of 1.8 m deep, were used. The shallow edges of the ponds supported a zone of vegetation, mostly cattails, the rest being open water. The ponds could be drained, so as to permit complete censuses of the fish present at the end of the experiments.

The first experiment tested for the effect of crowding on the food habits of the three species. The experimenters introduced 900 individuals of each species into one pond in June, and continued the observations into December. Three other ponds received introductions of 900 individuals of one species each. As these were young fish, few reached sexual maturity, and the experiment was not complicated by reproduction. Samples of fifteen specimens of each species under both conditions were removed from the ponds several times during the experiment for determination of stomach contents and growth. At the end of the experiment, the ponds were drained, and the fish were counted and weighed. Mortality was not excessive: *L. macrochirus* had 12% mortality both alone and with the other species; *L. gibbosus* had 3% mortality alone, and 7% with the other two; *L. cyanellus* had 17% mortality alone, and 16% with the others. Thus, crowding had little effect on survival. Growth of *macrochirus* was reduced approximately in proportion to the crowding: its mean dry weight alone was 3.6 g, and with the other species it was only 1.29 g. The other species were not as strongly affected by crowding: *gibbosus* weighed 1.38 g alone and 1.21 g in crowded conditions; *cyanellus* weighed 1.74 g alone and 1.34 g when with the others.

The most interesting result came from the examination of stomach contents, which consisted of crustaceans, insects, and other invertebrates. Most of the items could be classified according to the part of the pond from which they most probably came: the zone of vegetation around the edges, the pond bottom, and the open water, which supported a fauna of small crustaceans. In the ponds where each of the three species was alone, all had eaten the majority of their food in the vegetation: *macrochirus* 77%, *gibbosus* 77%, and *cyanellus* 64%. In the pond holding all three species, *cyanellus* continued to forage mainly in the vegetation (72%); *macrochirus* took 52% of its food in the open water and only 24% in the vegetation; and *gibbosus* took 76% of its food from the pond bottom, and only 11% in the vegetation. Clearly, without competition, all three species foraged mostly in the vegetation, but when in the presence of competitors, *macrochirus* and *gibbosus* changed their foraging habits. This was a rare experimental demonstration of niche partitioning under the force of interspecific competition. The superior ability of *L. cyanellus* to continue to forage in the preferred vegetation, even when in competition with the other species, was shown in another experiment. This time, the cattail part of a pond was separated from the open water by a nylon-mesh fence and was divided transversely into three sections of the same size. One of the sections received 500 *macrochirus*, one received 500 *cyanellus*, and the third received 250 of each of the two species. This time, crowding of individuals was the same in the one-species and two-species sectors. Survival and growth were of interest, as there was no choice of foraging habitat. Both species survived better together than alone, *macrochirus* 75% to 70%, and *cyanellus* 85% to 80%, indicating that intraspecific competition was stronger than interspecific competition. In growth, *cyanellus* also fared better, and the effect was greater in the mixed-species sector. Alone, its growth was 24% greater than that of *macrochirus* alone; together, it was 44% greater. Thus, *cyanellus* was the superior competitor, accounting for the fact that it was *macrochirus* that changed its foraging habitat when it had a chance.

Tadpoles and newts

Experimentation in the field is fraught with problems. Subtle differences between plots can cause large variations in both experimental results and controls, thus making statistical significance difficult to achieve. The more complex the design of the experiment, the worse this problem becomes, but there are some questions in community ecology that require complex designs for their solutions. In one series of such experiments, the problem was approached by having more control over the experimental conditions than could be obtained in nature. These experiments used amphibian communities in small ponds. Instead of natural ponds, or even ponds specially constructed as were those for the sunfish experiments, these experimenters at Duke University confined the communities in metal tanks, 1.52 m in diameter and 0.61 m deep, which held about 1 m^3 of water. To the water was added litter from the edge of a natural pond, rabbit chow, and many small crustaceans and aquatic insects, to produce a reasonably natural condition, but one over which the experimenters had control. The experimental communities consisted of several species of tadpoles; some had newts (*Notophthalmus viridescens*) added, and some also received the larvae of tiger salamanders (*Ambystoma tigrinum*). Some tadpoles feed by filtering the algae and other microorganisms out of the water; others feed by scraping the microorganisms that grow on solid surfaces, such as stems and leaves of aquatic plants, and in this case the sides of the tanks. Newts are predaceous and will feed on tadpoles unless the latter become too large. Tiger salamander larvae feed on both tadpoles and larval newts, but adult newts are toxic.

The experiment to be described was designed to test for the combined effects of competition among six species of tadpoles, as well as the effect of predation by newts and tiger salamander larvae on the tadpole species. Sampling natural ponds gave estimates of realistic densities of each species used. Eggs or hatchlings of the frog species were added in the sequence in which they normally breed, the earliest being the southern leopard frog, *Rana utricularia*, the spring peeper, *Pseudacris crucifer*, and the spadefoot, *Scaphiopus holbrooki*; toads, *Bufo terrestris*, were next, and two species of tree frogs, *Hyla chrysoscelis* and *H. gratiosa*, were the last to breed. The grand total was 1,200 tadpoles added to

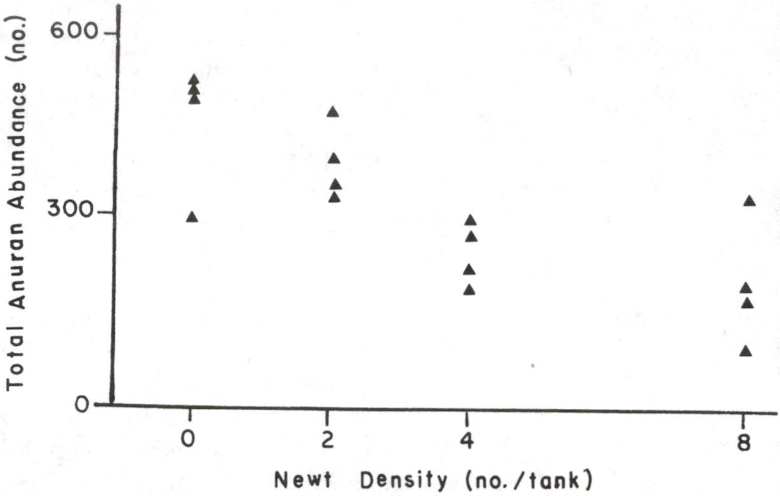

Figure 7.1. *Relation between final abundance of all censused frogs (including overwintering* Rana*) and initial density of predatory newts,* Notophthalmus, *in each of four tanks. (Redrawn from Morin 1983.)*

each tank. The effects of competition were measured in four tanks to which no predators were added. Four tanks were assigned to each of the following number of newts: two, four, and eight. Finally, three tanks received four tiger salamanders, and three tanks received four newts plus four tiger salamanders. All tanks were covered with tight-fitting screen lids to exclude all immigrant amphibians and insects.

Predation was most dramatic in the case of the tiger salamanders. Of the 7,200 tadpoles added to the six tanks, only 13 (0.18%) survived to metamorphose into frogs. Next, consider the effect of newts (Figure 7.1). Each symbol represents the total survival in one tank. Even without newts, fewer than half of the tadpoles survived to metamorphosis. Newts reduced that number, approximately in proportion to the number of newts present. Notice that even in the controlled conditions in these tanks, the overall survival was surprisingly variable with the different numbers of newts. Nevertheless, newt predation clearly had effects on the pond communities.

An even more dramatic effect of newt predation was on the relative abundances of the different species of tadpoles. In the tanks without predators, spadefoots (*Scaphiopus*) and toads (*Bufo*) were the most abundant at the end of the season. As the number of newts was increased, there was a marked increase in the relative importance of spring peepers (*P. crucifer*), with a marked decline in the importance of the competitive dominants. *Hyla gratiosa* reached its greatest relative abundance with intermediate numbers of newts, but it was never important. The other species declined in numerical importance with increasing numbers of newts. Predation had two noteworthy effects: a general reduction in numbers of tadpoles, and alteration of the relative abundances of the different species.

Table 7.1. *Effects of competition and predation on the survival of six species of anuran tadpoles*

	Percentage survival for given no. of newts			
Species of tadpole	0	2	4	8
Scaphiopus holbrooki	93.3	57.6	42.6	18.0
Pseudacris crucifer	4.0	27.6	27.1	30.3
Rana utricularia	62.5	57.6	24.0	12.0
Bufo terrestris	38.0	23.8	4.8	8.0
Hyla gratiosa	2.3	20.6	11.7	3.2
Hyla chrysoscelis	48.3	29.5	12.0	17.2

Source: Morin (1983).

Table 7.2. *Effects of competition and predation on the mean mass at metamorphosis for six species of anuran tadpoles*

	Mean mass (mg) for given no. of newts			
Species of tadpole	0	2	4	8
Scaphiopus holbrooki	466	552	577	650
Pseudacris crucifer	49	72	148	211
Rana utricularia	781	989	1637	2457
Bufo terrestris	80	92	88	135
Hyla gratiosa	524	724	1337	1685
Hyla chrysoscelis	167	273	372	402

Source: Morin (1983).

The effects of interspecific competition were assessed in four ways: survival, growth rate, size at metamorphosis, and duration of the larval period. The results for survival and size at metamorphosis are shown in Tables 7.1 and 7.2. The effect on survival has already been described in general. All species reached larger sizes with increased numbers of predators, but the effect was much greater for some species than for others. For all species, the growth rate (in mg/day) followed mass at metamorphosis. Similarly, the larval period decreased as the number of newts increased, but for *Scaphiopus, Pseudacris,* and *Bufo* the decreases were not statistically significant.

By looking at the two tables, we can see that the combination of differential survival and differential size suggests that the maximum mass of frogs coming out of the tanks might not be at the maximum number, and indeed that was the case. Intermediate amounts of predation resulted in an increase in total biomass. This experiment revealed a great deal about the interactions among a combination of competing herbivore species and their predators. The only problem in interpretation is in the degree to which the controlled conditions in the tanks affected the realism of the results, as far as understanding natural pond communities is concerned.

Streambank salamanders

A regular spacing of sizes of coexisting species, such as was shown for desert rodents in Figure 6.7, has been interpreted to reflect the effects of competition. It was claimed that the successive size differences (by at least a factor of 1.3 for the length of feeding structures) represented sufficient differences in food utilization to permit coexistence without excessive competition. Many examples were published, some of which have been shown to be of questionable statistical validity. The semiaquatic salamanders of the genus *Desmognathus* are abundant in the southern Appalachians and occur in "communities" of three to at least five species. They generally show the regular size differences among coexisting species. At the Coweeta Hydrologic Laboratory of the U.S. Forest Service, four species occur together, with the following ratios between the lengths of their heads: *D. quadramaculatustFS is 1.35 times as long as D. monticola, monticola* is 1.4 times as long as *D. ochrophaeus*, and *ochrophaeus* is 1.38 times as long as *D. aeneus*. The figures are for adult specimens of average size. All except *quadramaculatus* begin metamorphosed life at a very small size, with little or no difference between species. All except *aeneus* have larval stages in mountain streams and adjacent seepage areas. For *quadramaculatus*, the larval period lasts at least 3 years at Coweeta; *monticola* and *ochrophaeus* have larval periods up to 10 months; and *aeneus* lacks a larval stage.

After metamorphosis, they inhabit streams and stream banks in the forest. Classes at the University of North Carolina have studied the ecological distribution of these four species for a number of years. The students identified each specimen caught and recorded the distance it was found from surface water. The data for the first 13 years are shown in Table 7.3. There is much overlap in the distributions, but each species has a characteristic distribution with respect to surface water. In combination with the regular size differences, that information was interpreted as a classical example of niche partitioning.

Such an optimistic opinion did not last long. There is an elementary physiological consideration that conflicts with the easy ecological interpretation. These salamanders are lungless, and their respiration must be through the skin or the oral cavity (they can use throat muscles to pump air into and out of the oral cavity). The respiratory surface must be kept moist, or the gases will not diffuse properly through the membranes. Therefore, a terrestrial adult will lose water from its body in the process of keeping its skin moist, unless the atmosphere is saturated. The geometry of any body is such that for the same shape, a small body has a

Table 7.3. *Ecological distribution of four coexisting species of the salamander genus* Demognathus

Species		Distance from water (m)					
		0	<0.3	0.3–1.5	1.5–3.0	3.0–6.0	>6.0
D. quadramaculatus	no.	201	24	18	1	1	2
	%	81.3	9.7	7.3	0.4	0.4	0.8
D. monticola	no.	377	229	250	47	28	9
	%	40.1	24.4	26.6	5.0	3.0	0.9
D. ochrophaeus	no.	81	114	198	83	59	31
	%	14.3	20.1	35.0	14.7	10.4	5.5
D. aeneus	no.	0	3	19	21	43	76
	%	0	1.8	11.7	13.0	26.5	46.9

greater surface area relative to its volume than does a large body. Therefore, the smaller salamanders should be at a disadvantage, because they will lose water faster, all other things being equal. There should be no benefit for the more terrestrial species to be smaller, or for the smaller ones to be more terrestrial, and competition does not appear to be a satisfactory explanation for the direction of the evolution of the genus, which is thought by students of the subject to have started with a remote ancestor approximately like *quadramaculatus* or *monticola* and to have changed toward more terrestrial and smaller species.

That reasoning led to the examination of some predictions about what should happen to the different species when the adjacent one in the aquatic–terrestrial sequence is absent. Thus, with *monticola* absent, as it is from higher elevations, it would appear that *quadramaculatus* should be less aquatic with the streambank habitat open, and it should be larger where *monticola* is present, a greater difference being advantageous. Neither of those predictions was correct, and there were six more that were tested. Of the total of eight predictions, five were incorrect, and only three were valid – too poor a score to justify the classical interpretation of niche partitioning.

An alternative hypothesis to competition was needed. It was postulated that predation has been the driving ecological force in natural selection. Thus, if we start with the largest and most aquatic species, any origin of a new species in the series should put the smaller of the two species under pressure to become more terrestrial to avoid predation by the larger one. Predation by *quadramaculatus* on both *monticola* and *ochrophaeus* has been observed in the field. Thus, there were two hypotheses about the evolution of the species of *Desmognathus*, and some means of choosing between them was needed.

A set of experiments was conducted to answer the question. It was proposed to remove *ochrophaeus* from one set of plots, and *monticola* from another set, keeping a third set as unmanipulated controls to observe the natural changes in the populations. Removing either species had to result in a decrease in its population relative to the controls, or the experiment would be an obvious technical failure. Removing *monticola* should cause an increase in the abundance of *ochrophaeus* under either hypothesis (competition or predation), because *monticola* is the larger of the two. The experiment would have been a failure had that result not been obtained. Both of the requirements were met, as both species declined in abundance on the plots from which they had been removed, and *ochrophaeus* increased markedly on the plots from which *monticola* had been removed.

The critical experiment was the removal of *ochrophaeus*. If it had been competing with *monticola*, the latter should increase in numbers; if the relationship had been predation by *monticola*, that species should not gain by the removal of its prey, and it might decrease if *ochrophaeus* had been an important food item. Similar predictions would apply to the effect on *quadrama ulatus* of removing *ochrophaeus*, but the relationship might be weaker because of the greater difference in ecological distributions, as shown in Table 7.3. Removing *monticola* would involve the same predictions with respect to *quadramaculatus:* an increase in the latter if competition were the interaction, and no increase or a possible decrease if predation by *quadramaculatus* were important. There was no point in having plots from which *quadramaculatus* was removed, because both hypotheses would predict an increase in *monticola* and probably in *ochrophaeus*, and the considerable work would have accomplished nothing.

The results were mostly as predicted by the predation hypothesis. There was a significant increase in *ochrophaeus* populations where *monticola* had been removed; relative to the controls, there was a significant *decrease* in *monticola* where *ochrophaeus* was removed, and there was also a significant negative effect on *quadramaculatus* from removing *ochrophaeus*. The only effect that indicated any competition among these species was a weak and statistically marginal increase in *quadramaculatus* on the plots from which *monticola* had been removed.

This experiment has revealed the necessity of testing hypotheses derived from observations. Previous ecological interpretations of regular size differences and differences in ecological distributions had emphasized the theory that interspecific relationships in natural communities were competitive. This experiment has demonstrated that the theory lacked validity in this community.

References and suggested further reading

Ehrlich, P. R., & J. Roughgarden. 1987. *The Science of Ecology*. New York, Macmillan.

Hairston, N. G., Sr. 1987. *Community Ecology and Salamander Guilds*. Cambridge University Press.
Hairston, N. G., Sr. 1989. *Ecological Experiments. Purpose, Design, and Execution*. Cambridge University Press.
Morin, P. J. 1983. Predation, competition, and the composition of larval anuran guilds. *Ecological Monographs* 53:119–38.
Werner, E. E., & D. J. Hall. 1976. Niche shifts in sunfishes: Experimental evidence and significance. *Science* 191:404–6.
Werner, E. E., & D. J. Hall. 1977. Competition and habitat shift in two sunfishes (Centrarchidae). *Ecology* 58:869–76.

8

An unsolved problem
Population cycles

Changes in the abundances of animal species have been the basis for much ecological research. Some of these changes are readily explained as seasonal responses in reproduction and survival; others are relatable to fluctuations in the food supply. Two groups of mammals show remarkably regular fluctuations in abundance, with periods longer than 1 year and much longer than a generation. The first group contains the snowshoe hare, with an average time between peak abundances of about 10 years. This was first noticed in the records of the Hudson's Bay Company (Figure 8.1) and was brought to the attention of ecologists by the English ecologist Charles Elton. The example is famous in ecology textbooks because of the correlation with the population cycles of the hare's principal predator, the lynx. These figures were first interpreted to be confirmation of the mathematical prediction of mutual effects by predator and prey populations, as discussed later.

Much more is known about the population cycles of voles and lemmings. Collectively, these species belong to the Subfamily Microtinae of the Family Cricetidae. The genera involved are *Microtus*, *Clethrionomys*, *Lemmus*, and *Dicrostonyx*. They are short-tailed, ground-dwelling "meadow mouse" kinds of animals. They feed primarily on grasses and similar harsh vegetation and have grinding teeth adapted to that kind of food. The population histories of two species of microtines are shown in Figure 8.2. At their peaks of abundance, they are forty to seventy times as numerous as at their lows. The most striking features of these histories are that the number of years between peaks is nearly always either three or four and that the numbers of animals at the peaks are remarkably similar from one peak to another – much more similar than any random series would be. So far, only one species of microtine rodent has failed to show that pattern when adequately studied, and no other group of animals shows it.

The differences between highs and lows are brought about by

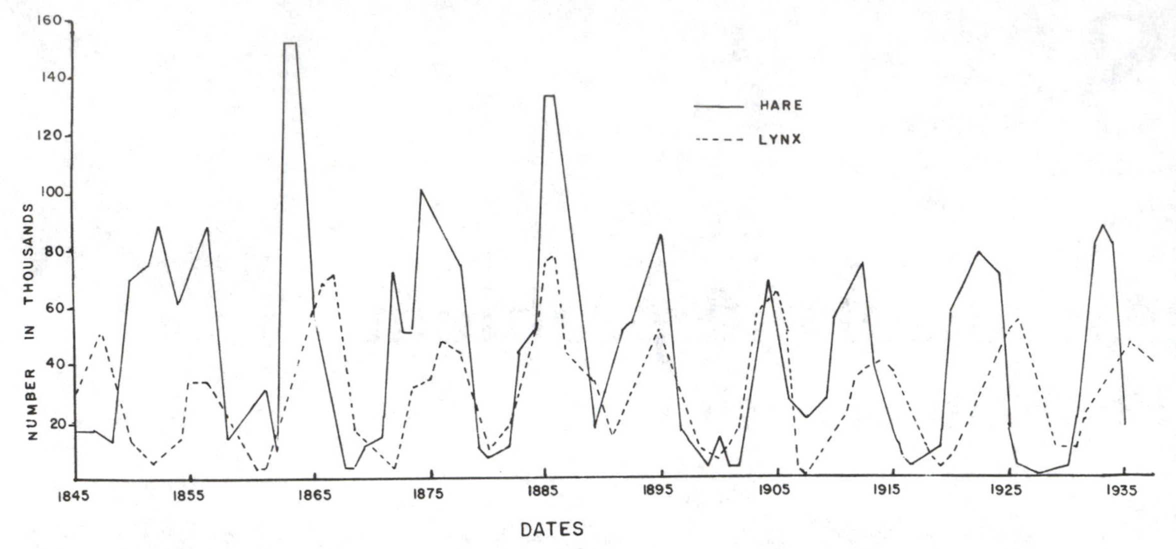

Figure 8.1. Changes in the abundances of lynx and snowshoe hare as indicated by the number of pelts received by the Hudson's Bay Company. (Redrawn from Odum 1971.)

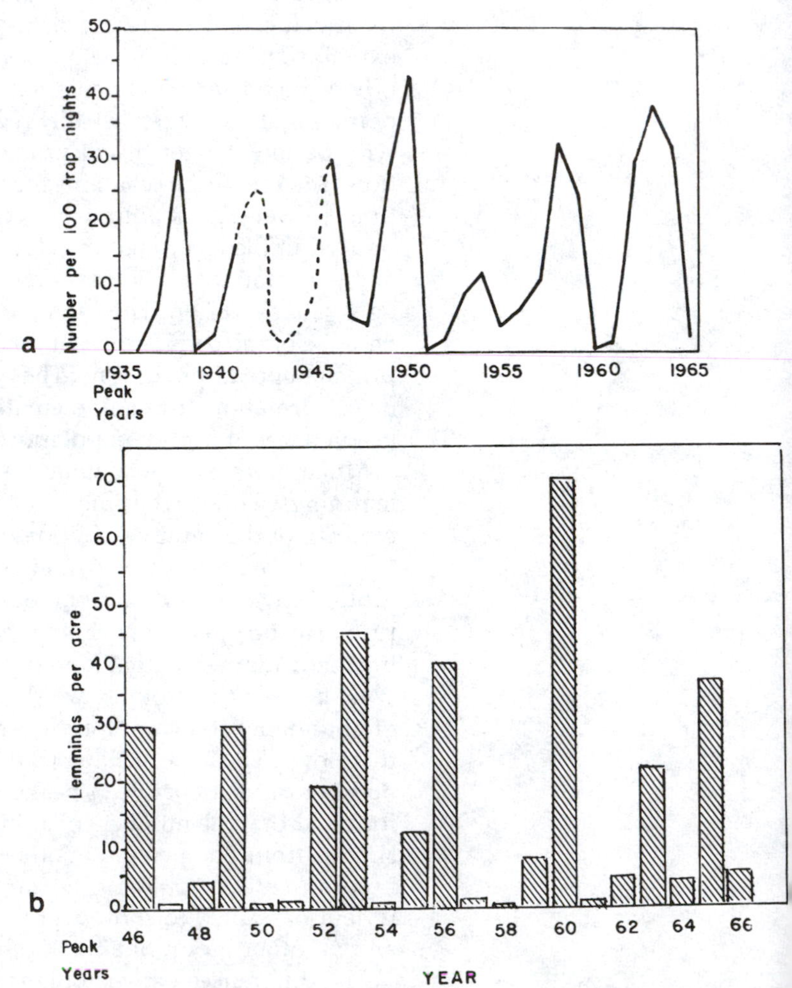

Figure 8.2. Long-term records of population densities of two small rodents: (a) autumn densities of the red-grey vole, Clethrionomys rufocanus; (b) summer densities of the brown lemming, Lemmus trimucronatus. (Redrawn from Krebs & Myers 1974.)

Table 8.1. *Intrinsic rates of natural increase for eight species of microtine rodents*

Species	r (per week)
Microtus agrestis	0.088[a]
M. pennsylvanicus	0.051
M. ochrogaster	0.070
M. californianus	0.066
Clethrionomys glariolus	0.104
C. rufocanus	0.146
Lemmus trimucronatus	0.083
Dicrostonyx groenlandicus	0.042

[a]All values except this one were estimated from maximum changes observed in the field.

marked changes in the intrinsic rate of natural increase (r), which was described in Chapter 5 and used again in the equations in Chapter 6. For the periods of population increase, r has been estimated for eight species, as shown in Table 8.1. Actually, only the first value in the table was obtained by a strictly proper method. It was obtained by determining survival and fecundity in the laboratory under uncrowded conditions. The remaining values were estimated from population changes, and hence are only approximations of r, being maximum estimates of $dN/N\,dt$, an important distinction when the rate of population growth is inversely related to the number of individuals present. When the population is increasing at the maximum rate, it is assumed that there is little influence of the number present at the start of the rapid-growth period. When the population is declining, r has a negative value, and these values can be estimated in the same way as the positive ones: by changes in the number present. These negative values are as dramatic as the positive ones: between –0.065 and –0.275 during crashes.

There are other observations that are related to the direct counts of numbers present. For reproduction, the duration of the breeding season is more important than is litter size, and the age at maturity increases as the population peak is approached. This reduces r and hence slows the growth of the population. Mortality, as might be expected, is higher when there is a declining population than during the expansion phase of the cycle. These observations show how the populations change, but they do not explain the regular cycles.

At least eight hypotheses have been advanced as explanations:

1. MATHEMATICAL ARTIFACTS. It has been postulated that the cycles are mathematical artifacts of the way in which the data

were analyzed. Nearly 40 years ago it was argued that certain methods of simplifying a large set of data could result in the appearance of a cycle when none actually existed. The method involved "smoothing" a graph of a series of numbers by calculating "running averages," that is, calculating the average of three or five successive observations, dropping the first number and adding the next in the series, calculating the average of those, and continuing through all of the observations. The process reduces the impression given by deviant values, because they are smoothed out by adjacent ones, and it makes the graph easier to understand. The mathematical argument against the reality of cycles was that the smoothing process also made regular "waves" in the graph, even when the original numbers used were a series taken from a table of random numbers. That is true, and as far as some alleged cyclical phenomena are concerned, it provided a sufficient explanation for them. It did not, however, provide an explanation for the cyclical behavior of microtine populations, because the range of numbers between peaks and lows was much greater than could be accounted for by the mathematical argument used.

2. FOOD. Food has been considered in two ways, the amount and the nutrient quality. As far as the amount of food is concerned, the hypothesis is that a population will grow until the food resource is exhausted, after which the population will decrease. The food will then recover, and the population will follow. This hypothesis requires that a high percentage of the food be consumed when the population reaches its peak. Actually, this is nearly always less than 5% (eleven of twelve cases where it has been measured). Thus, cyclical variations in the amount of food cannot provide a satisfactory explanation for the cycles. The alternate food hypothesis is that the nutrient quality changes cyclically, independent of the amount. Phosphorus has been studied in this regard, but shows no consistent relationship to the abundance of the rodents. In a second kind of test, bluegrass, *Poa pratensis*, was grown on soils differing in essential nutrients. The grasses obtained obviously were of different quality, but *Microtus pennsylvanicus* showed no preference between the two kinds.

3. PREDATOR–PREY RELATIONSHIPS. Under this hypothesis, as a population of the rodents increases, predator populations will follow, and eventually will become numerous enough to decimate the prey, after which the predators will starve, and the rodent population will recover. This is an ever-popular theory, as it was shown mathematically in the 1920s that such a situation is theoretically possible. In its simplest form, the mathematical theory could not work, because it did not provide for any kind of self-

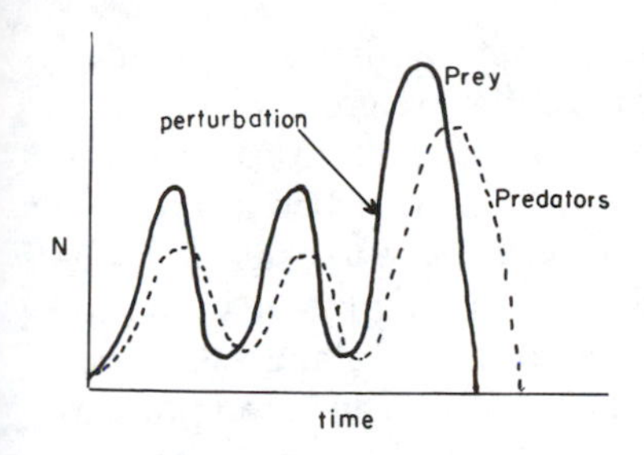

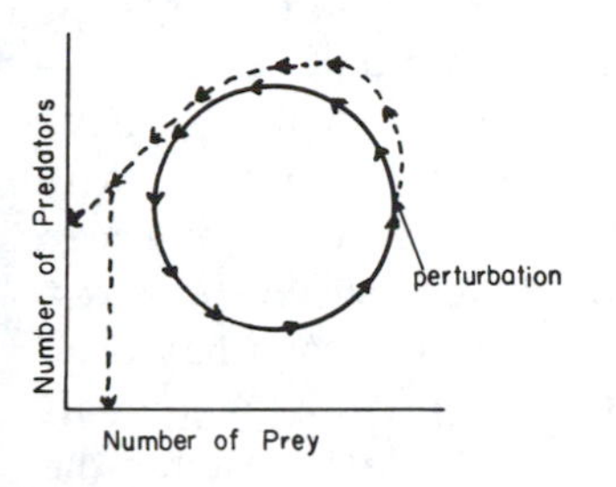

Figure 8.3. Two graphic representations of the Lotka–Volterra predator–prey model.

correcting mechanism to return to the cycle if accidental perturbations occurred (Figure 8.3). These would lead to wider fluctuations in the populations of both predator and prey, until the prey became extinct, followed by the extinction of the predators. The reason for the difficulty was that there was no limit to the prey population other than that imposed by predation. That problem was corrected mathematically by adding a term that reduces the growth of the prey population at high densities, allowing the system to return to the original cycle. The correctness of the new mathematical model has been shown experimentally among protozoans. In real populations of microtine rodents, however, predators have not been able to control population growth during the increase phase. There is evidence that predators can hold microtine populations to a low level, thus delaying the cycle. The trouble is that the cycles occur even where there are few predators. In most cases, they seem to follow the populations of microtines, rather than driving the cycle.

4. WEATHER. This is much too irregular to exercise complete control over the cycle. Weather might be able to synchronize cycles that are out of phase by speeding up a lagging cycle by providing favorable conditions (the faster cycle would not be affected), or by depressing an advanced cycle by imposing poor conditions. None of this has been demonstrated.

5. STRESS. Crowding causes increased adrenal and pituitary gland activity, which upsets the animals' physiology and brings on susceptibility to many causes of death. The effect has frequently been shown in laboratory populations, but only when they have been more crowded than field populations have ever been demonstrated to be. Field studies have always failed to confirm the physiological details. Animals brought in from the field during population declines nearly always remain healthy in the laboratory. The stress syndrome might possibly operate in conjunction with hypothesis no. 6.

6. AGGRESSIVE BEHAVIOR. Voles and lemmings are aggressive, and under some conditions they will not tolerate members of the same sex and age group in their vicinity, although we need to know more about this. Aggression would provide an ideal mechanism for keeping a population at one density, such as the salamander populations discussed in Chapters 5 and 6. The aggression could be combined with stress to give a cause-and-effect relationship. In a few studies, changes in aggressiveness have been found in different parts of the population cycle, but for a reason for the changes, we need to look to yet another hypothesis.

7. GENETIC CHANGES. The theory here is that a population has at least two kinds of individuals, differing genetically in aggressiveness. During the period around the peak of population density, natural selection would favor those that had aggressive tendencies and large size, as they would be able to chase the others away. During and after the decline, these aggressive individuals would be selected against in favor of the less aggressive but more actively reproducing ones, which would cause the population to increase. An important question here is, Can the genetic composition of the population change so rapidly? Electrophoretically detectable differences in serum proteins, which are gene products, have been monitored over the population cycle, with the result that changes in the frequencies of some alleles have been observed, but they have not been consistent between the sexes. The potential for such genetic mechanisms to work has been shown, but more recent research has involved removing animals to the laboratory at different parts of the cycle to test for the heritability of the relevant behavioral traits. The heritability was too low to give the claimed effect.

8. CHAOS. This term is the name of a relatively new development in mathematics, in which it has been discovered that seemingly simple mathematical models can give surprisingly complicated results when the constants are of appropriate magnitudes. Consider the equation for exponential growth

$$\frac{dN}{dt} = rN$$

and its integral

$$N_t = N_0 e^{rt} \qquad (8.1)$$

which tells us how to calculate the number expected after time t, N_0 being the number at time zero. We might, in an unsophisticated way, assume that the integral of the logistic (Equation 6.2) might be as follows:

$$N_t = N_0 e^{r([K-N]/K)t} \qquad (8.2)$$

It cannot be, because N in the exponent cannot be specified as to time. What can be done is to restrict the application to one time unit, from t to $(t + 1)$:

$$N_{(t+1)} = N_t e^{r([K-N_t]/K)t} \qquad (8.3)$$

This equation has interesting properties when solved repeatedly for successive time units. If r is large enough, some of the graphs of numbers against time will look surprisingly like the microtine cycles (Figure 8.4). The question now is, Is this at all realistic? If we remember that r is per time unit, we can see that what we must do

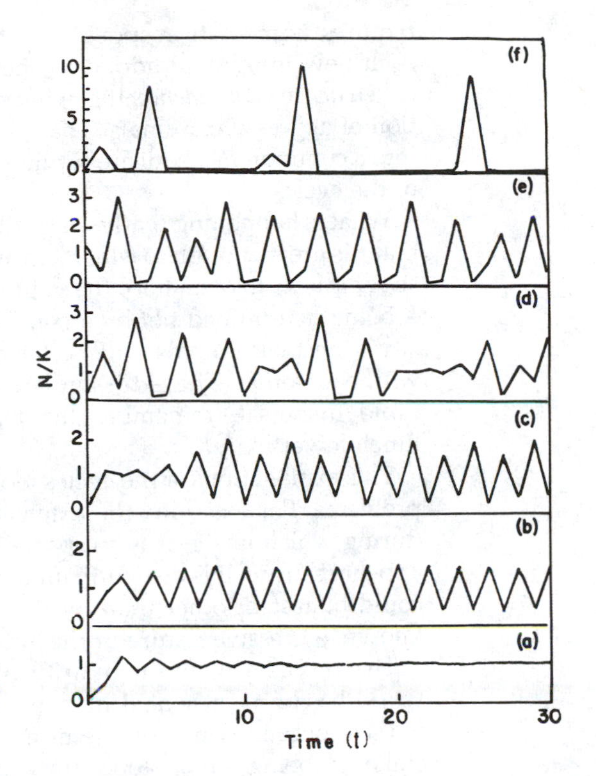

Figure 8.4. *Dynamic behavior of the population density* N_1/K_1 *as a function of time* t, *as described by the difference equation for various values of* r. *(a)* r = *1.8; (b)* r = *2.3; (c)* r = *2.6; (d)* r = *3.3* (N_0/K = *0.075); (e)* r = *3.3* (N_0/K = *1.5); (f)* r = *5.0* (N_0/K = *0.02). (Redrawn from May 1981.)*

Table 8.2. *Intrinsic rates of natural increase as in Table 8.1, recalculated to conform to the time scales in Figure 8.4*

Species	Recalculated *r*
Microtus agrestis	2.96–3.96
M. pennsylvanicus	1.71–2.30
M. ochrogaster	2.34–3.14
M. californianus	2.33–2.99
Clethrionomys glariolus	3.5 –4.77
C. rufocanus	4.93–6.6
Lemmus trimucronatus	2.81–3.76
Dicrostonyx groenlandicus	1.42–1.90

is to recalculate the values in Table 8.1, given in *r* per week, so that they correspond to the values for *r* in Figure 8.4. That can be done by noting the number of time units between the peak numbers on the graphs, equating that with 3 or 4 years, and multiplying *r* per week by the correct factor. That translates Table 8.1 into Table 8.2. The values overlap nicely with those required to produce the "cycles" shown in Figure 8.4c–e. Remember that the term $(K - N)/K$

requires immediate response by the population growth rate to each new individual added to the population. This property is called *density dependence*. It is what might be expected of a population of aggressive animals. That might bring stress and behavior into a common mechanism, leaving out all other suggested causes of the cycles.

What is happening mathematically in the graphs in Figure 8.4 is that r is great enough to cause a growing population to reproduce so rapidly as to overshoot the value of K by a substantial amount, K being determined not by a consumable resource but by a density such that animals will be far enough apart not to interfere with each other. The extra aggression will then cause great and rapid dispersal (remember the lemmings!), leaving a density much lower than K.

This explanation for the cycles looks plausible, but there are still problems. For example, the extension of r to such long periods, during which its effect is supposed to remain constant, does not appear realistic. It is also difficult to visualize how this hypothesis could be tested, other than by the observed values of r and the known aggressive nature of the animals.

The problem of the regularity of cycles in a highly variable world has been presented in some detail to show that ecology is not a science in which problems are easily solved. Nature has no stake in being understood by ecologists. Ingenuity and persistence will be required if we are to obtain real answers.

References and suggested further reading

Krebs, C. J., & J. H. Myers. 1974. Population cycles in small mammals. *Advances in Ecological Research* 8:267–399.

May, R. M. (ed.). 1981. *Theoretical Ecology: Principles and Applications* (2nd ed.). Sunderland, Mass., Sinauer Associates.

Odum, E. P. 1971. *Fundamentals of Ecology* (3rd ed.). Philadelphia, Saunders.

PART

IV

Geographic distributions of vertebrates

Explaining the distributions of animals and plants has been a challenge to biologists ever since Europeans began to visit remote places in the fifteenth century. Despite the progress made, some distributions remain puzzling, such as the lizard family Iguanidae, found throughout the Western Hemisphere, but also in Madagascar, Fiji, and Tonga, or the salamander genus *Hydromantes,* with three species in California, two in France and Italy, and perhaps four in Sardinia. Neither of these groups nor any of their fossils have been found outside those stated distributions.

The fascination of detective work in explaining distributions is not the only claim of the distributions of animals and plants to biological fame. Both Charles Darwin and Alfred Wallace, who published the principle of natural selection simultaneously in 1858, were stimulated toward that conclusion by their experiences with the distributions of animals on archipelagos: the Galapagos for Darwin and the East Indies for Wallace.

Some explanations appear to be obvious. Climate must be important, at least in preventing the spread of well-adapted species into areas with climates different from those to which they are adapted. The principle should not be pushed too far. There are no polar bears or walruses in the tropics, but is it the climate that keeps them in the Arctic? Darwin pointed out the problem in a famous passage in *The Origin of Species:*

> When we travel from south to north, or from a damp region to a dry, we invariably see some species gradually getting rarer and rarer, and finally disappearing, and the change of climate being conspicuous, we are tempted to attribute the whole effect to its direct action. But this is a false view; we forget that each species, even where it most abounds, is constantly suffering enormous destruction at some period of its life, from enemies or competitors for the same place and food; and if these enemies or competitors be in the least degree favoured by any slight change of climate, they will increase in numbers; and as each area is already stocked

full with inhabitants, the other species must decrease. When we travel southward and see a species decreasing in numbers, we may feel sure that the cause lies quite as much in other species being favoured, as in this one being hurt.

In Part IV, the effects of each of a number of factors are described, as well as the interactions among some of them. Besides the climate and other physical factors, barriers to dispersal are important determinants of distributions. These can be bodies of water for terrestrial animals, and can range from oceans to large rivers. Mountain ranges and deserts must also be considered, and of course a narrow strip of land can be an impossible barrier to fishes or other aquatic species. As Darwin pointed out, other species can have profound effects on the one under consideration, and their effects can interact with those of climate. These other species can be predators, parasites, or competitors, or, in the case of food specialists, they can be the food species.

In Chapter 9, the discussion begins with a consideration of the distributions of vertebrates in a minute fraction of the globe. The state of North Carolina is used as an example, but any comparable area with an equally varied topography would serve as well. As many workers, from Wallace to the present, have shown, there are general patterns in the worldwide distributions of animals and plants, and an exploration of those patterns will suggest explanations that may or may not be valid. The patterns that have been observed are described in Chapter 10, with a discussion of the influences of climate and of obvious barriers. Chapter 11 covers the effects of other species, and Chapter 12 considers the theories that have been advanced to explain what is known of distributions.

9

Distributions of North Carolina vertebrates

The vertebrate fauna of North Carolina has been chosen for description in this chapter so as to illustrate on an easily comprehended scale some of the factors that have influenced the worldwide distributions of animals and plants. In describing those distributions, marine species have been omitted. The reason for that omission is primarily the problem of deciding on a boundary. Species that breed regularly in the saltwater sounds within the outer banks would not cause problems, but those that remain for varying distances outside would not be cleanly separable from true oceanic forms that are known to occur off the coast. There is no simple way of separating those living both within and outside the outer banks from those living just outside.

In following the representation of the various groups, it should be understood that the world totals are only estimates by the best authorities. They could be in error by as much as 10%. Most of the major groups of vertebrates are well represented, as described in the following listing.

There are only about 50 species of Agnatha in the world, of which 3 (6%) occur in North Carolina.

All of the Chondrichthyes are marine, although freshwater species occur in other parts of the world. North Carolina has 200 species of freshwater Osteichthyes, which is 2.9% of the world fauna of 6,580 species.

Two of the three orders of Amphibia are represented, and the differences are instructive, as is described later. The Order Caudata is the best represented of all North Carolina vertebrate groups. Of the total of nine families, seven are represented. Except for the southeastern states, no other area in the world has more than four families. There are 49 species, which constitute 12.9% of the 378 recognized species worldwide. In contrast, only five of the eighteen families of Anura (frogs) are present, and only 29 species out of a total of 2,700 (1.1%). All members of the third order, the

Table 9.1. *Number of species in vertebrate faunas: temperate and tropical islands versus continental areas of comparable size*

	North Carolina	Great Britain	New Zealand	Cuba	Costa Rica
Location	Continent	Contin. isl.	Oceanic isl.	Oceanic isl.	Continent
Climate	Temperate	Temperate	Temperate	Tropical	Tropical
Area (km^2)	84,625	143,201	160,000	70,796	31,496
Amphibia					
Caudata	49	3	0	0	35
Anura	29	4	3	24	120
Reptilia					
Lacertilia	10	3	6	37	68
Serpentes	37	3	0	15	127
Mammalia	31	46	2	25	203
of which, bats	(14)	(12)	(2)	(23)	(102)

Gymnophonia, are confined to the tropics. All three recognized orders of reptiles are represented in North Carolina. The one species of Crocodilia constitutes 4.8% of the 21 in the world. Turtles (Order Chelonia) are well represented by 19 of the 240 species in the world (7.9%). Lizards and snakes, belonging to two suborders of Squamata, are unequally represented. North Carolina has only 10 species of Lacertilia (lizards), or 0.3% of the 3,000 in the world. The state has three times that representation of snakes (Serpentes): 1.1%, or 37 species out of 3,260.

North Carolina has ten of the nineteen orders of Mammalia, and of the 4,060 species in the world, 81 (2%) are present.

Birds (Class Aves) are divided into twenty-seven orders, of which nineteen are or have been represented, one of them (parrots) being extinct locally. So many species of birds migrate to and from breeding areas that species nesting locally are customarily listed separately. There are (or were) 167 nesting species present, 1.9% of the 8,700 in the world. If we count all species ever recorded, there are 323, or 3.7%, in North Carolina.

Thus, there are 765 species of vertebrates exclusive of marine fishes and mammals – 2.6% of the world terrestrial and freshwater fauna. This is twenty-nine times as many as the proportion of land area involved. North Carolina covers 84,629 km^2, or 0.091% of the 92,547,114 km^2 of the world.

Three factors are most important in giving that state such a large fraction of the world's fauna. The first factor is its warm temperate climate. Compared with deserts and with cold climates, such as Antarctica, Greenland, northern North America, and northern Eurasia, or major mountain ranges, such as the Himalaya, Andes, Alps, and Caucasus, North Carolina is greatly favored. The climate is not as favorable as those in many tropical areas, however.

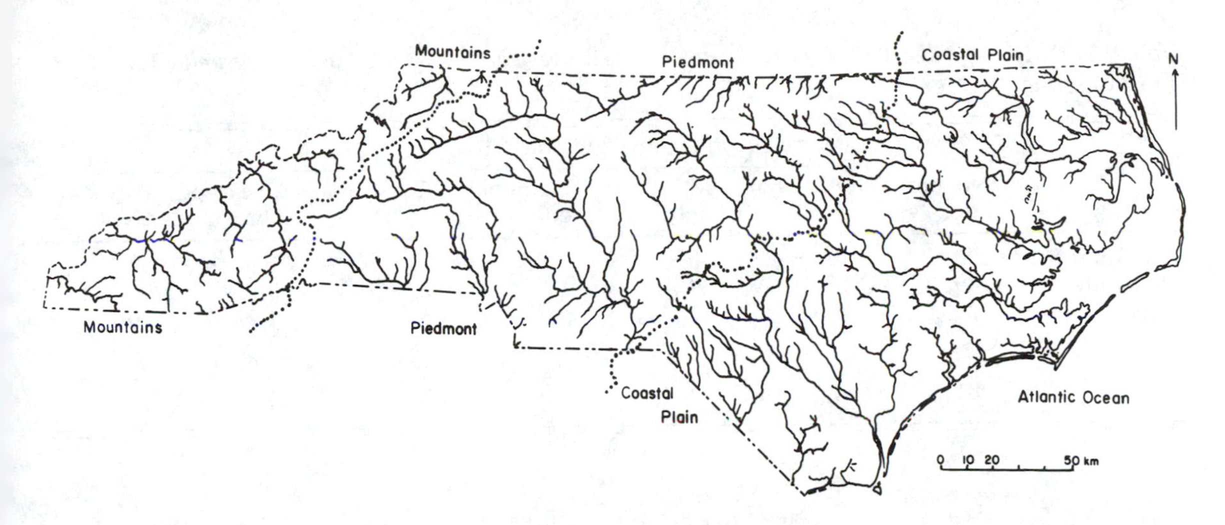

Figure 9.1. Map of North Carolina, showing all major drainages. Note that nearly all streams in the mountains flow westward into the Mississippi drainage system. (Adapted from Clay, Orr, & Stuart 1975.)

The second factor is that it is part of a continent. This means that it is accessible to land animals that are able to extend their distributions, in contrast to the situation on islands, even those close to continents. These first two factors can be visualized from an examination of Table 9.1. Few amphibians or reptiles have been able to reach an island as close to a continent as Great Britain, which until quite recently, in geologic terms, was part of Europe. Being an oceanic island completely inhibits colonization by salamanders, and very nearly inhibits colonization by nonflying mammals. Comparing the two tropical locations with temperate ones with more than twice as much area shows what a difference climate makes to all of the vertebrate groups listed.

The third factor favoring the large variety of North Carolina vertebrates is the varied topography. It is conventional to refer to three physiographic regions in the state: the mountains, the Piedmont, and the Coastal Plain (Figure 9.1). The different classes and orders of vertebrates are distributed differently over these three regions (Table 9.2). It is obvious that the Piedmont, though distinctive to humans, is not a separate region as far as its vertebrate fauna is concerned. Individual species either do not make distinctions among any of the three regions or treat the Piedmont as indistinguishable from either the mountains or the Coastal Plain. Thus, despite the physiographic partition into three regions, from the standpoint of the distributions of vertebrates there are two, with an overlapping area between them. There are some distinct patterns among the vertebrate groups: Frogs and reptiles are much alike. More species are confined to the Coastal Plain than anywhere else. Both rely on the environment for maintenance of body temperature, and the Coastal Plain is the warmest part of the state. Few species of either are confined to the mountains.

Birds and mammals are alike in tending to be found throughout the state, probably because they are endotherms. Thus, these spe-

Table 9.2. *Percentage distributions of different groups of vertebrates among the physiographic regions of North Carolina*

		Percentage in the given region					
Group	No. of spp.	All (whole state)	Mtns. only	Piedmont only	Coast only	All except mtns.	All except coast
Freshwater fish	200	19	30	7	14	22	9
Salamanders	49	18	49	2	22	0	8
Frogs	29	40	7	0	50	4	0
Reptiles	67	32	4	3	43	13	3
Birds	167	49	17	0	19	9	5
Mammals	81	42	23	0	23	4	5

cies can adjust to a wider range of outside temperatures than can the other classes, which are ectotherms.

Salamanders are unique among vertebrates in having almost half of their species confined to the mountains. They are also different from other amphibians and reptiles in having all but one family confined to the North Temperate zone. The inference from that distribution is that they originated and underwent most of their evolution there, and thus are better adapted to cool conditions than are frogs and reptiles. The mountains not only are cool but also are wetter than the rest of the state, the high elevations, especially, receiving at least half again as much rainfall as lower elevations or the rest of the state. The combination of cool and moist conditions is important to the respiration of most of the species, those belonging to the Family Plethodontidae. All members of that family lack lungs; they respire through their skins and mouth cavities. Any respiratory surface must be kept wet to function properly in the transfer of gases, and small animals like salamanders require a wet atmosphere so that they will not dry out because of exhaustion of their internal supply of water. Moreover, many Plethodontidae are exclusively terrestrial, never going to water, even to breed.

Like salamanders, freshwater fishes have few species distributed throughout the state. Two-thirds of the species either are confined to the mountains or are excluded from the mountains. The reason for this pattern is in the distribution of watersheds (Figure 9.1). The map shows that almost all of the streams in the mountains flow west or northwest, into the Mississippi drainage, and then into the Gulf of Mexico. Those in the Piedmont and Coastal Plain flow into the Atlantic Ocean. The difficulties of crossing from one drainage to another, or moving around Florida, are obvious. Thus, mixing of the fish fauna is severely inhibited. The only means by which a strictly freshwater species could spread (aside from human introductions) between the Gulf and

Atlantic drainages would be through stream capture. That is the process in which a stream with a high gradient, and therefore able to cut rapidly into its headwater terrain, erodes into the area of a slow-flowing stream belonging to a different watershed and cuts off the headwaters of the slow-flowing stream. As the headwaters of the streams of the Atlantic drainage are on the steep escarpment on the eastern and southeastern faces of the mountains, capture of streams from the Gulf drainage is the more likely.

Summary

This chapter has introduced some of the kinds of data that are used to explore the determinants of animal and plant distributions. The ones illustrated involve both intrinsic properties of the organisms and the varying properties of the area under discussion. The first intrinsic property comprises the respective abilities of the different groups to disperse. Such abilities are highest among birds and bats, and still high for the rest of the mammals. Lizards and snakes can disperse easily, but are more temperature-limited than are mammals. Amphibians tend to be moisture-limited in addition. Fishes are the most limited in their ability to disperse. The second intrinsic property is illustrated by salamanders, the evolutionary history of the group having been important in determining their current distribution.

The regional climate has been shown to be important to the distribution of frogs and reptiles. On a very local scale, the presence of a strongly competitive species can be a determinant of distribution, as was described in Chapter 6.

Finally, a physical barrier has been shown to be important in determining the distribution of fishes. It should be apparent that there are few absolutes in biogeography. For nearly all generalizations, there are exceptions, as can be seen by examining the two tables. This should be remembered in reading the succeeding chapters, as it is not feasible to cite all exceptions to the rules that are described there.

References and suggested further reading

Clay, J. W., D. M. Orr, & A. W. Stuart (eds.). 1975. *North Carolina Atlas: Portrait of a Changing Southern State.* Chapel Hill, University of North Carolina Press.

Martof, B. S., W. M. Palmer, J. R. Bailey, & J. R. Harrison III. 1980. *Amphibians and Reptiles of the Carolinas and Virginia.* Chapel Hill, University of North Carolina Press.

Mehinick, E. F. 1991. *The Freshwater Fishes of North Carolina.* Raleigh, North Carolina Wildlife Resources Commission.

Potter, E. F., J. F. Parnell, & R. P. Teulings. 1980. *Birds of the Carolinas.* Chapel Hill, University of North Carolina Press.

Webster, W. D., J. F. Parnell, & W. C. Biggs, Jr. 1985. *Mammals of the Carolinas, Virginia, and Maryland.* Chapel Hill, University of North Carolina Press.

10

Global pattern and distribution of climates

Worldwide pattern

Interest in the geographic distributions of animals and plants has gone through several cycles, depending on the prevalent theory of how those distributions came about. The first analyses, in the middle of the nineteenth century, were on a very broad scale, with the biogeographic realms corresponding to the continental landmasses (Figure 10.1). The animals and plants on a given continent clearly are more similar to one another than they are to the animals and plants of any other continent. In the nineteenth century, interest centered largely on the locations of the boundaries between the realms. Of the realms, none corresponded exactly to the physical limits of any continent. The extension of the Palearctic Realm onto North Africa and the extension of the Ethiopian onto Arabia are easily explained by the desert barriers. The extension of the Neotropical Realm onto the southern end of North America is due to a clear climate change at that location. The Oriental Realm is separated from the Palearctic by a desert in the northwest and the Himalaya Mountains across its northern edge. Only in China is there any question about the boundary. There was much early discussion about where the limits of the Oriental and Australian realms are. Alfred Wallace placed the line between the East Indian islands Bali (Oriental) and Lombok (Australian) and extended the line to put Borneo in the Oriental and Sulawesi (Celebes) in the Australian. Weber argued that the line should be farther east and included Sulawesi and Timor in the Oriental Realm. Wallace based his division on the location of the deepest water; Weber based his on the location where the faunas of the two realms are most nearly balanced. That kind of argument reinforces the earlier observation that absolutes are very rare in biogeography. Scme groups of animals are distributed in a way that will support one person's idea; other groups have different distributions. The Palearctic and Nearctic realms have so many

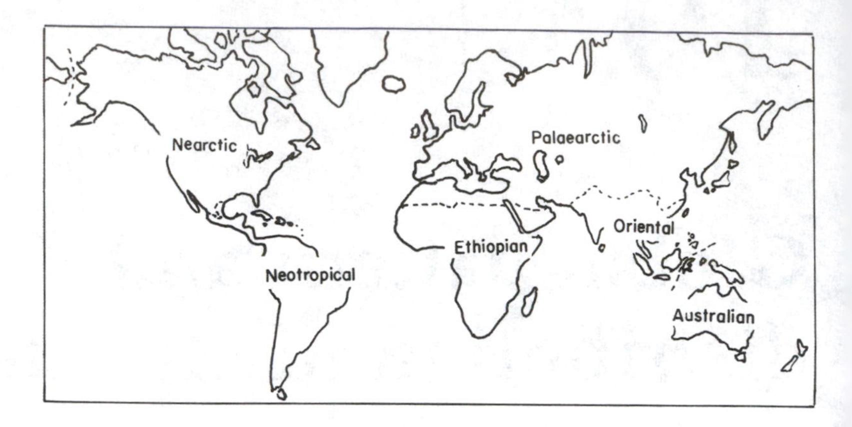

Figure 10.1. The faunal realms on a Mercator projection.

animal and plant groups in common that they are frequently combined as the Holarctic.

Recognition of all of these realms depends on the easy observation that each one contains certain groups that are found nowhere else and some that are almost exclusively found there, and there are some groups that are conspicuous by their absence. It is conventional to analyze the information by using families or subfamilies as the units that have been called "groups" in the foregoing discussion. Those that are confined to the different zoogeographic realms are shown in Table 10.1. The Neotropical Realm has the most endemic families, and the Palearctic Realm the fewest. That is partly due to the relative isolation of South America and partly due to the tropical climate of most of it. Notice that the Ethiopian Realm (also largely tropical) is second in the number of endemic families of vertebrates, and the Nearctic has the second fewest. The reason for the greater diversity in the tropics frequently is given as the upheaval in the northern realms caused by the advances and retreats of glaciers not long ago, during the Pleistocene epoch, in contrast to the long period of climatic stability in the tropics.

Something can be learned by examining the number of families of vertebrates that are in common between two adjacent realms, but are found nowhere else (Figure 10.2). The most striking result is that in contrast to all other combinations, there are more families in common between the Palearctic and Nearctic realms than are restricted to either of the two. That is the basis for regarding them as one large realm. The larger numbers of families restricted to the Nearctic and Neotropical combined, and to the Ethiopian and Oriental combined, are reflections of the very large numbers of families restricted to each of the three tropical realms, rather than to special affinities, as the fractions represented by the combinations are only about one-fourth of the totals. The separate realms appear to be real, with the possible exception of the Nearctic and Palearctic. One might be more convinced of the accuracy of that

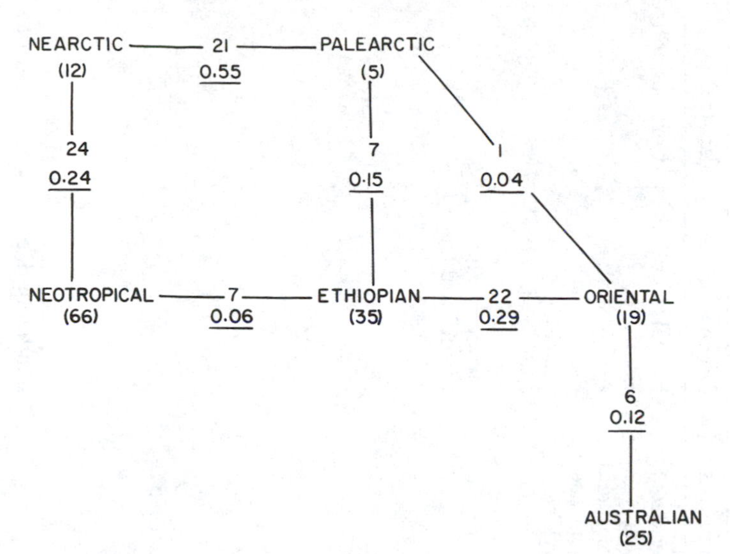

Figure 10.2. Numbers of families of vertebrates confined to each of the zoogeographic realms (in parentheses), numbers of families confined to each pair of adjacent realms (numbers interrupting the connecting lines), and proportion of the total represented by the shared families (underlined).

statement if such a large number of families did not occur in three or more realms. Of 384 families of vertebrates, exclusive of marine or quasi-marine families, Figure 10.2 accounts for only 181 (47%). These are not distributed proportionally among the classes of vertebrates: 57 (90%) of 63 families of primary freshwater fishes, 86 (79%) of 109 families of mammals, 12 (54%) of 22 families of amphibians, 73 (53%) of 138 families of birds, and 19 (36%) of 52 families of reptiles. The remaining 203 families are distributed more widely. It is not easy to account for the differences among the classes, as separate theories would be required for most of the comparisons. One could always think of something, but the arguments would not be convincing. It must be stated that taxonomists working with the different classes probably would disagree with details of the analysis, such as exactly how many families to recognize, but even if we could obtain perfect agreement, the pattern described would not be changed.

Patterns within continents

Within the realms, the patterns of distribution are nearly always described in terms of the dominant vegetation, because that is what is conspicuously observable, although the distributions of many animal species come close to being congruent with the patterns of vegetation. The first attempt to rationalize these distributions made use of the obvious correlation with climate, as shown in both latitudinal and altitudinal zonation. It was first claimed that the distributions were determined by average cumulative temperature. What seemed to work best for limiting the northern distribution was the number of days with a mean temperature of 6°C or greater during the period of reproduction and growth; the

Table 10.1. *Names of vertebrate families confined to each of the zoogeographic realms*

Class	Neotropical	Nearctic	Palearctic	Ethiopian	Oriental	Australian
Osteichthyes[a]	Hemiodontidae	Amiidae		Polypteridae	Homalopteridae	Ceratodontidae
	Xiphostomidae	Hiodontidae		Kneriidae	Sisoridae	
	Gasteropelecidae	Percopsidae		Phractolasemidae	Amblycepidae	
	Rhamphichthyidae	Aphredoderidae		Pantodontidae	Pangasidae	
	Sternarchidae	Centrarchidae		Mormyridae	Chacidae	
	Gymnotidae			Gymnarchidae	Cranoglandidae	
	Electrophoridae			Mochochidae	Pristolepidae	
	Diplomystidae			Malopteruridae	Luciocephalidae	
	Ageneiosidae				Chaudhuriidae	
	Pimelodidae					
	Helogeneidae					
	Hypopthalmidae					
	Cetopsidae					
	Pygidiidae					
	Bunocephalidae					
	Callichthyidae					
	Loricariidae					
	Astroblepidae					
Amphibia	Rhinophrynidae	Ambystomidae	Hynobiidae	Phrynomeridae		
		Dicamptodontidae				
		Amphiumidae				
		Sirenidae				
Reptilia	Dermatemydidae	Anniellidae			Lanthanotidae	Pygopodiadae
Aves	Rheidae					
	Tinamidae					
	Anhimidae					
	Opisthocomidae					
	Cariamidae					
	Psophidae					
	Eurypygidae					
	Thinocoridae					
	Chionididae					
	Nyctibiidae					
	Steatornithidae					
	Momotidae					
	Todidae					

	Bucconidae			Struthionidae	Irenidae	Apterygidae
	Galbulidae			Sagittariidae		Casuariidae
	Ramphastidae			Scopidae		Dromaeidae
	Rhinocryptidae			Mesoenatidae		Rhynochetidae
	Conopophagidae			Coliidae		Aegothelidae
	Formicariidae			Philepittidae		Xenicidae
	Furnariidae			Vangidae		Menuridae
	Pipridae					Atricornithidae
	Cottingidae					Cracticidae
	Phytotomidae					Grallinidae
	Tersinidae					Callaeidae
						Ptilonorhynchidae
						Paradisaeidae
Mammalia	Caenolestidae	Aplodontidae	Spalacidae	Tenricidae	Cynocephalidae	Tachyglossidae
	Solenodontidae	Antelopcarpridae	Glirinae	Potamogalidae	Myzopodidae	Ornithorhynchidae
	Noctilionidae		Seleviniidae	Chrysochloridae	Tupaiidae	Dasyuridae
	Desmodidae		Dipodidae	Macroscelicidae	Tarsiidae	Notoryctidae
	Natulidae			Lemuridae	Platacanthomyidae	Peramelidae
	Furipteridae			Indridae	Plantanistidae	Phalangeridae
	Thyropteridae			Daubentomiidae		Phascolomidae
	Cebidae			Anomaluridae		Macropodidae
	Callithricidae			Pedetidae		Mystacopidae
	Caviidae			Nesomyninae		
	Hydrochoeridae			Lophiomyninae		
	Dinomyidae			Thryonomidae		
	Dasyproctidae			Petromyidae		
	Chinchillidae			Bathyergidae		
	Capromyidae			Orycteropodidae		
	Octodontidae			Procaviidae		
	Ctenomyidae			Hippopomatidae		
	Abrocomidae			Giraffidae		

[a]For fishes, only primary freshwater families are given, namely, those that cannot tolerate salt water.
Source: Mostly from Darlington (1957).

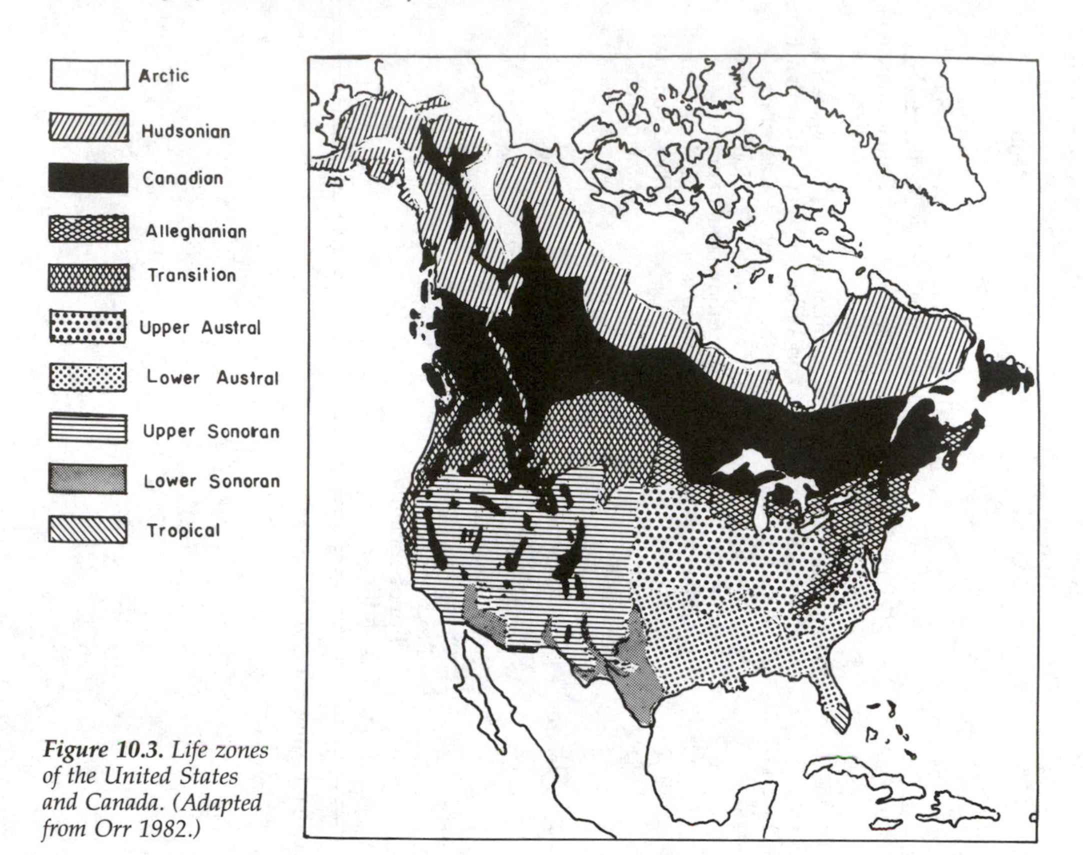

Figure 10.3. *Life zones of the United States and Canada. (Adapted from Orr 1982.)*

southern limit was thought to be determined by the average temperature for a limited period in the hottest part of the year. Those limits created a series of zones (Figure 10.3) that correspond reasonably well with observable distributions of plants and animals. Of course, the necessary weather data were not completely available in the early years of the twentieth century, when the theory was popular, and maps of the zones were drawn to fit the observable vegetation in many places, especially in western North America, where the terrain interrupted the smoothly drawn zones of the East. Even away from the western mountains, it is obvious that temperature alone cannot account for the distributions, an observation even clearer in the West. The extra factor that must be considered is rainfall. The same temperature combination is correlated with deciduous forest and red-eyed vireo in the East, but with grassland and antelope on the Great Plains, and with desert shrubs and kangaroo rats still farther west. The higher elevations do support coniferous forests, as do the colder climates, but there is no deciduous forest zone in the West, most of the native hardwoods being evergreen. In order to understand the correlation between climate and the distributions, it is necessary to learn something about how climate is brought about. The following simple facts apply:

1. Warm air rises.
2. Cold air sinks.
3. Warm air holds more moisture than does cold air.
4. As warm air rises, it cools and loses moisture, which condenses as rain.
5. As cool air sinks, it becomes warm and takes up moisture, thus drying out the land below.
6. The sun never shines directly from above, except in the area between the Tropic of Cancer in the Northern Hemisphere and the Tropic of Capricorn in the Southern Hemisphere. These lie at latitudes 23°27′ north and south of the equator, respectively. The belt between the tropics is thus warmed maximally by the sun, especially near the equator.

Air at the equator is heated by the sun and rises; as it rises, it cools, and rain falls. Air at and near the poles is cold, and it sinks. The sinking air at the poles cannot pile up, so it flows toward the equator. The rising air at the equator cannot rise indefinitely, so it flows poleward at high altitudes. At locations between the tropic and 30°, the poleward-moving air has become cold enough to sink, and having lost its moisture as rain, it is very dry. As it sinks, it becomes warmer and takes up more moisture from the ground beneath. It is at this latitude that the great warm deserts of the world are found (Figure 10.4). In the days of sailing ships, this zone was a difficult one to cross because of the baffling light winds. Ships becalmed there with horses aboard were in desperate straits for water, and the horses naturally suffered most. They would become frantic and kick and damage the ship; hence the term "horse latitudes." The cold deserts, such as the Gobi, have a different explanation. Moist air is prevented from reaching them by high mountains, by arid intervening areas, or by cold offshore currents that cool the air and prevent it from taking up moisture that might otherwise fall as rain. The Peruvian Desert is of that type.

The air cannot pile up at 30°, and so it flows at the surface both toward the equator and toward the pole. Between the equator and 30°, a circulating zone forms, named a Hadley cell. North of 30° there is a zone where the air tends to rise, with resulting instability that initiates storms (Figure 10.5). Now, we complicate the picture by taking into consideration the rotation of the earth. The movement of a point on its surface is fastest at the equator (1,040 mph, or 1,679 kph) and nil at the pole. At 30°, a point moves at 875 mph, or 1,410 kph. At 60°, the speed is 500 mph, or 806 kph. This means that air flowing toward the equator from either north or south will surely be left behind. A person between 30° and the equator will be rushing into that air and will feel a wind from the northeast or the southeast, depending on the hemisphere. This is the zone of the trade winds, so called because they are very dependable, and

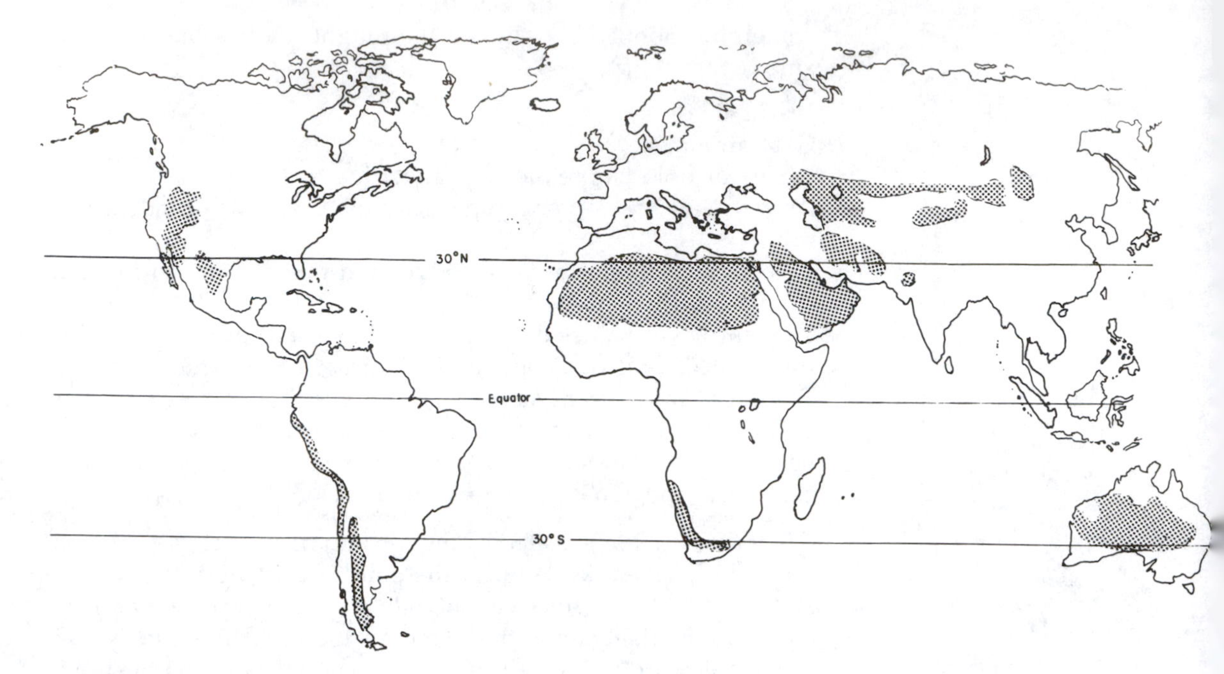

Figure 10.4. Locations of the major deserts of the world.

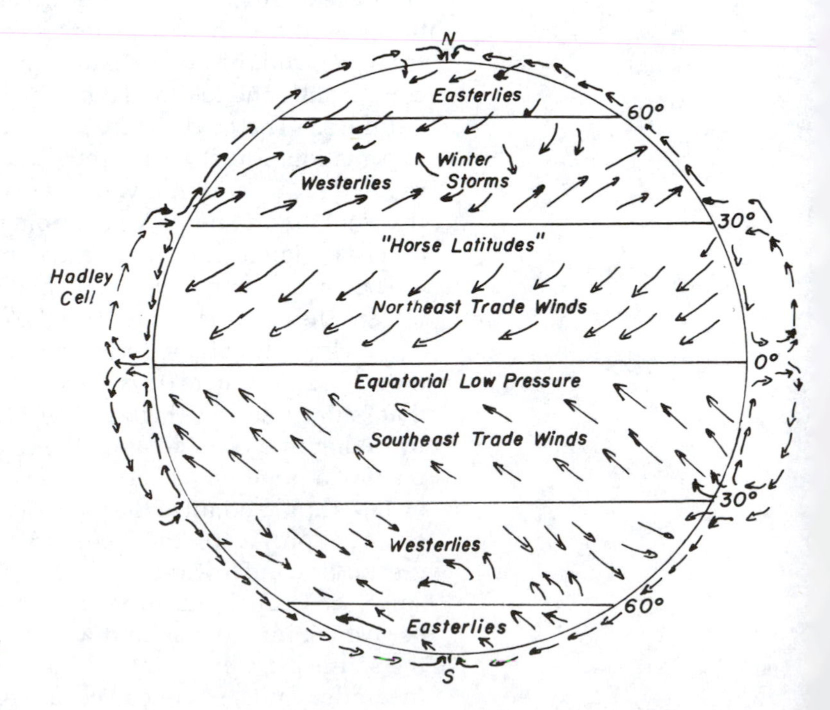

Figure 10.5. Diagram of the global wind patterns.

sailing ships used them as much as possible. Similarly, air flowing from the pole will also appear to come from the east. Conversely, air flowing from the horse latitudes poleward will tend to come from the west, as will the storms generated in latitudes north of 30°. These are the latitudes of the westerlies.

Inasmuch as more than 70% of the earth's surface is covered with water, and the moisture in the air can come only from that source, the winds dictate that land in the Trade Winds Belt will receive more rain if it is on the eastern side of a continent or an island, because the air will be forced up by the land, causing it to become cooler and to lose its moisture as rain. Conversely, in mid-latitudes, the western side of a continent will tend to be wetter. This neat pattern is disrupted by the rotation of the earth, which leaves the water behind at the equator, causing a westward current. This in turn forces the water next to the coast poleward, and eventually it moves in an east or northeast direction, because the water flowing westward at the equator is replaced from north or south, and from water rising from the bottom on the eastern side of the oceanic basin. The net result is a vast circulation in each hemisphere for each ocean. The gyre in the center of the North Atlantic, the location of the Sargasso Sea, is mentioned in connection with the migration of eels (Chapter 13). It is surrounded by the equatorial current to the south and the Gulf Stream to the west and the north. This circulation of water has an important impact on climate, bringing warmth to northern Europe and to Alaska and northwest Canada in the Pacific. These favored lands are much more habitable than are areas at corresponding latitudes on the western sides of the respective oceans (Labrador and Siberia). The upwelling water on the eastern sides of the basins is cold, and air flowing over it will not take up much water. Therefore, the western side of the continent is less wet than would otherwise be the case. The west coasts of South America, Africa, and Australia below the tropics are dry.

The positions of marine basins make a further difference in the climate over the nearby land. The Gulf of Mexico extends into the region of the westerlies, providing moisture feeding into the storms and bringing more rain into central North America than would be the case without it.

In the early twentieth century, it was realized that although there were general correlations between the kinds of animals and plants and the climate, those correlations were not close enough to be useful to biogeographers. The study of interspecific interactions was becoming important in ecology, and the life-zone concept was replaced by the theory of biomes. The biotic zones were considered to consist of many interacting species of plants and animals, the interactions consisting of competition and predation. Each was thought to maintain its integrity by competing as a group with the adjacent biomes (Figure 10.6). The competition is

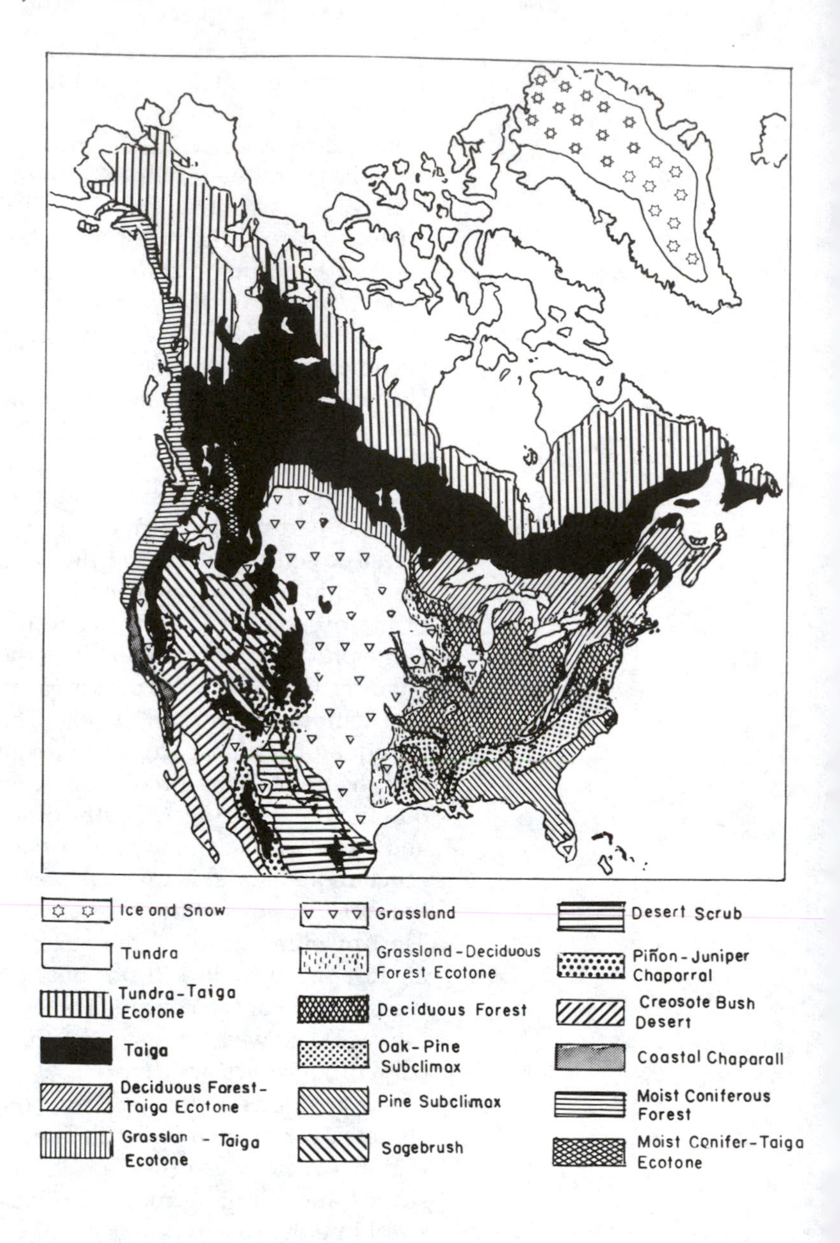

Figure 10.6. *Biomes of North America. (Adapted from Pitelka 1941.)*

expressed over a zone where the biota is a mixture of the two biomes, called an ecotone. Some of the ecotones are narrow, and others are wide, depending on the interpretation. For example, the Piedmont of the southeastern United States is interpreted as being within the Eastern Deciduous Forest Biome, according to some authors, but according to others it is an ecotone between the deciduous forest and the Pine Subclimax of the Coastal Plain. The distribution of vertebrates within a small part of this general pattern was described in Chapter 9.

References and suggested further reading

Darlington, P. J., Jr. 1957. *Zoogeography: The Geographical Distribution of Animals*. New York, Wiley.
Odum, E. P. 1971. *Fundamentals of Ecology* (3rd ed.), Chapter 14. Philadelphia, Saunders.
Orr, R. T. 1982. *Vertebrate Biology* (5th ed.). Philadelphia, Saunders.
Pitelka, F. A. 1941. Distribution of birds in relation to major biotic communities. *American Midland Naturalist* 25:113–37.

11

Vertebrate distribution
The importance of other species

In Chapters 9 and 10 it was shown that climate has a strong influence in determining the distributions of vertebrates, as do barriers to dispersal. The quotation from Darwin in the introduction to this part of the book, however, reveals that it would be incorrect to attempt to explain distributions on the basis of the physical world alone. Other species of organisms also can be shown to influence distributions by three different routes: through competition, through predation and parasitism, and by providing a necessary resource. These effects are the subject of this chapter.

Competition

The effect of competition can be appreciated with least effort by looking at the side of a high mountain (Figure 11.1). The mountainside above Zermatt in the Swiss Alps is dark, because it is covered with conifers, larch and spruce. Those trees do not continue upward to the bare rock. Above them, there is a zone of low-growing plants, grasses, forbs, and tiny shrubs, that are collectively known as alpine tundra. The boundary between the two vegetation zones is abrupt, as is the boundary between the conifers and the broadleaf trees in France at the Château de Menton lower in the Alps (Figure 11.2). Clearly, climate has a strong influence, but if it were the only determinant of the distributions, one would expect to see a gradual change from one vegetation type to another. The effects of interspecific interactions prompted ecologists to change from classifying the distributions by life zones to classifying them by biomes.

Determination of the altitudinal distributions of vertebrates requires much more effort than simple observation of a mountainside, but when the necessary observations have been made, sometimes it has been found that they have the same kinds of distributions as the vegetation types. It should be remembered that in the Black Mountains of North Carolina and the Great

Figure 11.1. *Photograph of the Alps near Zermatt, Switzerland, to show vegetational zonation.*

Figure 11.2. *Photograph of the French Alps and the Chateau de Menton, to show lower limit of conifer forest.*

Smoky Mountains of North Carolina and Tennessee, the salamander species *Plethodon jordani* is confined to high elevations, and the forms of *P. glutinosus* are confined to low ones, with a narrow zone, 70–100 vertical meters wide, where they are able to

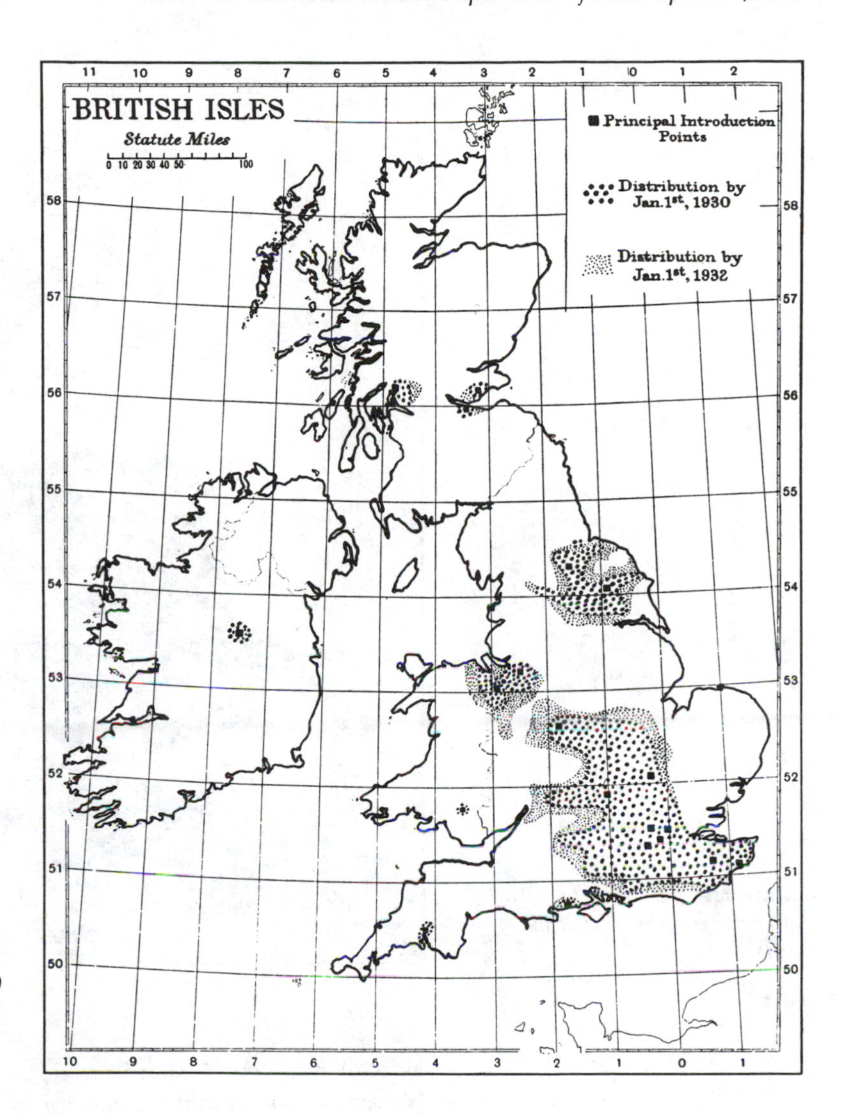

Figure 11.3. Range of the gray squirrel in the British Isles in 1930 and the approximate area colonized from 1930 to 1932. (From Middleton 1932.)

coexist (see Figure 6.2a). As described in Chapter 6, experimental removal of each species from plots in this zone resulted in either an increase in the total numbers of the other or an increase in the proportion of young individuals of the other, showing that reproduction had increased and would eventually give a significant increase in numbers. Those observations and behavioral experiments showed that the narrowness of the zone of overlap is caused by intense interspecific aggression as an expression of competition between the two species.

There are some records in which the effects of competition have been followed over historical times. Many of these examples, especially from oceanic islands, have resulted in the extinction of local species when exotic ones have been introduced by humans.

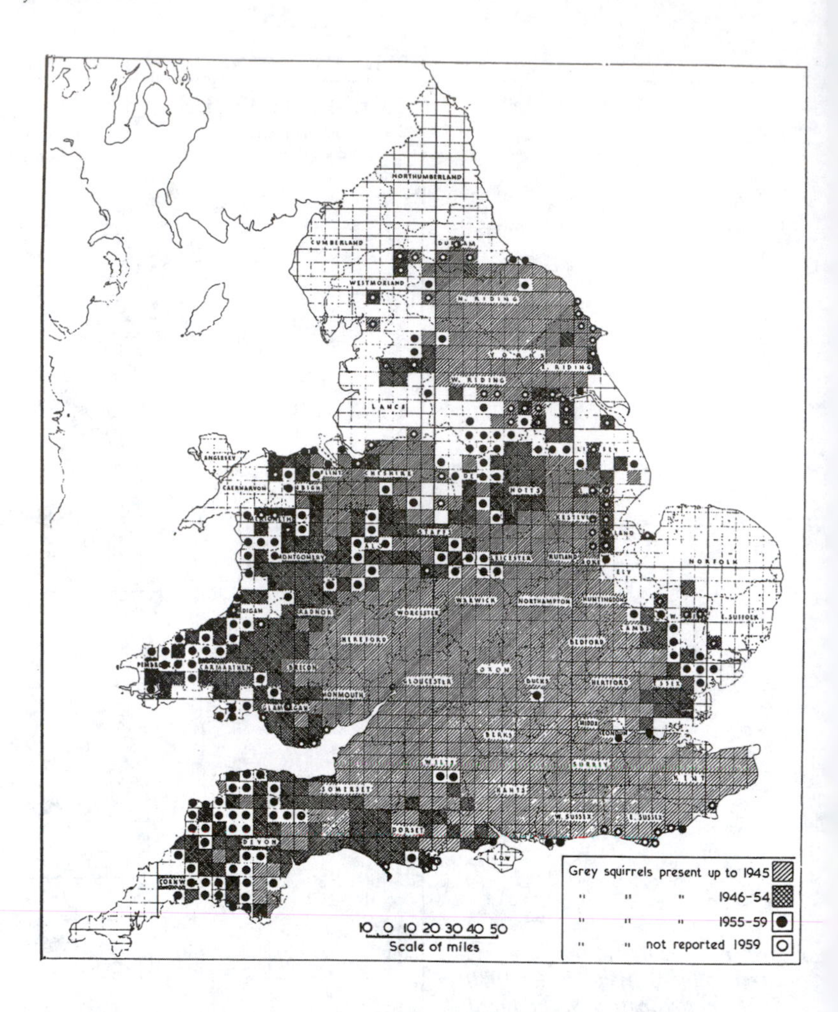

Figure 11.4. *Spread of the gray squirrel, 1945–59. (From Lloyd 1962.)*

Hawaii and New Zealand provide repeated examples of such folly, carried out in many cases for esthetic reasons rather than for any practical purpose. In both places, the damage has been compounded by further introductions in attempts to reduce the effects of the original ones. The best example is treated later with the topic of predation.

One example of the effects of competition with good distributional data is that of the introduction of the American gray squirrel, *Sciurus carolinensis,* into Great Britain. At some time in the eighteenth century, the gray squirrel began to be introduced into private parks in several parts of the British Isles. Further introductions were made in 1830 in northern Wales, in 1880 into Cheshire, the adjacent part of England, and in 1890 into Bedfordshire, north of London (Figure 11.3). By the early 1930s it was beginning to be observed that the native red squirrel, *Sciurus vulgaris,* could not be

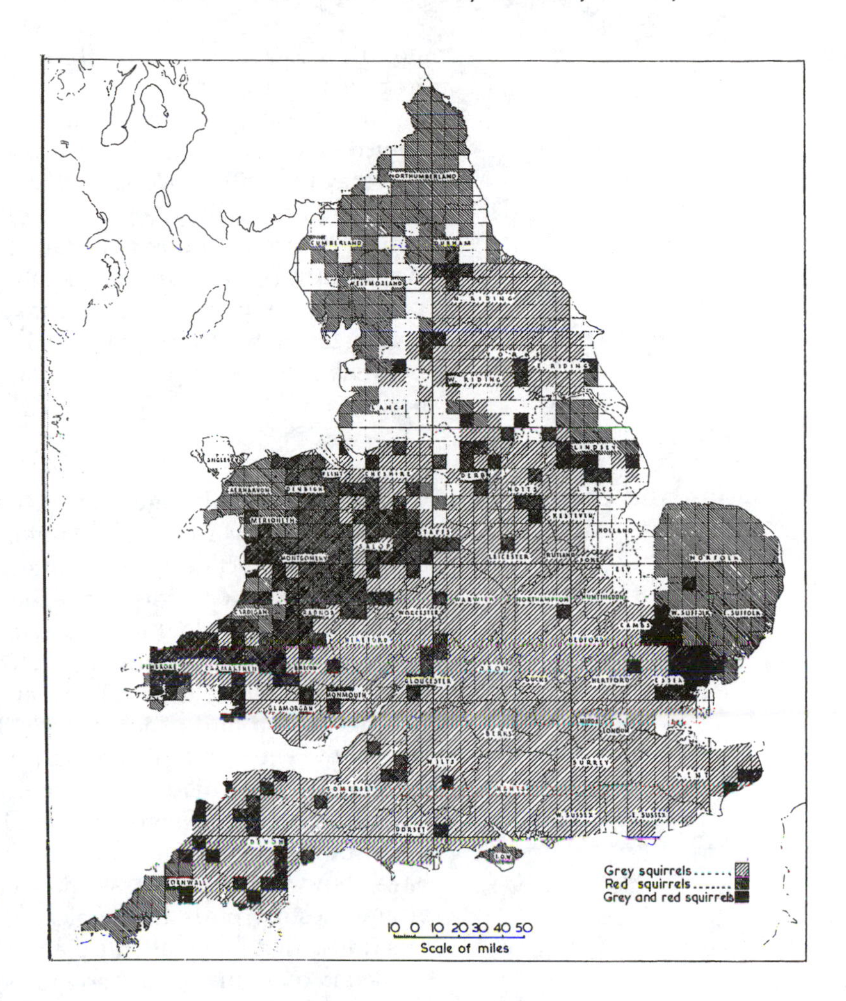

Figure 11.5. Distributions of gray and red squirrels in 1959. (From Lloyd 1962.)

found in the areas into which the gray squirrel had expanded its range (Figure 11.4). The rather slow spread of gray squirrels up to 1931 was quite unlike the rate at which they extended their range in the next 28 years. By 1959 they had been found over most of England (Figure 11.5). The spread was not uniform, and in a few localities the species apparently lost a little ground. Nevertheless, over most of its new range, the red squirrel had disappeared, and although there is some disagreement that that was a direct effect of the gray squirrel, it is difficult to reach any other conclusion. Moreover, a very recent study (Kenward and Holm 1993) has shown that both species can thrive on hazelnuts, but only the gray squirrel thrives on acorns. There is competition for hazelnuts, which ripen first. Thus, the gray squirrels have a food refuge in acorns, and red squirrels cannot persist in woods with more than 14% oak canopy. Exacerbating this disadvantage has been a decline in hazel production through coppicing (repeated pruning to give new growth).

An interesting feature of this example is that initially the gray squirrel made little headway. It is typical of many introductions that they have little success at first, and then after a long period of just hanging on, they begin to multiply. In fact, a number of attempts at introductions have failed, in some cases several times. Apparently, some fortuitous combination of genetic material, or perhaps a favorable mutation, takes place that makes the new organisms better adapted to local conditions, and they can then multiply much more effectively that they had before the change occurred.

Predation and parasitism

Competition is only one way in which other species affect the distributions of vertebrates. Many introductions by humans have caused the extinction or decimation of local species. Rats, *Rattus norvegicus* and *R. rattus,* have been introduced inadvertently throughout the world. In addition to quickly becoming pests, they are predators on local animals, such as the local coot on Tristan da Cunha. Lord Howe Island, between Australia and New Zealand, was declared a bird sanctuary in 1879, but a plague of rats has caused the extinction of the local bird fauna. Many different attempts have been made to control them, some through biological control programs. Mongooses have been introduced as predators to control the rats, but they have rarely been successful. They have, however, been disastrous in their effects on the local ground-nesting birds in Jamaica and elsewhere. Feral domestic cats (those that have taken up a wild existence) have caused the extinction of a number of endemic species of island birds. Direct predation by humans has caused the extinction of many species, but the best-documented story is that of the passenger pigeon, *Ectopistes migratorius*. This bird was present at the time of the European colonization in numbers that are difficult to grasp. In 1806, Alexander Wilson observed a flock that was moving at "a mile a minute" (probably exaggerated), with three birds per square yard, several tiers thick, and at least a mile wide. It required 4 h to pass his observation point. His estimate was that there were at least 2,230,272,000 birds in the flock. Also early in the nineteenth century, John J. Audubon reported a flock that contained 1,115,130,000 and estimated that as a small part of the total. The bird had two fatal traits: It was very good to eat, and the entire population apparently nested in the same area, which had to be where the acorn crop was abundant. This second trait may have been a behavioral necessity for successful nesting. It allowed market hunters to converge on the nesting area, and with the arrival of railroads and better transportation, they could come from great

Figure 11.6. *Photograph of Lake Manyara, Tanzania, in the African Rift Valley. The western wall of the valley is shown.*

distances. In a number of successive years, the activity of the hunters completely disrupted breeding by the pigeons, and the lack of replacements doomed the species. The last specimen died in the Cincinnati Zoo in 1914.

Another example that has been claimed is that the arrival of humans in North America at the end of the last glacial advance exposed the large game animals, mammoths, mastodonts, horses, ground sloths, and many other species, to a socially organized and much more efficient predator than any they had ever experienced. The hypothesis is that the early humans swept over the Western Hemisphere in a rapidly moving wave and exterminated these species in a short time. That hypothesis is not universally accepted, but the fact remains that the species in question did become extinct at the end of the last major glacial, after successfully experiencing a number of such glacial advances during the Pleistocene. It is difficult to attribute their extinction to any effect of climate, as some of the critics of the hypothesis attempt to do.

These examples are to some extent artificial, all being related to human interference, but from that basis it can be imagined how the connection of North and South America in the Pliocene, after having been separated during most of the Tertiary, could have led to a rapid loss of species of primitive placental mammals in South America, as the more highly evolved forms moved in from the Holarctic.

One of the more interesting effects of parasitism is in the indirect limitation of humans themselves. The Lake Manyara Game Reserve in Tanzania is located at the foot of the western escarpment of the African Great Rift Valley (Figure 11.6). It is immediately adjacent to the grazing lands of the Masai, a pastoral people who have steadfastly retained their way of life, and for whom the possession of many cattle is the greatest mark of status. There is no geographic barrier to their encroaching on the Lake Manyara Reserve, as they have at the Amboseli Reserve at the foot of Mount Kilimanjaro. What stops the Masai at Lake Manyara is the presence of a very dense population of tsetse flies. These insects transmit protozoan parasites of the genus *Trypanosoma,* which, in various parts of Africa, cause lethal sicknesses of both people and cattle. The form at Lake Manyara transmits *T. brucei,* which is fatal to cattle, and any encroachment by the Masai would lead to the destruction of their culture. The game animals are immune to *T. brucei.*

Food species

If a species has a restricted diet on which it must subsist, its distribution obviously is restricted to that of its food species. Such a restriction is not common among vertebrates, and most such examples are found among herbivores. One possible example among carnivores is provided by the saber-toothed cats of the genus *Smilodon,* commonly called "saber-toothed tigers" (Figure 11.7). The huge canine teeth of these animals apparently had a single adaptive function: stabbing through the thick skins of elephants and other very large mammals. As described earlier, these large herbivorous mammals became extinct over a short period at the end of the Pleistocene. Without such prey, *Smilodon* presumably was unable to capture the smaller, faster species, and it also became extinct. That scenario is inferential, of course, and perhaps should not be taken too literally.

Among the herbivorous mammals, the giant panda, *Ailuropoda melanoleuca,* and the koala, *Phascolarctus cinereus,* are the most interesting. The giant panda subsists only on bamboo, which has the trait of "mast cropping" to an extreme degree. There are many species of bamboo, and each species has a characteristic period for flowering and producing seeds. After sprouting, all individuals of that species grow for a certain number of years, up to at least 60 years for some species. They are easily transplanted, but when they are, every individual, regardless of where it is, blooms at once, sets its seed, and dies. Recently, the pandas were in trouble because the bamboo species on which they depend for their food had reached the stage of flowering and dying, and the process of the sprouting of the new seeds and growth of new plants takes too long for them to wait. They are no longer able to disperse to areas where other species of bamboo could be found, because those areas are now occupied by people.

The marsupial koalas feed exclusively on the leaves and bark of eucalyptus trees, but only a very limited number of species of

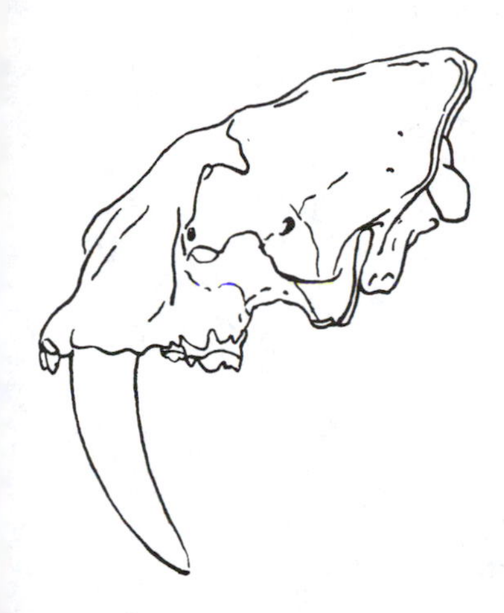

***Figure* 11.7.** *Skull of the extinct saber-toothed cat,* Smilodon.

Eucalyptus will serve as food. There are hundreds of species of *Eucalyptus* in Australia, and intensive research was necessary to discover which trees are satisfactory food trees and why. First, it was well known that in the state of Victoria the animals regularly feed on the manna gum, *E. viminalis,* but that at irregular intervals they leave that species and select one of a very few others. After a brief period, they return to the manna gum. Investigation revealed that the young leaves of the manna gum contain cyanogenic glycoside, and when these leaves are chewed it combines with an enzyme to produce prussic acid, a deadly poison. At intervals, especially in winter, most of the leaves, even of mature trees, develop the property. The few species to which the kaolas can resort at such times all produce an essential oil, phellandrine. Many species have high concentrations of cineol, another essential oil, and koalas always reject those trees. When fed on them in zoos, koalas die. They are able to metabolize phellandrine, and in Victoria, at the southeast corner of Australia, they can exist only where the manna gum and one or more of a few other species are found together. A bizarre situation was revealed when it was discovered that in Queensland, in the northeast of Australia, the koalas prefer one of four species, all of which are rich in cineol, but lack phellandrine. The Queensland koalas reject the manna gum absolutely. It appears that there is the potential, if not the necessity, for speciation between the two forms of the koala.

Emphasis on the interactions among species, as described in Chapters 6 and 7, and earlier in this chapter, gave rise to a new way of thinking about animal distributions. Instead of considering distributions as static, determined by such factors as barriers to dispersal and climates, it began to be recognized that the presence of a species at any given place is also the result of competitive exclusion, predation, and the ability to consume the resources that are found there. Although this has not been formalized for distributions generally, especially those on large areas, such as continents and very large islands, a formal theory has been proposed to account for certain regularities in the faunas of smaller islands.

The model was created by Robert MacArthur and Edward O. Wilson in 1963, but some of the ideas used were published by P. J. Darlington in 1957. The first such regularity is an increase in the number of species with increasing area of their island. The second is a decrease in the number of species on an island with increasing distance between that island and a much larger area that could provide colonists. When applied to islands, the theory proposes that, other things being equal, there are two factors that should operate to determine the number of species present. The first is that the rate of immigration of new species should decrease with increasing distance between the island and a mainland source. We

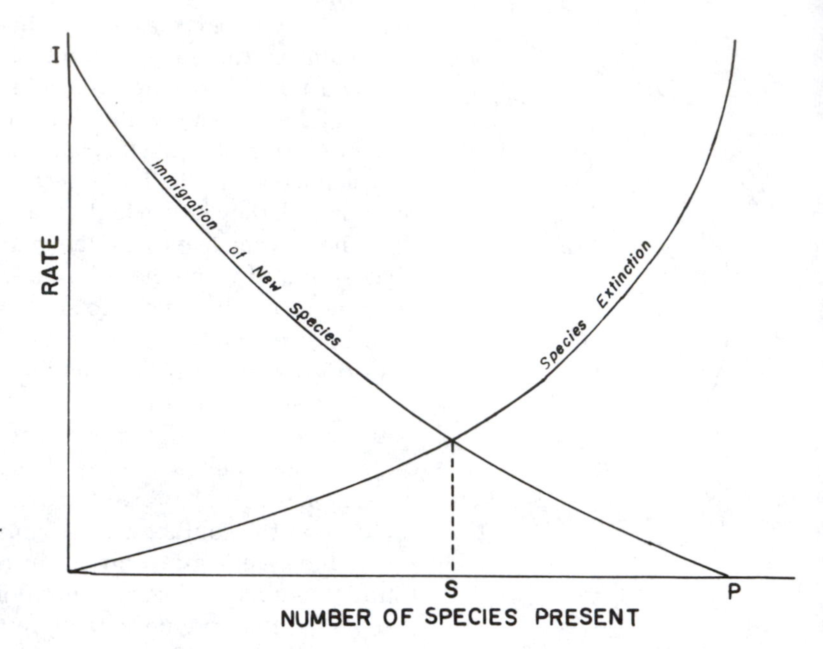

Figure 11.8. *Equilibrium model for the fauna of a single island. (Redrawn from MacArthur & Wilson 1963.)*

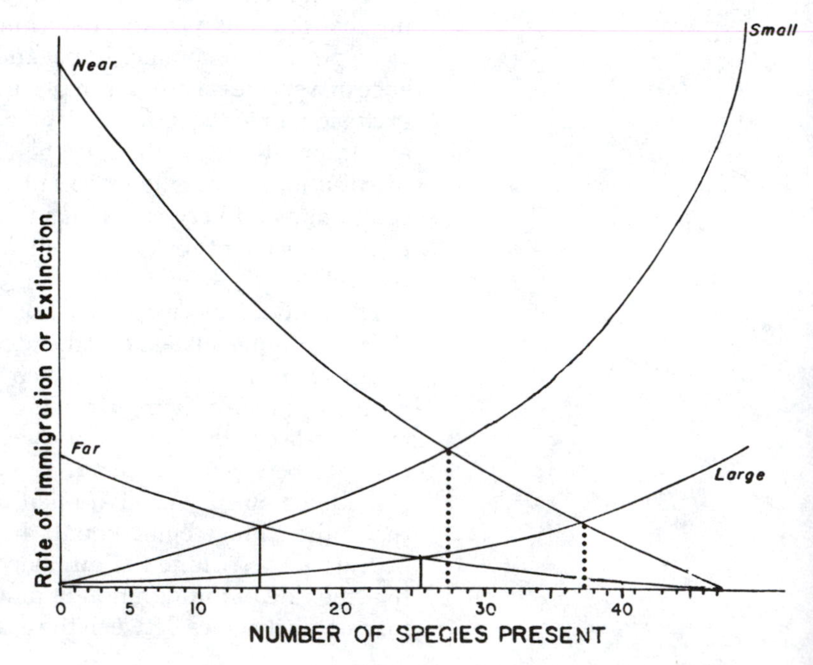

Figure 11.9. *Equilibrium models for a large island both near to and far from a source fauna, and a small island similarly placed, to show the effects of size and distance on the predicted faunas. (Adapted from MacArthur & Wilson 1963.)*

can thus draw a graph plotting the rate of immigration as a descending function of the number of species already present. When the line reaches the total number of species available to immigrate, the rate will be zero. For an island near a source, the curve must be steeper than for one far away, simply because each species has a better chance of reaching the nearer island. The line will be a curve because the species are not equal in colonizing ability, and those most able to reach the island will do so first, leaving a decreasing rate of immigration to those with poor dispersal abilities. On the island, however, another set of factors will come into play. Competition and predation will tend to eliminate some of the arriving species. In the imaginary situation that all available species have arrived, these factors will eliminate some of them, and the rate of elimination will vary among species, with the most susceptible losing most rapidly, and the resistant ones at a slower rate. A line describing this process will be a curve beginning at the total number of species available and descending toward zero as the number of species declines. If we put the two lines on the same graph, we shall find that they will cross at the point where the rate of extinction just balances the rate of immigration, and that point will lie directly above the theoretical number of species that should be present if we have measured the two rates correctly (Figure 11.8). As might be imagined, no one has made such measurements for vertebrates, but we can make use of the regularities that have already been stated. For islands at differing distances from the continental source, the rate of immigration will begin at a higher level for one closer to the source than for one farther away. We can thus draw a series of curves descending toward the total number of species available, each curve representing a more distant island. Also, for a small island, the rate of extinction of species is sure to be greater than for a large island, and we can draw another series of curves, each one representing a larger island (Figure 11.9). Thus far, we are sure of the relative positions of the curves and do not need to be concerned about the absolute values. What is interesting about the series of curves is that they give us a prediction that can be confirmed or rejected.

The prediction can be made as follows: If we create any arbitrary scale on the horizontal axis for the number of species and project the expected number of species from the intersections of the near curve with the curves for the smallest and the largest islands, and do the same for the far curve, we find that the expected numbers predict that the effect of size will be greater for far islands than for near ones. MacArthur and Wilson published a graph showing the relationships between the number of bird species present and the sizes of a number of islands in the tropical Pacific (Figure 11.10). The numbers rise more slowly for "near" than for "far" islands, confirming the prediction.

It has not been feasible to test the theory with experiments on vertebrates, but an experiment on insects and spiders on small mangrove islets in the Bay of Florida gave most interesting results. The number of species on each was established by repeated surveys, and then plastic tents were placed over the islets, which were fumigated with methyl bromide, which was successful in eliminating the fauna. Each islet was surveyed again fifteen times during the year after the fumigation. For most islets, approximately the original numbers of species had returned after about 5 months. The interesting feature was that for all posttreatment surveys, an average of only 58% of the species were the same as in the baseline survey. Moreover, there was a large amount of turnover of species among the posttreatment surveys. The dynamic nature of island faunas was amply confirmed.

There are a few examples among vertebrates of spontaneous extensions of distributions in the absence of deliberate introductions by humans. The horned lark, *Eremophila alpestris,* a small, ground-living bird, is a typical prairie species that nests in open areas. With the spread of agriculture in North America, more and more open land appeared in the eastern United States, replacing the original forest. These birds were first recorded as winter visitors, but they increased their nesting distribution: north-central Indiana in 1880, Ohio in 1890, north-central New York in 1895, Massachusetts in 1903, West Virginia in 1908, Virginia in the late 1920s, central North Carolina in 1937, and Georgia in 1940. By that time, farming had become unprofitable in much of the Piedmont of the Southeast. Flying over that area in winter, one is impressed by the numerous green areas. These are pine forests, a regular stage in the succession from plowed fields to hardwood. Each green area was once a field, showing what a large proportion of the land had once been cleared, and hence suitable habitat for horned lark nesting. In recent years, they have not been recorded nesting, and it is difficult to find them in winter.

Two species of mammals, the opossum, *Didelphis marsupialis,* and the armadillo, *Dasypus novemcinctus,* are extending their ranges northward in North America. The causes of these extensions are not known, but the opossums thrive in proximity to humans, and the expansion of people into suburbs may encourage them.

The most spectacular example of a range extension is that of the cattle egret, *Bubulcus ibis.* This is a common species in the Old World tropics; it is well adapted to accompany large grazing herbivores, feeding on the insects that they flush. It was unknown in the Western Hemisphere until near the end of the nineteenth century (1882), when it was observed in Surinam. There is no evidence that these birds reached South America by any means other than their own. It is generally supposed that they were blown from Africa by a hurricane. It was only after 30 years that cattle

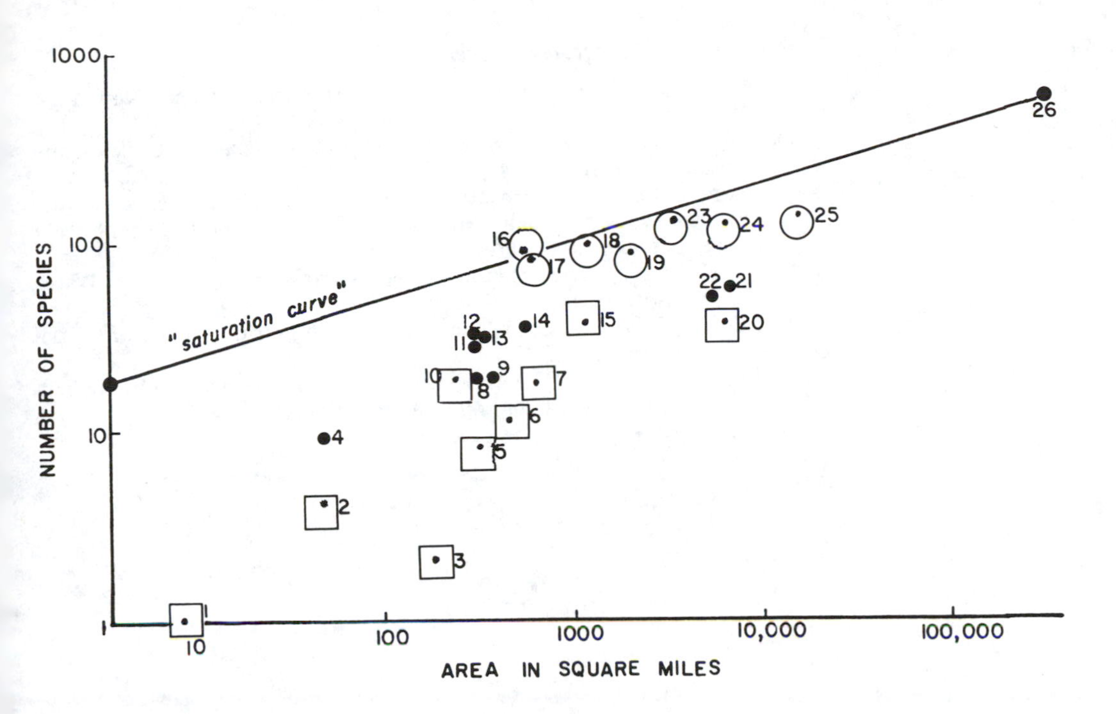

Figure 11.10. *Numbers of land and freshwater bird species on various islands in the Moluccas, Melanesia, Micronesia, and Polynesia. (Redrawn from MacArthur & Wilson 1963.)*

egrets had spread to Guyana, the adjacent country (1912), and another 30 years were required for them to reach Venezuela (1943). There was a report from Florida in 1942, but proof was not available until 1952. Interestingly, the intervening islands were not colonized until 1954–7. They spread rapidly from Florida and by 1959 had been observed from Minnesota to Nova Scotia. They have been seen as far north as the Northwest Territories and are thoroughly established throughout North America. Like the gray squirrel in Great Britain, they scarcely expanded their distribution in the Western Hemisphere for 67 years, but over the next 24 years they expanded over both North and South America.

Summary

Recognition that competition and predation are important in community ecology has turned the attention of some zoogeographers to the effects of other species on the distributions of many animals. Introduction by humans of many species into areas in which they were not indigenous has had powerful effects on the local faunas, and it is likely that spontaneous changes in distributions in the past have also had important effects. This point of view is one of a dynamic interaction among the various species in an area, in contrast to the previous view that distributions were static and depended on climate and barriers to movement. A formal theory of how these dynamic processes operate on islands has been confirmed in several ways.

References and suggested further reading

Elton, C. 1958. *The Ecology of Invasions by Animals and Plants*. New York, Wiley.

Hairston, N. G., Sr. 1987. *Community Ecology and Salamander Guilds*. Cambridge University Press.

Kenward, R. E., & J. L. Holm. 1993. On the replacement of the red squirrel in Britain: A phytotoxic explanation. *Proceedings of the Royal Society. Series B. Biological Sciences* 251:187–94.

Lloyd, H. G. 1962. The distribution of squirrels in England and Wales, 1959. *Journal of Animal Ecology* 31:157–65.

MacArthur, R. H., & E. O. Wilson. 1963. An equilibrium theory of insular zoogeography. *Evolution* 17:373–87.

Middleton, A. D. 1932. The grey squirrel (*Sciurus carolinensis*) in the British Isles, 1930–1932. *Journal of Animal Ecology* 1:166–7.

12

Principles of global distribution

The explanations of vertebrate distributions described in Chapters 9 and 11 are all piecemeal, restricted to specific examples or to a small fraction of the earth's surface. Those explanations, however, are sufficient to explain the distributions of more than 93% of the nonmarine families of vertebrates. The explainable distributions are of those families confined to one continent or to two or three adjacent continents, connected or separated by shallow water. There are gaps in the distributions of some of these families, but the gaps can be explained by competition or predation in the unoccupied area, or by a change in climate in the area no longer occupied. For example, both of the salaman er genera *Plethodon* and *Aneides* are found in eastern as well as western North America, but they are absent from the Great Plains, which lack their forest habitat. The glacial periods of the Pleistocene were wetter than is the current interglacial, and their distributions probably were continuous during those periods.

The remaining general problem in biogeography is that of explaining disjunct distributions of groups that include related species that are found in widely separated areas. Some examples are the tapirs in the Oriental and Neotropical realms, the pipid frogs in Africa and South America, and the paddlefishes in the eastern United States and China.

There are two general theories that attempt to explain the overall distributions of families of animals and plants, and there is a modification for one of them. These two theories are antithetical, and it is necessary here to explain them and to try to understand how they can be so different. The difference hinges on the importance of continental drift, which was a theory held in low regard by most biologists until the reality of plate tectonics was demonstrated in the 1960s. As one theory assumes that the continents were in approximately their current positions before the distributions were determined, and the other assumes that the continents drifted apart after the families had become recognizably distinct from one another, timing is the essential factor. To return to the pipid frogs, if the Family Pipidae became distinct from other frog families after the formation of the South Atlantic Ocean, any theory must require their presence at some previous time in North America, as the ability of any amphibians to tolerate salt water is so limited as to be negligible. If, on the other hand, the family became distinct before South America and Africa drifted apart, and their total distribution at that time included parts of that combined continent, their current distribution needs no further explanation.

The first theory depends more strongly on the dispersal of animals after the families became distinct from one another, but neither theory can afford to ignore dispersal, because it is most unlikely that the original distinct population covered an area large enough to be included in two or more breakaway continents. Thus, in discussing the two theories, it is important, first, to know the relative dispersal abilities of the different groups and, second, to have an estimate of the amount of time that was available to each of them for dispersal.

It is obvious that flying animals have a great advantage, and the example of the cattle egret crossing the barrier of the South Atlantic Ocean has been described. There are other general traits of some groups that help them. Endotherms have an advantage because they can overcome a greater variety of climatic challenges than can ectotherms. At the other extreme, Amphibia are at a special disadvantage, because nearly all of them have delicate skins used in respiration that cannot be allowed to dry out, and because almost none of them can tolerate exposure to seawater.

A different aspect of dispersal is the question of how much time the different groups have had to disperse from their area of origin. Given enough time, almost any group could, in principle, become distributed over a great area, if not worldwide. The question is how much time each group has had. In discussing this problem, we must decide in advance the size of the group under consideration. Obviously, if our taxonomic arrangements mean anything biologically, the genera within a family of vertebrates have, on the average, been in existence for a shorter time than have the families within the order. Thus the genera have had less time to disperse than have the families to which they belong. In what follows, the comparisons to be made are among families, as is usually the case in biogeographic discussions. It must be remembered that, like other taxonomic categories above the level of species, families are delimited in less than uniform ways among the different classes of vertebrates. They do, however, appear to be more uniform than some other categories, and we must accept some arbitrary limits in order to make any comparisons.

We then must make use of such information as we have about

Table 12.1. *Worldwide distributions (numbers and percentages) of nonmarine families of tetrapod vertebrates*

No. of continents reached	Amphibia		Reptilia: Testudinata		Reptilia: Squamata		Mammalia: Nonbats		Mammalia: Bats		Aves	
	No.	%	No.	%	No.	%	No.	%	No.	%	No.	%
1	25	67	1	12.5	6	23	41	46	3	21	47	30
2	6	16	3	37.5	7	27	34	39	5	36	35	22
3	3	8	1	12.5	6	23	5	6	3	21	15	9
4	1	3	3	37.5	2	8	8	9	0	0	20	13
5	2	5	0	0	5	19	0	0	3	21	40	25
Totals	37		8		26		93		14		157	
Probable age of families ($\times 10^6$ yr)	180		200		65–100		60–75		60		60	

Note: Eurasia is counted as one continent, and Antarctica is omitted.

the time of origin of the families in the different classes of vertebrates. The nearest that we can come is to say that the families of birds and most placental mammals can be no older than the Tertiary period, and that many of them undoubtedly are young. The families of reptiles must vary greatly in age, because the orders arose at such different times: no later than Triassic for turtles (200 million years ago), but probably Cretaceous or Tertiary (no more than 65 million years ago) for snakes. The families of Amphibia probably are older than most of the reptile families, but the fossil record is too poor for a definitive decision.

We must finally adjust the information for the different numbers of families in the different classes. All figures, therefore, are also given as percentages of the total number of nonmarine families in the class (Table 12.1). In studying this table, we can see the importance of both the ability to disperse and the time available to disperse. The flying vertebrates, birds and bats, have distributed their families more widely than other groups, even though they had a shorter time available. On the other hand, the rest of the mammals, despite their endothermy, still have more of their families confined to one continent than do the reptiles, which have had much longer to disperse. The Amphibia, as anticipated, have dispersed little as a class; two-thirds of their families are confined to single continents.

Another piece of evidence regarding the ability to disperse comes from island faunas. Islands are informative because of the difficulty most terrestrial and freshwater animals have in crossing even a short stretch of salt water. There are basically three different kinds of islands, in addition to the differences in island sizes already discussed. There are the islands of the continental shelf – those that have had land connections to a continent at times when sea levels have been low, as during glacial maxima. Britain, Newfoundland, Ceylon, and the East Indies are examples. These islands have had some reductions in their total numbers of species relative to the adjacent continents, but still have populations of all kinds of vertebrates, regardless of their difficulty in crossing seawater barriers. Those of the second type are the true oceanic islands that have never been connected to any continent. All of them are volcanic in origin, and they are separated from continents by deep oceans. The Galapagos Islands off Ecuador, the Hawaiian Islands, and all of the islands of Micronesia and Polynesia are examples. Finally, there are the islands that are now separated from continents by deep water, but which at one time were parts of continents, from which they became separated by the processes of plate tectonics. Examples are Madagascar, New Zealand, and the West Indies. The differences in the faunas of these three types of islands reflect their locations and origins. No true oceanic island more than 1,000 km from a continent has any nonflying vertebrate that has not been introduced by humans. Hawaii is the largest such island, and it is 3,300 km from North America. The Galapagos, 965 km from South America, have small rodents, tortoises, snakes, and lizards. Bermuda, much smaller, is the same distance from North America, and has only a species of lizard. Lord Howe Island, 500 km from Australia, has a tortoise and lizards. Reptiles have thus been most successful in crossing oceanic barriers, which, in conjunction with their greater age, accounts for their wider distribution than mammals.

The large islands that were once parts of continents require individual consideration. New Zealand is easiest to describe. It became separated from Australia in the middle to late Cretaceous, after marsupials and even some primitive placental mammals were widespread, although neither has any representative there. The native vertebrates are a frog whose only possible relative is in western North America, the unique lizardlike reptile *Sphenodon*, endemic lizards, and a number of kinds of flightless birds. The ancestors of the frog, like those of *Sphenodon*, must have been present before the separation from Australia. Tailed frogs are unknown as fossils, but the ancestral forms must have been present in Australia and Asia at some time. Rhynchocephalia, the group of reptiles to which *Sphenodon* belongs, were widely distributed in the Triassic and Jurassic. The lizards possibly could have been on New Zealand since before the separation from Australia, but lizards have reached a number of oceanic islands never connected with continents, and they could have colonized the islands while they were closer to Australia than at present. Flightless birds are part of a separate problem in zoogeography, and they are discussed later in this chapter.

Madagascar became separated from Africa during or shortly after mid-Cretaceous. It may seem surprising that it has a vertebrate fauna that is so much more extensive than that of New Zealand. It is best known for its lemurs (primitive Primates), but there is a large and varied group of endemic Insectivora, plus small carnivores related to an African group, as well as rodents. Most surprising is the presence in the Pleistocene of a pygmy hippopotamus, different from the one now in Africa. Madagascar also has turtles, snakes, lizards, and frogs. An interesting peculiarity, to be considered later, is that one of its lizard families is South American, but not African, in affinity. Until 2,000 years ago it had large flightless birds, whose consideration is also deferred until later in the chapter. Its deep-water separation from Africa is about 100 km, a distance that one might expect to deter the hippo and most of the small mammals. The reptiles, frogs, and insectivores could have been present before separation from Africa, but the more recently evolved mammals must have reached Madagascar on their own. North and northwest of Madagascar there is a group of small islands, the Comoros, that could have acted as steppingstones for colonization.

The Greater Antilles, including Cuba, Hispaniola, Jamaica, and Puerto Rico, belong in a somewhat different category from New Zealand and Madagascar. There is no evidence that they were ever attached to a continent, but they are not on a continental shelf, being surrounded by deep water, and they are not completely volcanic in origin. Cuba is ten times as large as Hawaii, and Hispaniola is eight times as large as that largest of volcanic islands. One hypothesis is that the Greater Antilles were part of a series of islands between South America and what is now Mexico until near the end of the Cretaceous, when the westward movement of both North America and South America left the intervening part of the Pacific Tectonic Plate with its Caribbean Arc behind, in effect moving the Antilles eastward to their current locations. Compared with those of mainland tropical areas, Antilles Amphibia are represented by few frogs and no Gymnophonia or salamanders. The reptile fauna is extensive, and as might be expected on the basis of other islands already described, lizards are more diverse than snakes, with nonmarine turtles limited to two genera, one of them extinct. As the Greater Antilles have never been far from continents, birds and bats are numerous and require no comment. The mammals, although many species are extinct, are interesting in their relationships to North and South America. There are no marsupials, living or fossil, although they must have been present at some time, because they have been on both continents since the Cretaceous. As insectivores have reached South America only very recently, and Cuba has an endemic genus, *Solenodon,* it and a fossil form must have arrived from the north, but as the insectivores arose in the Cretaceous, they may have been on the Antilles for a long time. There are two extinct small ground sloths, belonging to the Order Xenarthra, of South American origin. There are four families of rodents related to South American families, as well as the rice rat *Orizomys*, a form of northern origin. Like Madagascar, but unlike New Zealand, the Greater Antilles have many vertebrates that appear to have arrived by dispersal over at least narrow water barriers.

The theories

As stated at the beginning of this chapter, theories of biogeography attempt to explain disjunct distributions of related species of organisms – vertebrates in this case. The first question that arises is whether or not the species are descended from a common ancestor not so remote as to be meaningless. Thus, it is generally agreed that all amniotes are descended from Paleozoic amphibians, but that is irrelevant to the question of whether the large flightless birds had a common ancestral group that was present over Gondwanaland before that continent drifted into its current pieces or whether the several species are from different avian stocks and became flightless independently, subsequent to the drifting. That process is known as convergence, and it can be almost impossible to resolve the question whether two forms are convergent or closely related.

The older theory, as far as biogeography is concerned, was proposed by the paleontologist W. D. Matthew in 1915. He was an expert on fossil mammals, and he assumed, as was almost universally accepted at that time, that the continents, though varying in how high above the oceans they rose, had always been in their current positions. He tried to account for the observation that many disjunct distributions involved the presence of primitive forms on two or three of the southern continents. "Primitive" is the accepted term for organisms that retain characters in much the state possessed by ancestral forms. Thus, the large flightless birds, which were thought to be primitive, are found in Africa (ostriches), Australia and New Guinea (emus and cassowaries), and South America (rheas), as well as the kiwis and the recently extinct moas of New Zealand and the recently extinct elephant birds of Madagascar. Other examples are the lungfishes in Australia, Africa, and South America and the marsupials in Australia and South America (with a single species, the opossum, in North America). The disjunct groups were not necessarily primitive, such as the tapirs mentioned earlier. Until Matthew's theory, there had been much speculation about these distributions, with many authors theorizing that there had been "land bridges" across oceans at some time in the past and that the bridges had subsequently dropped to the level of the ocean bottom. Matthew realized that such geologic phenomena were so unlikely as to be virtually impossible, and he provided an alternative theory that made such phenomena unnecessary. He proposed that most evolutionary advances had taken place in the largest landmass, namely, the Holarctic Realm, which has been a single supercontinent for

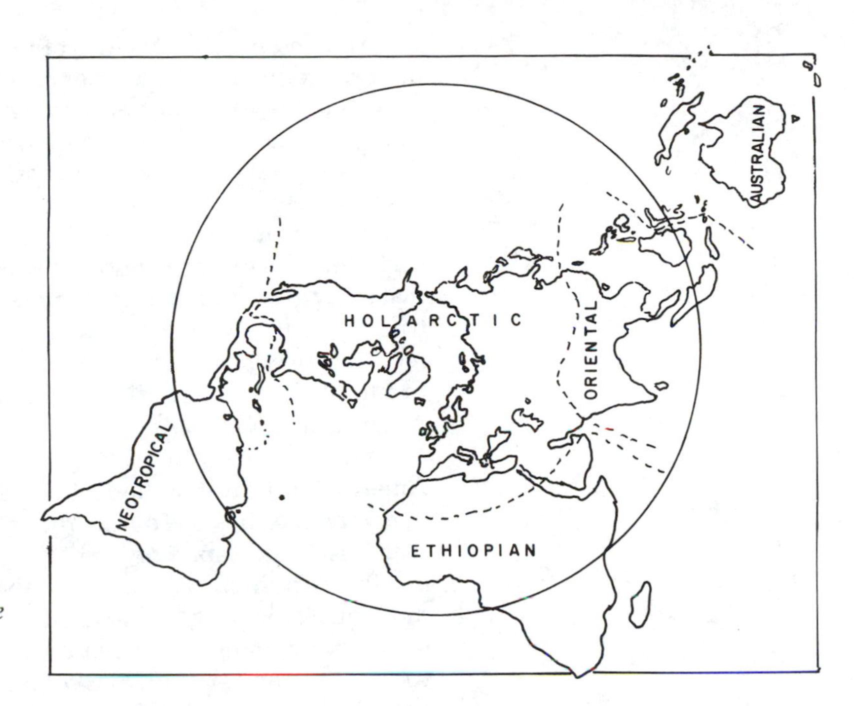

Figure 12.1. *The faunal realms on a North Polar projection. Compare with Figure 10.1. (Adapted from Matthew 1939.)*

much of its geologic history. Viewing the earth from above the North Pole, Holarctica appears to be at the center of the world, with the southern continents radiating out away from there (Figure 12.1). He argued that populations would have been larger on large land areas, and that the chances of a favorable mutation would have been greatest in such large populations. Moreover, the climate had changed much more in Holarctica than in the tropics, and that should have increased natural selection in favor of these new mutations. Thus, one should expect that better-adapted forms would have arisen in Holarctica, and their populations would have spread outward from there, outcompeting and replacing their existing, more primitive relatives as they enlarged their distributions. The evidence presented was partly to show that the current distributions of more advanced types are Holarctic and those of more primitive types are in the southern continents, as is true for the Family Canidae (dogs). The true wolves and foxes, regarded by Matthew as advanced, are in North America, Asia, and Europe, whereas the primitive African hunting dogs and the genus *Cyon* (India and the East Indies) are peripheral in distribution. Similarly, fossil forms ancestral to the primitive forms are known from the Miocene of North America. More impressive are the distributions and fossil records of the tapirs and of the Camelidae. As mentioned earlier, the tapirs, large, semiaquatic hoofed mammals with elongated snouts, are present in the Neotropical and Oriental realms, and are absent from the rest of the world. Fossil ancestors of the tapirs are known from both North

America and Europe. A series of fossil camels and camel relatives has been found in North America in every epoch from the Eocene to the Pleistocene. None are there now, but they are living in Africa and Asia, and the llamas (vicuñas and huanaco, close relatives) are present in the Andes of South America. There are no fossils elsewhere until the Pliocene. The remaining orders of mammals were also reviewed, and most of them either confirm Matthew's theory or cannot be used to refute it. He recognized that a few groups of mammals gave him trouble. That is especially true of the hystricomorph rodents, which include the porcupines, members of which are present in profusion in South America, and to a limited extent in Africa and Eurasia. The porcupine of North America has been present only since the late Pliocene, when it arrived from the south, and there are no fossils known from North America for the rest of the Tertiary, despite an excellent record for many rodents. It is one of the frustrating features of biogeography that fossils and especially geologic strata of the appropriate age are absent from many regions of the earth where they would be most interesting. The presence of fossils provides excellent evidence, but nothing can be made of their absence, unless there are rocks of the appropriate age containing many fossils, rocks that should contain the relevant ones if the animals had been present when the strata were formed. Until the demonstration of plate tectonics, the only proposal for change in Matthew's theory was the suggestion that the location of the major advances in adaptive types had been in the Old World tropics, adjacent to the Holarctic.

Matthew's theory brought rationality to a chaotic field and put an end to fanciful theories about land bridges that rose and disappeared at convenient times. The theory that most groups arose in Eurasia or North America was very influential and was responsible for a number of expeditions to central Asia to look for human ancestors ("missing links"). It was also responsible for the long-standing rejection of Raymond Dart's discovery of *Australopithecus* as being important in human ancestry, because it had been found in southern Africa. Matthew knew about the theory of continental drift, but he argued against it on the grounds that no mechanism was known that could drive continents through the solid crust of the earth. That such mechanisms exist is now known, and they have been used to document the alternative theory for the disjunct distributions that stimulated the original theory.

The discovery of the nature of the Mid-Atlantic Ridge in the 1960s brought about a revolution in geology. The ridge is a mountain range rising from the ocean floor in the middle of the Atlantic, and extending from Iceland through the entire length of the ocean. In only a few places does the range reach the surface of the ocean, notably at Iceland and Ascension Island. It is active volcanically, and magma rises at the center, mostly as lava. The process apparently is due to drag by the nonsolid mantle, pulling the crust

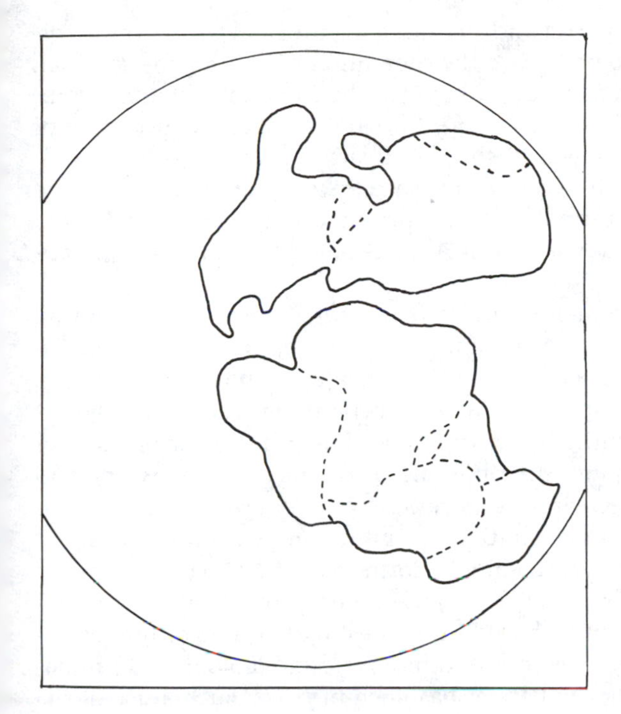

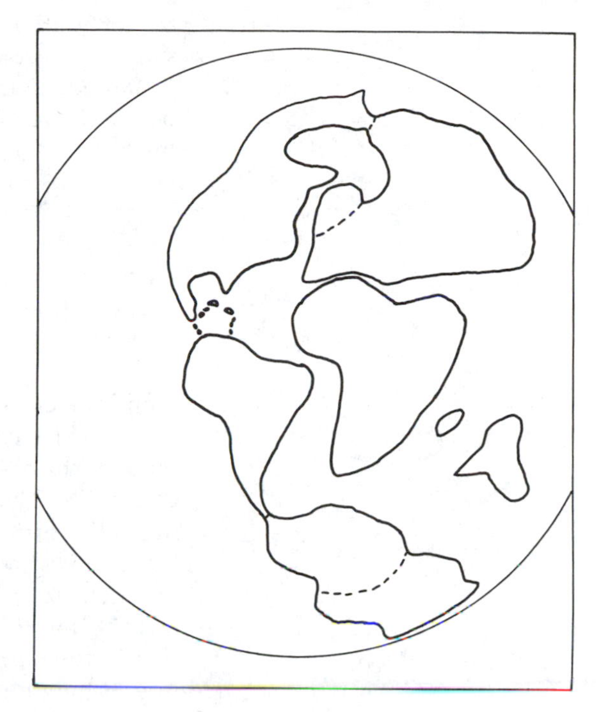

Figure 12.2 (above left). *Outlines of the continents in the late Jurassic.*

Figure 12.3 (above right). *Outlines of the continents in the late Cretaceous.*

apart. It was soon discovered that similar ridges are present under all oceans and that the same kind of process is under way on the continent of Africa, causing the great rift system where lakes Malawi, Tanganyika, and Turkana are located. The continents are surface features of great crustal plates that are moving in known directions at rates that can be estimated. Space does not permit a description of the details here, but it has been possible to reconstruct the major features of the earth's geography at many periods in the geologic past. Some of those reconstructions were described in the introduction to Part I of this book. Those were reconstructions of a past too distant to be relevant to current biogeography. The later periods are of interest here.

About 200 million years ago, early in the Mesozoic, all of the continents were part of a single supercontinent, Pangaea (see Figure I.4). During the early Jurassic period, rifts appeared that began to separate a northern group, called Laurasia, from a southern group, Gondwanaland (see Figure I.5). There may have been some gaps within Laurasia and Gondwanaland, but that does not imply true separation except between the two, which was complete by the end of the Jurassic, 130–140 million years ago (Figure 12.2). During the Cretaceous, first the South Atlantic and then the North Atlantic became completely open; Africa became completely separate from the rest of the other elements of the former Gondwanaland. India and Madagascar separated from Africa, Antarctica, and Australia (Figure 12.3). The timing of the separation of South America and Australia from Antarctica is controversial. Some geologists place the separation of South America from Antarctica in the late Cretaceous, at the same time as the formation of the Antilles Arc, and for the same reason: the formation of the Scotia Arc, which pushed eastward between the two continents in the same way. Others place the separation as late as the Eocene, 49 million years ago. The same range of times for the separation of Australia from Antarctica can be found among authorities.

These disagreements make it frustrating for anyone attempting an unbiased analysis of biogeography, because the period involved was precisely the period when most bird and mammal families were arising. The later the separation, the greater the extent to which continental drift can be viewed as the major cause of the current disjunct distributions in the former Gondwanaland continents. There can arise a serious problem in circular reasoning, if the current distribution of a group on two sides of a barrier is used in estimating the time of formation of that barrier, because an interpretation regarding any other group with a similar distribution might be based on that estimated time of the formation of the barrier. Without an independent geologic determination of how long the barrier has been in place, such a use of the origin of the barrier in explaining a disjunct distribution is not warranted.

Everyone who discusses biogeography laments the imperfection of the fossil record. As already noted, the absence of a fossil record for a family in critical places and for critical times cannot be used as evidence, unless there is a good record of other forms in general at that place and time. Thus, the discovery of an important fossil can reverse conclusions that had been made on the tacit assumption that such a fossil did not exist. The recent discovery of a fossil marsupial of Eocene age in Antarctica demonstrates the possibility that marsupials spread from South America to Australia by that route, a route that previously had been questionable because of uncertainty about the timing of the separation of the continents. Later, it will be claimed that all of the mammal families, except for the monotremes and marsupials, have distributions that can be better explained by Matthew's theory than by continental drift. It must be admitted, however, that for the entire Cretaceous, the only fossil record of living families of mammals in any Gondwanaland continent is a jaw and a few teeth of late Cretaceous age in Peru. The best-preserved fossils are marsupials, but some may be from an extinct order of placentals. There are no fossil mammals in the Cretaceous of India, Madagascar, New Zealand, Australia, or Antarctica. As there are few fossil vertebrates of Cretaceous age from those continents, at least those that might be useful in judging the importance of the absence of mammals, that absence is weakened as evidence. The abundance of fossil

Table 12.2. *Earliest fossil occurrences of the nonmarine orders (and families) of vertebrates*

	Vertebrate class				
Period and epoch	Osteichthyes	Amphibia	Reptilia	Aves	Mammalia
Quaternary					
Pleistocene					(5)
Tertiary					
Pliocene	(1)	(1)		(4)	(5)
Miocene	(1)	(1)	(3)	4 (28)	1 (27)
Oligocene	(1)	(3)		5 (13)	1 (28)
Eocene	6 (7)	(1)	(7)	14 (26)	5 (27)
Paleocene		(3)	(3)	2	8 (5)
Cretaceous					
Late	2 (2)	(3)	(11)	2 (1)	1
Middle	1 (1)	(3)	(4)		1 (1)
Early		(1)	(2)		
Jurassic					
Late		2 (3)			
Middle	1 (1)		1 (1)		
Early		(1)			
Triassic					
Late			1		
Middle	1 (1)		1		
Early		1 (1)			

Source: Data summarized from Darlington (1957) (Osteichthyes), Goin, Goin, & Zug (1978) (Amphibia and Reptilia), Olson (1985) (Aves), and Nowak (1991) (Mammalia).

vertebrates throughout the Tertiary, especially mammals, does suggest that the apparent absence of most families from the Gondwanaland continents in the Cretaceous is not entirely irrelevant.

With those caveats in mind, we can begin to estimate the extent to which continental drift can be used to explain vertebrate distributions and the extent to which an explanation must be based on the current distribution of the continents. The age of a group of related species can be estimated only from the fossil record, and that can give only the minimum age, especially for families and lower categories. A new discovery of a fossil can always lengthen the estimated age of the group, but it cannot shorten that age without showing that the classification of the group had been incorrect, an unlikely possibility. The ages of the oldest fossils that can be assigned to living orders and families of nonmarine vertebrates are shown in Table 12.2. Comparison of this table with Figures I.4, I.5, 12.2, and 12.3 will give an idea of the possibilities. Bird and mammal orders and families are generally much younger than those of the other classes, and, as expected, families tend to be younger than orders in the same class. The immediate significance of the table for biogeography is the likelihood that most of the families of birds and mammals are too young to have had their distributions affected by continental drift. That conclusion is much stronger for placental mammals than for birds, because their skeletons, and especially their teeth, are more easily fossilized than are the light and toothless skeletons of modern birds. The only unequivocal example of a Gondwanaland distribution of a bird family is the penguins (Spheniscidae), for which there is a fossil record, none of it outside of the southern landmasses. Unfortunately, the fossils are no older than the Miocene. Thus, it is still possible that a new discovery could upset the conclusion that continental drift accounts for the present distribution of the family. An example of just such an upset is provided by the frog family Pipidae. They are now found only in Africa and South America, and the origin of frog families throughout the Cretaceous, including the Pipidae, would appear to solidify the conclusion that they are a Gondwanaland group, separated now by continental drift. Unfortunately, Cretaceous and Paleocene fossils have been found in North America and the Middle East, as well as in Africa and Argentina. As these frogs are rated primitive by herpetologists, Matthew's theory seems to fit them very well.

Table 12.3 lists the families whose present distributions in the southern continents appear to fit the theory that continental drift is responsible. Two groups are discussed in detail, because so many texts and papers have been written about them. They are the flightless Apteriges, or ratite birds, and the marsupials.

The ratites, so named from the shape of the sternum (breast bone) which lacks the keel that serves as the place for attachment of the flight muscles of flying birds, are nearly all very large, and except for the former presence of the ostrich in Asia, their present and recent distributions are in Africa, South America, Australia–New Guinea, Madagascar, and New Zealand. They are extinct on the last two islands, but only recently so (the kiwi of New Zealand is not so large and may have had an independent origin). Besides their size and the absence or near-absence of wings, they share a peculiar palate that is different from those of all but one family (Tinamidae) of flying birds. This palate, named paleognathous ("old jaw"), is claimed to be evidence that all of these birds are closely related and are descended from an ancestor with a general Gondwanaland distribution. They are not, however, the only giant flightless birds that ever lived. Such forms are known from the Tertiary of most continents, and the question arises whether or not the ratites are similar simply because they are flightless, and represent independent origins from different stocks. Many birds on oceanic islands that originally had no mammalian predators have become flightless, including especially rails and geese. The best evidence favoring the hypothesis that the ratites are des-

Table 12.3. *Families whose distributions support the theory of continental drift*

Osteichthyes	Reptilia
Lepidosirenidae	Pelomedusidae
Characidae	Chelidae
Nandidae	Iguanidae
Amphibia	Amphisbaenidae
Caeciliidae	Aves
Leiopelmidae (? + Acaphidae)	Apteriges?
Leptodactylidae	Spheniscidae
	Captionidae
	Mammalia
	Marsupialia

cended from a common ancestor living prior to the breakup of Gondwanaland is the common possession of the paleognathous palate. Recently, fossils clearly related to the tinamous (Tinamidae), and having both a paleognathous palate and the ability to fly, have been found from the Early Tertiary, thus providing a possible common ancestor.

The difficulty with that possibility is that the late A. C. Wilson of the University of California extracted DNA from mummified remains of moas (New Zealand ratites) and found it to be too different from that of living ratites for the necessary close relationship. That finding demonstrates that the crucial paleognathous palate must have been evolved independently in moas, and thus potentially in the other forms as well. That some or all of the ostrich, rheas, emus, and cassowaries could easily have evolved independently is shown by the many features they share with some newly hatched shorebirds. The retention of juvenile traits in adult animals is called neoteny and is known in many examples, including *Homo sapiens*, which retains the rounded head and hairlessness of many baby mammals. Among different birds, flightlessness has evolved many times, both on oceanic islands where there have been no ground-living predators and on continents, where the advantage of large size in escaping or defying predators has meant a size too large for flight. These have all lost the keel on the sternum, and that feature cannot be used as evidence for common ancestry. Tempting though it is to speculate about the current distributions of these very similar birds, final solution of the puzzle of those distributions awaits the discovery of some crucial fossils.

The other prime example of vertebrates postulated to have had their distributions determined by continental drift concerns the marsupials. Australia is, of course, the location of the greatest abundance and diversity of these primitive mammals, but there are also many of them, belonging to two families, in South America. The recently discovered fossil marsupial of Eocene age in Antarctica provides support for the continental drift theory. There are none in Africa, Madagascar, or India, all of which became separated from Australia and Antarctica before the critical period in the middle Cretaceous. The time of separation of Africa from South America may be a problem, but without the necessary fossils from South America dating to earlier than the latest Cretaceous, that question cannot be resolved. The principal evidence against the theory lies in the existence of many marsupial fossils of middle and late Cretaceous age in North America, and a few of Tertiary age in Europe. In the absence of comparably old fossils from South America, these have been used to argue for a North American origin for the order, but as has been pointed out, dependence on the absence of fossils is risky. By the Paleocene, South America had a varied marsupial fauna, consisting of three families, in contrast to Europe and North America, where all marsupial fossils of whatever age belong to a single family. That indirect evidence has been used to argue for a South American origin of the order. Incidentally, if that is the case, the marsupials may have evolved from reptiles independently of the placental mammals. An abundance of South American fossils of Jurassic and Cretaceous ages would be necessary to establish such a contention.

Be that as it may, there remains the strong probability that Australia was colonized by marsupials from South America via Antarctica. The subsequent rifting apart of those continents would account for the present distributions and would constitute a triumph for *vicariance biogeography*, as the theory based on continental drift has become known.

It appears that both of the two global theories of biogeography are needed. Matthew's theory is needed for placental mammals and nearly all birds, and vicariance is needed for marsupials and at least some of the older classes, but neither is needed for most vertebrates.

References and suggested further reading

Condie, K. C. 1989. *Plate Tectonics and Crustal Evolution* (3rd ed.). Elmsford, N.Y., Pergamon Press.

Darlington, P. J., Jr. 1957. *Zoogeography: The Geographical Distribution of Animals*. New York, Wiley.

Feduccia, J. A. 1980. *The Age of Birds*. Cambridge, Mass., Harvard University Press.

Goin, C. J., O. B. Goin, & G. R. Zug. 1978. *Introduction to Herpetology* (3rd ed.). San Francisco, Freeman.

Lillegraven, J. A., Z. Kielan-Jaworowska, & W. A. Clemens (eds.). 1979. *Mesozoic Mammals. The First Two-Thirds of Mammalian History*. Berkeley, University of California Press.

Matthew, W. D. 1939. *Climate and Evolution*. Special Publications of the New York Academy of Sciences.

Nowak, R. M. 1991. *Walker's Mammals of the World.* (5th ed.). Baltimore, Johns Hopkins University Press.
Olson, S. L. 1985. The fossil record of birds. Pp. 79– 252 in *Avian Biology,* Vol. 8, D. S. Farner, J. R. King, & K. C. Parkes (eds.). New York, Academic Press.
Pielou, E. C. 1979. *Biogeography.* New York, Wiley. Veevers, J. J. (ed.). 1984. *Phanerozoic Earth History of Australia.* Oxford University Press.
Veevers, J. J. (ed). 1984. *Phanerozoic Earth History of Australia.* Oxford University Press.

Veevers gives the separation of Australia– Antarctica–New Zealand as 95 million years ago; Condie gives the New Zealand separation as 80 million years ago and the Australian–Antarctica separation as mid-Tertiary (about 30 million years ago).

PART

V

Migration

Migration is the most conspicuous activity engaged in by vertebrates, and it is fascinating to us because the many different capabilities demonstrated by representatives of the vertebrate classes are so far outside our experience.

The ability of small birds to migrate was so unbelievable that up through the eighteenth century most people were convinced that when swallows disappeared in the autumn and reappeared in the spring, they had spent the winter like frogs and turtles, buried in the mud at the bottoms of ponds. Indeed, so firm was that conviction that supposed eyewitness sightings were widely circulated, as shown by the following account:

In 1878, Dr. Elliott Coues, one of America's most distinguished ornithologists, listed the titles of no fewer than 182 papers dealing with the hibernations of swallows. The following quotation from his discussion shows how strong some of the supporting evidence appeared to be: "Dr. Wallerius, the distinguished Swedish chemist, wrote in 1748 that he had more than once seen swallows assembling on a reed, till they were all immersed; and that he had likewise seen a swallow caught in a net under water, revived in a warm room, where it fluttered about and then died. Mr Klein, secretary of Danzick (Danzig), procured many sworn affidavits of such occurrences. The mother of the Countess Lehndorf said she had seen a bundle of swallows brought from Lake Frith-Hoff, which were revived in a room and flew about. Williams writing of the swallows of Vermont in 1794 says that at Danby in that state . . . a man digging up roots of the pond lily found several swallows enclosed in the mud; alive but in a torpid state. About the year 1760, two men digging in the salt marsh, in Cambridge, in Massachusetts, on the bank of the Charles River, about two feet below the surface of the ground, they dug up a swallow, wholly surrounded and covered with mud. The swallow was in a torpid state, but being held in their hands, it revived in about half an hour. The place where this swallow was dug up was every day covered with salt water, which at every high tide, was four or five feet deep. The time when this swallow was found was the latter part of the month of February." (Lincoln, 1939, p. 7)

Aristotle recorded that cranes migrated from the plains of Scythia (southern Russia) to the marshes in the upper Nile, but he also believed that not only swallows but also storks, doves, and kites hibernated.

Not all movements of populations are migrations in the sense intended here, although past movements of human populations have been called migrations. As used here, a migration involves the return of a population to the same area (large or small) from which it departed, after an absence of at least a season. The definition does not cover such animals as the crossbill, a bird that is specialized for opening the cones of conifers, especially those of spruce and white pine. These trees, like oaks, do not set seeds in a regular manner, year after year, but in different years in an irregular manner. This is thought to be an effective means of avoiding a buildup of the populations of seed-eating animals, which might otherwise become so numerous as to consume virtually all of the seeds that the trees could produce. By producing seeds at unpredictable intervals, the trees allow the populations of seed eaters to decline between crops. The crossbills wander between nesting seasons, searching for areas where the conifers are setting seeds, and settle there to nest. They do not return to the same area again unless seeds are being set, and that is ordinarily several years later, or they may not find it again. These wanderings are not migrations.

All vertebrate classes have migratory species. Among the Agnatha, the sea lamprey mates and lays its eggs in freshwater streams. After a period of development and growth of 3–7 years in the silt and sandbars, the lampreys follow the streams to the sea (now also to the Great Lakes), where they feed parasitically on larger fish. After 1 year or at most 2 years, they return to the streams to breed.

Many Osteichthyes follow a fairly similar life history, with a much shorter life in fresh water. The most famous are the various species of salmon, which differ in the duration of the marine stage. Most migrate to the sea after a short season of development in the streams where they hatched; others may remain in fresh water for 1 or 2 years. These species are anadromous. The Atlantic eel is catadromous. Its breeding takes place in the Sargasso Sea, in an area that overlaps with the breeding of the European eel. On hatching, the larval eels (elvers) begin a migration to the Gulf Stream, which they follow along the North American coast, some leaving the group for the mouth of each river as they pass. The individuals of European origin stay in the Gulf Stream for another year. The young eels move up the streams, where they grow for several years. They then return to the Sargasso Sea, where they breed and die.

Why should some fish be anadromous and others catadromous? A recent study pointed out that there is a correlation between the number of species following each kind of life history and the relative productivity of oceanic and fresh waters at different latitudes. Production in the ocean increases with latitude, and there are many more anadromous species than catadromous ones in these cooler waters. In tropical oceans, production is less than in streams, and there the proportion of catadromous species is greater.

Among the Chondrichthyes, a number of species have regular migrations. The common dogfish, *Squalus acanthias*, a small shark, migrates from its winter habitat off the coast of the southeastern United States to give birth to its young off the coast of New England and eastern Canada.

Amphibia have many species that migrate. Frogs and many salamanders migrate to bodies of water to mate and lay eggs. These migrations may involve short distances of 20–50 m, but they are regular, and for most of the species the same body of water is visited by the same individuals.

Migration is less common among the reptiles than among the other classes of vertebrates, but sea turtles undertake spectacular migrations over great distances, returning to the same beaches for egg laying after long intervals. One location of such beaches is on Ascension Island (12 km × 9.5 km) in the middle of the South Atlantic. The turtles that nest there spend the remainder of their time off the coast of South America, at least 2,100 km away. To find Ascension Island they must be accurate to less than 10 minutes of arc.

A few species of large mammals migrate considerable distances. The caribou spend the winter inside the edge of the northern coniferous forest (taiga) and migrate to summer calving grounds on the northern tundra. The wildebeest of the Serengeti migrate over a regular course, following the rains and the best grazing. Observing the migration of the gray whales up and down the California coast has become a popular pastime. They mate and give birth in warm lagoons on the coast of Baja California and return to waters off Alaska to feed. The ability of fur seals to locate the Pribilof Islands in the Bering Sea after a prolonged absence of several years, and no previous experience, is an unsolved mystery.

It is the birds that really excite our imagination. Some of their migrations are quite short, as is the case of the mountain-nesting Carolina junco, which moves down to the Piedmont for the winter, a distance of no more than a few kilometers. Many sandpipers and some plovers nest on the tundra near the Arctic Ocean and spend the winter in South America. The species that migrates the farthest is the Arctic Tern, which nests in northern North America, and then crosses the Atlantic to Europe, follows the coast to Africa, and crosses back to southern South America, to spend the northern winter in the southern ocean.

How the various species of vertebrates achieve these feats of navigation and endurance is the subject of later chapters.

References and suggested further reading

Carr, A. 1986. *The Sea Turtle. So Excellent a Fishe* (2nd ed.). Austin, University of Texas Press.

Lincoln, F. C. 1939. *The Migration of American Birds*. New York, Doubleday, Doran.

Schmidt-Koenig, K., & W. T. Keeton (eds.). 1978. *Animal Migration, Navigation, and Homing*. Berlin, Springer-Verlag.

Storm, R. M. (ed.). 1967. *Animal Orientation and Navigation*. Corvallis, Oregon State University Press.

Twitty, V. C. 1967. *Of Scientists and Salamanders*. San Francisco, Freeman.

13

Migration of vertebrates other than birds

The accuracy of both the ability to return to the same place and the timing of migration is so remarkable that much effort has been spent in discovering the behavioral and physiological bases of such abilities. In this chapter we describe some experiments that have demonstrated the sensory mechanisms that selected species of vertebrates use in locating their goals. Most of such information is known about birds, which are deferred to the next chapter.

The best work on fishes has been carried out with Pacific salmon (*Oncorhynchus* species). The eggs of these species are laid in gravel banks of swift streams. After hatching, the young salmon remain in the stream for a period of a few weeks or more, some up to 2 years, depending on the species. They then descend the stream to the ocean, where they remain for several years, the time again depending on the species. In the ocean, they feed and grow to maturity. When their period of growth is complete, they are able to return to the river from which they came, even though they may have been thousands of kilometers away, and they find the exact forks of the smaller streams to return to the same part of the stream in which they hatched. The first test of exact homing was carried out in British Columbia. It consisted in marking 469,326 young sockeye salmon (*Oncorhynchus nerka*) in a tributary of the Fraser River and releasing them. After they had migrated to the ocean and back, 11,000 were recovered from the same stream, and not a single marked fish was recovered from any other stream. In tests that involved removing young salmon at various times after hatching, and transplanting them to other streams after marking them, it was shown that the ability to find the home stream develops within a very short time, no more than 1 week in some species.

The fishes have chemoreceptors for both taste and volatile substances (Chapters 1 and 2). During the period when they are still in the stream where they hatched, the chemical composition of the water becomes fixed ("imprinted") in their nervous systems, and that allows them to recognize that particular set of chemicals again, even in the dilution of the river up which they migrate as adults. The experiment demonstrating that it is the chemical sense that is involved was carried out by capturing a number of fishes in each of two forks of a stream, marking them accordingly, and dividing each group into two. One of the two groups from each stream had their nasal capsules blocked with absorbent cotton; the other remained unmodified. All specimens were then released 800 m downstream from the fork. The unchanged fishes selected the same fork from which they had been captured; the fishes with olfactory pits blocked chose the two streams randomly. Clearly, the normal fishes responded to the exact chemical compositions of the respective streams. The imprinting of the chemical composition of their home water was used to stock a large hatchery at the University of Washington. Salmon eggs were removed from a natural stream and allowed to develop, hatch, and grow at the hatchery. These fishes returned to the hatchery, instead of to the stream where they had been laid as eggs.

These and similar experiments explain the return of the adult salmon to the stream of their origin, but their ability to find the correct river from thousands of kilometers away in the ocean cannot be explained by chemical memory. It was suggested that they found the river by random search until the chemical could be identified, but the probability of success would be too low for the observed frequency of return (2.3% for the Fraser River experiment), even discounting the mortality between marking and recapture years later. Thus, an explanation for their homing ability in the open ocean is needed.

For homing under such conditions to succeed, there must be both the ability to maintain a constant direction and the ability to sense errors and to correct them, as is the case in true navigation. In the case of migrating salmon, it is possible that direction finding is sufficient, because that would place them at the coast, and by swimming along it they would come to the estuary containing the chemicals from their home stream. They would need to turn in the correct direction along the coast, but ocean currents are sufficiently consistent that instinct for the correct choice could have evolved from short periods or distances in the past.

Fishes have been shown to have abilities that could allow them to maintain a constant direction. The first is the ability to see the sun on a clear day. Combined with a sense of the time of day, that would provide a means of keeping a constant direction. Related to this sun-compass orientation is the ability to react to polarized light. Light from the sky is polarized because of the effect of the atmosphere on sunlight. Humans cannot perceive the effect, but many animals can, including insects and many vertebrates. As the polarization is related to the position of the sun in the sky, that would provide information about direction, even if the sun were obscured in a partly cloudy sky, or before sunrise or after sunset.

Laboratory experiments on salmon and other species of fishes have demonstrated that they do orient in appropriate directions under open sky, or under a polarizing lens. Salmon migrating in the open ocean have not been tested, but Quinn (1980) conducted experiments on the use of celestial cues by sockeye salmon migrating through two lakes in the Pacific Northwest. As part of the experimental setup, he also tested for a different means of orientation, namely, perception of the magnetic field of the earth, an ability that would permit direction finding day or night, or even under cloudy skies. Both Lake Washington and Chilko Lake are elongate in the north–south direction, and the salmon migrate along the length of the lakes. Those in Lake Washington migrate out of the tributary stream at night, whereas those in Chilko Lake migrate from the outlet stream into the lake during daylight. Thus, they go in opposite directions relative to the flow of the streams. The experiments were carried out in cross-shaped tanks (78 × 78 cm), in each arm of which was a trap. Fishes were introduced from a circular container in the middle of the tank, thirty at a time, after which the container was removed by remote control. The fishes were left under the experimental condition for 45 min, after which the traps were closed, and the numbers in each arm and in the center were tallied. No fish was used more than once. The control experiments were performed with the clear sky visible and with the normal magnetic field present. The sky was blocked out at night with black plastic to test for celestial orientation; during the day, a thick translucent plastic cover allowed the position of the sun to be seen, but interfered with the pattern of polarization. To test for orientation on the magnetic field of the earth, a coil was constructed that shifted the field so that magnetic north was rotated to west. The possible effect of the presence of the coil without the current was tested by removing it and leaving the other conditions unchanged; that had no effect. The various combinations of covers and magnetic rotation showed that in all but two of fifteen tests, the fishes were significantly nonrandom in distribution among the arms of the tank. For the Lake Washington fishes, the activated coil produced a rotation in orientation toward the west, whether the cover was on or not. With the coil off or taken away, the fishes oriented in the direction expected from their normal migratory direction. Thus, for night-migrating fishes, the magnetic field is the factor determining orientation. For the day-migrating Chilko Lake fishes, celestial cues were stronger than magnetic ones, but with the tank covered, the magnetic field became the important cue for orientation. In these experiments, polarization of the light was not allowed to be a factor, but in other experiments, two species of teleosts, the goldfish, *Carassius auratus*, and the halfbeak, *Dermogenys pusilus*, have been shown to orient using only polarized light.

Sharks and rays have also been shown to respond to magnetic fields. As they have the ampullae of Lorenzini, which allow them to respond to weak electric currents (Chapters 1 and 2), the ability is not surprising.

Migrating fishes therefore have at least three different cues by which they can maintain direction. It is important that the salmon experiments demonstrated that both sensory mechanisms are used in orientation, an ability that would be important under conditions where the primary cue was blocked out.

Little is known of the routes taken during open-ocean migrations, but a few observations on European eels, *Anguilla anguilla*, are of interest. Eels were followed by radio tracking from a research vessel at the beginning of their migration to the breeding area in the Sargasso Sea. Those from northern Europe originally headed in a northwesterly direction, indicating that their normal route takes them around Scotland, thus reflecting the geologic recency of the opening of the English Channel. Only eels released beyond the 200-m depth at the edge of the continental shelf moved toward the west or southwest, the correct direction. Those released there in daylight promptly dove to below 400 m; night-released eels remained at shallow depths. Eels have been shown experimentally to be sensitive to the earth's magnetic field, and that may be the means by which they maintain a direction below the photic zone of the ocean.

Study of the mechanisms by which Amphibia orient during their migrations was pioneered by Twitty (1959, 1967) and Ferguson, Landreth, and Turnipseed (1965). Early experiments with red-bellied newts, *Taricha rivularis*, showed that they were able to return to a 350-m stretch of stream with good accuracy after being moved to a downstream location 1,600 m away 1–5 years previously. Removals to a different stream 2,170 m away and beyond a ridge 350 m high gave excellent returns over the next 5 years. By placing fences across the direct route between the two streams, Twitty showed that the newts were not following the foreign stream down to the confluence with the home stream and going back up the latter. As blinded newts could still return, he showed that sight was not necessary. Most newts that had their olfactory nerves cut did not return, the exceptions being those that on dissection proved to have regenerated the nerves. The conclusion that their orientation is by sense of smell was based partly on the reaction of male newts to sponges soaked in water from a tank where females had been kept. When the sponges were tied to rocks in the stream, males approached from far downstream and clustered around the sponge. The conclusion that the chemical

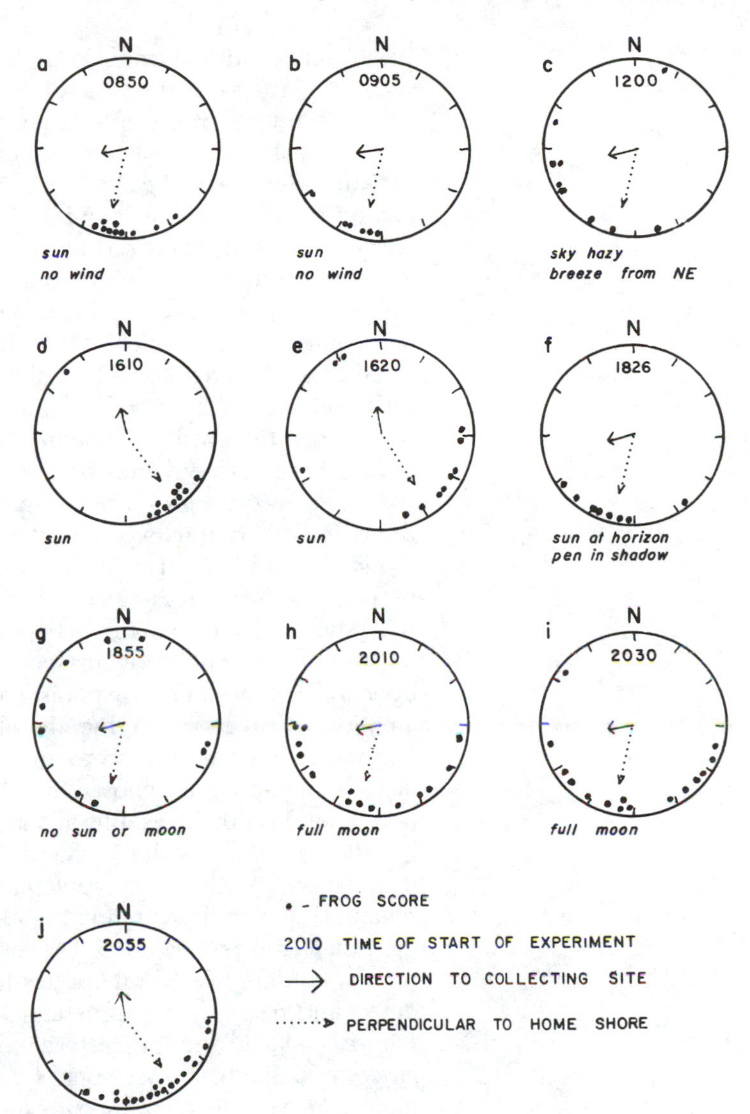

Figure 13.1. Results of experiments to test the solar and lunar orientations of cricket frogs to a direction perpendicular to their home shorelines. All frogs except those in parts a and b were transported and held in total darkness prior to testing. (Redrawn from Ferguson et al. 1965.)

sense is used in homing has not fared well in subsequent experiments on salamander migration, however.

Frogs of a number of species captured along the shoreline of a pond and transported to a distant circular pen with a 2-m-high plastic fence set in water, with only the sky visible, swim in the direction they would have taken to reach the shore had they been dropped into the water of their home pond (Figure 13.1). Ferguson attributed the response to a shoreward escape from predatory fishes. As the figure shows, they were able to orient in that way under a full moon, or after the sun had dropped below the plastic fence, but not under complete overcast. They oriented correctly regardless of the time of day, showing that they adjusted their swimming direction relative to the sun, which was moving across the sky. The possession of an internal "sun clock" is necessary in order to use the sky to maintain a constant direction. Correct orientation after the sun is no longer visible suggests that the frogs were using polarized light, as described earlier for salmon. Amphibians that leave their aquatic breeding sites after their reproductive season is complete face the problem of finding them in subsequent years. Fowler's toads, *Bufo terrestris fowleri,* when captured crossing a road on a breeding migration and transported to a distant terrestrial pen like the aquatic one described earlier, maintained their original orientation under a sky with bright stars but no moon. They could be diverted from their orientation by playing a recording of a breeding chorus of toads from another direction outside the experimental pen.

Tadpoles and larval salamanders swim toward the deeper parts of their ponds when disturbed, and experiments in artificial pools have shown that they swim in the same direction at different times on sunny days, but on overcast days they swim in random directions to the borders of the pools. During metamorphosis, the orientation is reversed, giving the shoreward movement already described. Experiments on a number of amphibian species, both larval and adult, have shown that their perception of the sky does not depend on the eyes. Blinded animals will orient correctly unless the roof of the skull is covered with opaque plastic film. When it is covered with transparent plastic, the correct orientation is retained. The ability extends to polarized light. It is thought that the pineal complex is the light-sensing organ.

Salamanders of several species have been found to be able to detect and respond to magnetic fields. Both the cave salamander, *Eurycea lucifuga,* and the Eastern red-spotted newt, *Notophthalmus viridescens,* have shown such sensitivity in laboratory experiments. *Eurycea* were trained to move back and forth in a box-shaped corridor by alternating which end was moistened by dripping water. One group was trained in the normal magnetic field; the other was trained with its box within a coil that rotated the magnetic field 90°. The two boxes could be combined in a cross shape, with the central area the location for introduction of the salamanders. The whole was inside a large coil that could dictate the orientation of the magnetic field inside the box. The two groups of salamanders were placed together for 60 min and then released at the center of the cross. Whichever way the field was oriented, the two groups moved into significantly different arms of the cross. The normally aquatic *Notophthalmus* were trained in aquaria each of which had an artificial shoreline at one end. Newts normally leave a pond when the water temperature rises above 30°C. The aquaria were placed in the normal magnetic field, or in one that had been rotated 90°, 180°, or 270°. After they had been trained to go to the artificial shore when the water was heated, they were placed, one at a time, in a circular tank, without light cues from the outside; following their release at the center of the tank, their directions of movement toward the edges were recorded. Their responses were as expected, as they moved in the magnetic direction to which they had been trained. *Notophthalmus* showed an additional response to a vertical component of the magnetic field, one of which we are not usually aware. We think of a compass as simply pointing to the North Magnetic Pole, but if the needle were free to point in any direction, including up and down, it would point not only north but also downward at an angle from the horizontal (69° at Ithaca, New York, the location of the experiments). By appropriate manipulation of the experimental coil, this vertical component can be reversed without changing the horizontal one. The responses of the newts to that reversal depended on the conditions in which they had been kept before the experiment (Phillips 1986). Those that had been kept at a relatively constant temperature of 18°C before it was raised to the stimulating 33°C on the day of testing oriented in the expected direction in fields of normal vertical polarity, but those exposed to reversed vertical polarity oriented in the opposite direction, showing that they were reacting to the vertical component of the magnetic field. In contrast, when newts that had been kept at fluctuating temperatures of 3–5°C at night and 25–27°C during the day were then stimulated with the higher testing temperature, they oriented in the experimental tank toward the pond from which they had been captured, 20 km away, regardless of the direction of the vertical component of the magnetic field. That response implies the ability to acquire information about map location while being held in the training tank before the test, and thus heading for the home pond represents true navigation.

Sea turtles are the best-known migratory reptiles, and much effort has been expended in trying to understand their life histories and their demonstrated abilities to locate and return to the same beaches after absences of several years, and from distances of thousands of kilometers. Archie Carr spent most of his scientific life on the problem. He marked thousands of turtles to find out where they went after hatching, where they grew to maturity, and where they fed between their visits to home beaches to lay eggs. Newly hatched sea turtles leave their nests on tropical or warm-temperate beaches and immediately head for the sea, even when they cannot see it. Apparently they distinguish the lighter horizon in the seaward direction. Once in the water, they swim directly away from shore until they are far beyond sight of land. For a long time, what happened to them after that was unknown, but it appeared that they reached the Gulf Stream, and accounts collected from fishermen showed that they were associated with drifting lines of sargassum weed, which come together where converging currents cause downwelling. To shorten a long account, it appears that the little turtles follow the North Atlantic Gyre around to the vicinity of the Azores (Figure 13.2), by which time they have grown to several times their initial length. Considerably larger turtles, still immature, are known from the area near Cape Canaveral, but there is a missing size class between them and those seen near the Azores. Apparently the small turtles go around the gyre not once but several times, growing to the size seen at Canaveral (Carr 1986).

Once in the gyre, they do not require a special orientation mechanism, but two problems remain: the constant heading away from land that they can no longer see, and the ability to find the nesting beach. It is not known that they return to their hatching beach to nest, but once having nested, they always return to nest at a distance of less than a few hundred meters from the previous nest site. The first problem was solved by Lohmann (1992). At first, knowing that sea turtles' eyes are so nearsighted out of water that stellar orientation is impossible, he tested their ability to react to the magnetic field of the earth. The experimental setup was a plastic satellite dish filled with water, with a pivot in the center to which a baby turtle could be attached with a monofilament line. The turtle was free to swim in any direction, and that direction could be recorded automatically. By surrounding the whole apparatus with a coil, and covering everything with opaque plastic, he could control the magnetic field and determine if the turtle oriented to it. Although there was a peculiar circling by all of the test animals at intervals, there was a consistent heading between the circlings of slightly north of east, the direction that would take them to the Gulf Stream from the Florida beach where the tests were conducted. Trials in a large pen offshore, with the turtle tethered to a floating buoy, were consistent with the laboratory results until a day of dead calm, when the turtle appeared lost, and wandered aimlessly. It developed that rather than orienting by the magnetic field, they were using the direction from which the waves were coming, which is consistent enough to give them reliable guidance. On the few days when the wind and the waves were from the west, they swam in the wrong direction: back toward shore.

Despite all of the research effort, long-distance navigation by the adults remains a mystery. A theory was once proposed that the turtles feeding off the coast of Brazil could follow a hypothetical chemical stream coming along the ocean current from Ascension Island, where they nest. The probable dilution of this chemical was calculated to be only 100 to 1,000 after the 2,250-km drift. That was before orientation by using the magnetic field was known in other vertebrates, and the theory is no longer given much credence. It is true that turtles can sense chemicals in water. A. K.

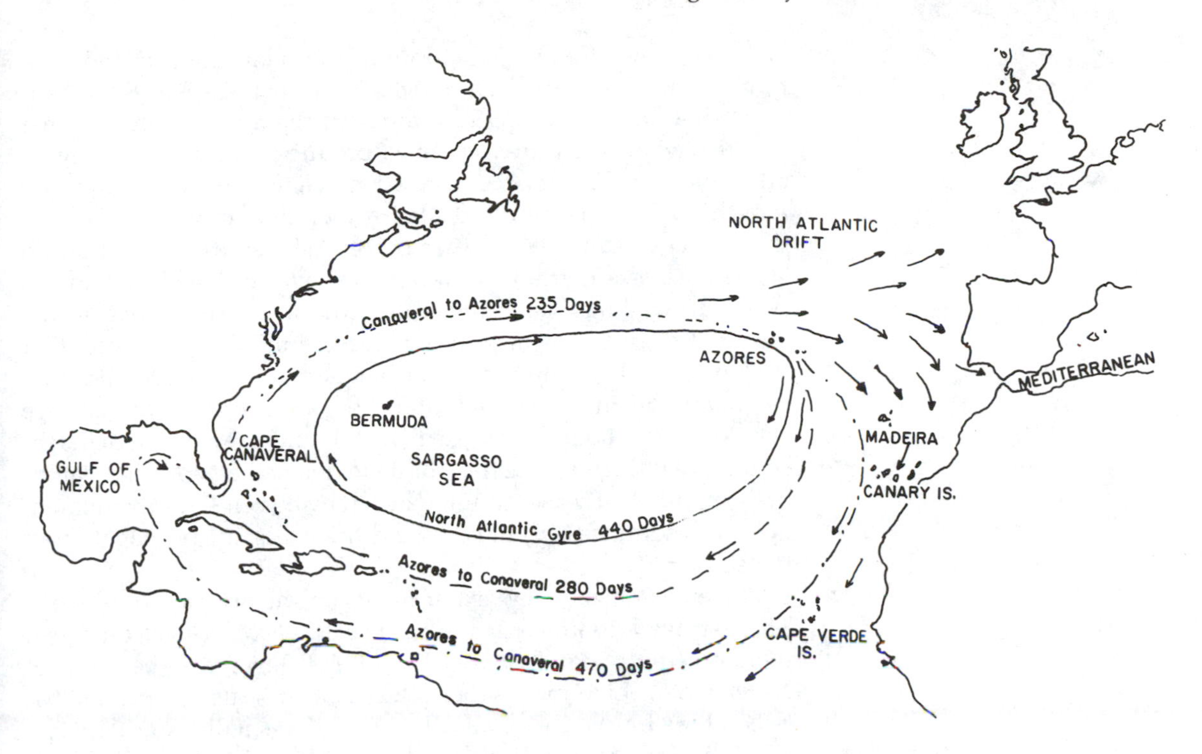

Figure 13.2. *Three transatlantic routes that turtles might follow in the main Atlantic currents, with calculated travel times. (Redrawn from Carr 1986.)*

Harris (pers. commun.) once followed a snapping turtle around a pond, using scuba gear. The animal wandered without noticeable orientation until it stopped, turned, and swam directly to a dead fish. It would not be wise to extrapolate that observation to transatlantic navigation, however.

Other reptiles make less spectacular migrations. A different kind of migration is made by prairie rattlesnakes, *Crotalus viridis*, at the northern edge of the distribution in Wyoming. Winters are harsh and long and must be spent in "dens," usually crevices or piles of large rocks. Small-mammal populations are rarely found in the vicinity of the dens, and summer migrations must be made to find these prey animals. It is important that the den be found before winter, and it has been discovered that they secrete skin lipids that make long-lasting trails that they follow back to their winter quarters. The lipids have the necessary properties of nonvolatility and persistence to last through the summer. Pregnant females do not migrate as far, but move to specific large flat rocks where they give birth to their young near the end of summer. The neonates follow their mother's trail back to the den, which they would not be able to find otherwise, as the mother does not guide them.

Migratory mammals include bats, whales, seals, and large hoofed species. The marine species undertake the longest migrations. Whales give birth in warm inshore waters, such as the lagoons around Baja California for gray whales (*Eschrichtius gibbosus*), and the waters around the Hawaiian Islands for humpbacks (*Megaptera novaeangliae*). From these kinds of locations they migrate to high latitudes in the appropriate summer months, where the waters are most productive and where they do all of their feeding for the year. Gray whales spend the summer months in the waters around Alaska, including the Bering Sea and the Arctic Ocean. Many other baleen whales (those that strain small crustaceans from the water) migrate to Antarctic regions, where upwellings make the waters the most productive in the world. The behavior of gray whales in migration suggests that they use familiar landmarks in orientation, as they are observed frequently raising their heads above the surface. All cetaceans use echolocation, and it is supposed (without more evidence than reasoning) that they can use this ability to detect the depth and nature of the bottoms of seas through which they travel. The low frequency of the sounds they make would favor distant propagation of the sound.

There have been a few experiments on the orientation of seals. The tests were conducted on harbor seals, *Phoca vitulina*, on one of their breeding islands, Sable Island, 160 km from the coast of Nova Scotia. Like most other seals, harbor seals are pelagic between breeding seasons, and they find this small island, a vegetated sandbar, for breeding. By the use of two kinds of blindfolds and some surgery it was established that whereas normal control seals, on release in an unfamiliar part of the island out of sight of the sea, headed directly for the coast by the shortest route, whether the surf was clearly audible from that direction (south) or from the north, seals prevented from seeing any polarization of light headed in the direction of the sound of the surf, as did blindfolded seals. Deaf seals went either due north or due south. Seals deprived of the sense of smell behaved like controls. As far as long-distance orientation is concerned, the experiments suggest the use of polarized light and possibly the magnetic field of the earth.

Small rodents, which do not migrate, are able to home to some extent, but do not do so uniformly. When displaced from their home ranges, *Apodemus* species and *Clethrionomys* orient in the correct direction using visual cues, and *Apodemus sylvaticus* has been shown to be sensitive to a magnetic field.

The ungulates of the Serengeti apparently migrate in direct response to distant rain, which can be seen from afar, and which stimulates growth of savanna vegetation. The rainfall appears in approximate sequence in different areas. The species involved are the wildebeest, *Connochaetes taurinus*, a large antelope, the zebra, *Equus burchelli*, and Thomson's gazelle, *Gazella thomsoni*. The other species present, including carnivores, do not migrate. The migration of the caribou of North America, *Rangifer tarandus*, has been described. The animals apparently follow well-established trails and may use scent glands on their feet and lower legs to mark their trails.

Bats migrate, and they use echolocation, as described in Chapter 2, but the high frequency of the ultrasonic tones used makes for short distance and accurate location, rather than for migratory direction finding. Their ability to return to a cave after displacement declines with distance moved.

Summary

Bird migration has attracted the most attention from biologists, but at least some representatives of all classes migrate, some of them over great distances. The methods by which they find their way have been elucidated by numerous experiments. The spectacular ability of salmon to return to the small stream in which they hatched after several years in the ocean depends on their memory of the chemical composition of the water of that stream. In laboratory experiments, their use of daylight celestial cues has been demonstrated, as has their orientation to the magnetic field of the earth. Amphibians of many species have been shown to use celestial cues, including the sun, stars, and polarized light from the open sky. One of them, the eastern red-spotted newt, also has shown highly complex orientation to the magnetic field, especially to the vertical component, which overrides the horizontal component under some conditions, and which they ignore under others. Sea turtles make some remarkable migrations; they are able to find Ascension Island in the South Atlantic, and they follow the Gulf Stream and the North Atlantic Gyre for several complete circuits while growing to adulthood. Newly hatched turtles orient to the brightest part of the sky in order to go directly to the sea, and then orient to the incoming waves to reach the Gulf Stream from Florida beaches. They can respond to the magnetic field of the earth, but whether or not they use that ability on long migrations is not known. The best-known migration among whales is that of the gray whale along the west coast of North America. They remain close to shore and apparently use landmarks on shore for orientation. Seals can use polarized light, the magnetic field of the earth, and the sound of surf to find their way. Large hoofed mammals, such as caribou, may use scent marking of trails between winter feeding areas and calving grounds.

References and suggested further reading

Carr, A. 1986. Rips, FADS, and little loggerheads. *Bioscience* 36:92–100.

Ferguson, D. E., H. F. Landreth, & M. R. Turnipseed. 1965. Astronomical orientation of the southern cricket frog, *Acris gryllus*. *Copeia* 1965:58–66.

Hasler, A. D. 1966. *Underwater Guideposts: Homing of Salmon*. Madison, University of Wisconsin Press.

Lohmann, K. J. 1992. How sea turtles migrate. *Scientific American* 1992(1):100–6.

Phillips, J. B. 1986. Two magnetoreception pathways in a migratory salamander. *Science* 233:765–7.

Quinn, T. P. 1980. Evidence for celestial and magnetic compass orientation in lake migrating sockeye salmon fry. *Journal of Comparative Physiology* 137:243–8.

Twitty, V. C. 1959. Migration and speciation in newts. *Science* 139:1735–43.

Twitty, V. C. 1967. *Of Scientists and Salamanders*. San Francisco, Freeman.

14

Bird migration
Patterns and capabilities

Bird migrations are among the most fascinating phenomena in nature. The fact that we cannot observe the complete process has been a challenge for centuries and has prompted the development of a number of different methods for studying migration. The earliest observations doubtless were that certain species could be observed in the north only in summer, and in the south only in winter, which merely served to stimulate curiosity.

Reasons for migrating

From the standpoint of survival, bird migration is expensive. It has been estimated that the survival rate of white-throated sparrows, *Zonotrichia albicollis*, is high at their wintering sites as well as during the nesting season, but that overall annual survival is close to the generally accepted 50% for adult passerines, and therefore most of the mortality occurs during migration (H. Wiley and W. Piper, pers. commun.). There are strikingly similar independent data: Dolnik (1990) reported that 1.5 billion (U.S. billion) birds passed his study area east of the Caspian Sea during autumn migration and that half that number passed northbound in spring. The survival in Africa is unknown, but Moreau (1972) estimated that the known 2,000 individuals of one of three falcon species, *Falco concolor*, plus their broods, consume 1.75 million migrants on the Mediterranean–Sahara route. It should be noted that Moreau was preoccupied with competition, rather than predation, and did not discuss survival during the migrants' stay in Africa.

These high costs of migrating must mean either that there would be higher costs of not migrating or that the gains from migrating are greater than the costs in mortality. The obvious way in which to overcome the cost in mortality is to increase reproduction, and that is the primary, if not sole, occupation of migrants during the summer. The winter ranges of migrants in and near the tropics already contain more resident species of birds than there are migrants, and they appear to maintain their populations. The question, therefore, is why the migrants do not follow the same course. Clearly, reproductive success in the tropics must be lower than that on their summer nesting areas, and it is well known that nest predation is high in the tropics, and brood sizes are low.

Despite the potential competition from resident tropical species during the winter, and there is a little evidence that the residents are superior competitors, the advantage to migrants lies in nesting where they could not maintain themselves in winter.

Migration by birds seems most closely related to the present climatic regime, as it is difficult to account for under conditions such as prevailed during the late Cretaceous and early Tertiary periods, namely, generally equable and warm over most of the earth. Whether or not it evolved earlier in the Cenozoic, bird migration must have had powerful selective advantages during the periods after the major glaciations of the Pleistocene, as areas newly vacated by the ice gradually became available for colonization by increasing numbers of species. Those nesting on tundra would have moved north first, followed by coniferous forest species, those two habitats being farthest removed from the tropics.

Observational methods

Banding

The first specific attempt to learn about the movements of individual birds was banding – the placing of metal rings on the legs. The first known record was the banding of a heron in Turkey and its recapture in Germany several years later in 1710. In North America, the artist-naturalist John James Audubon tied silver threads to the legs of a brood of phoebes in 1803 and observed two of them nesting in the same area the next year. It was not until early in the twentieth century that bird banding became organized on a large scale, with the North American effort coordinated by the United States Biological Survey (now the U.S. Fish and Wildlife Service), which also coordinates efforts in Canada and Mexico by treaties. As early as the 1930s, more than 300,000 birds were being banded each year, with recoveries of around 20,000. Numbers over the 5 years 1987–92 averaged 1.1–1.2 million banded, with about 65,000 recovered. Each band has a unique number, and that number should be reported to the Bird Banding Laboratory of the U.S. Fish and Wildlife Service, Laurel, MD 20708. Killing, capture, or even possession of wild birds is strictly regulated. Permission must be obtained, and it is not readily given. Extensive banding programs have been carried out elsewhere, especially in Europe, and some of the results are described later. The information obtained through bird banding has been invaluable in identifying principal migration routes and, in many instances, in determining the rates of migration, the lengths of stopovers during migration, and changes in the physical condition of free-flying migratory birds. None of that information is obtainable in any other way.

Direct watching

Migrating birds can be observed by direct watching with the help of a 20-power telescope. Many species migrate during the day, and such species as ducks and geese can be seen from the ground. Even nocturnal migrants can be seen when passing across the face of the moon. The method has been supplemented by projecting a light vertically at the sky. All of these methods give valuable information about the numbers of birds migrating and the directions of their flights. When wind is blowing, the difference between their headings and the direction of travel can sometimes be determined.

On ten calm spring nights in Piedmont South Carolina, Gauthreaux (1978) was able to make counts, and he calculated that the average number of birds passing along a 1-mile (1.6-km) front was 10,900 per hour; on ten calm autumn nights, the average was 13,250. The maximum observed was 32,400, and the minimum was 3,047.

Listening

The calls of migrating geese are familiar to many people living along the migratory route, but it is less well known that many species of nocturnal migrants call at intervals, and these calls can be heard even when conditions are not suitable for watching. These calls provided the first clue to the rather low altitudes at which some migrants fly.

Radio tracking

The space-flight program required maximum efficiency in the weight to be carried, and that requirement was met by miniaturization, the reduction of size and weight of any instrument or part to be propelled into space. The implications for many branches of science were great, and the possibilities for observations of migrating birds were quickly realized. If a sufficiently small radio transmitter could be attached to a bird without hindering its ability to fly, a directional antenna and receiver could be used to locate the bird. With the receiver mounted on an automobile or light airplane, the flying bird could be followed. As described later in this chapter, the method has been used extensively to determine flight paths and speeds for several migrating birds.

Radar

The advantages of using radar to observe migrating birds has been recognized for more than 30 years, and its use has been enhanced by the increasing sophistication of radar technology. The earliest sets used were surveillance types, which detected the presence of birds or flocks under conditions that made other observations impossible, either because of clouds or fog or because the birds were at greater distances than other methods could locate them. Mass migrations were detected, and the direction of movement could be determined by time-exposure photographs of the screen. The subsequent development of tracking radar allows monitoring individual birds or flocks of birds over considerable distances, thus obtaining flight speeds and directions. Radar also reveals the height at which the birds are flying. Its uses are described in some detail in later chapters.

Experimental methods

A great many experiments have been conducted in the attempt to understand different aspects of migration, and to test different hypotheses about direction finding, navigation, physiological problems, and learning. The methods and experimental techniques are described in later chapters covering the specific questions that have been raised.

Migration routes

North American flyways

It is commonly assumed that migration is simply a matter of flying in north–south directions, and although the general trends are along those lines, a number of geographic factors have caused many species of birds to converge into certain common routes, with the result that concentrations of migrants have been recognized in many parts of the world. In North America, four great geographic features appear to have been responsible for concentrating migrants into flyways: the two coasts, the Mississippi River, and the Front Range of the Rocky Mountains. For daytime migrants, these features provide easily recognized landmarks for orientation (Figure 14.1). The two coasts and the Mississippi River valley also provide suitable habitat for rest stops for ducks, geese, and shorebirds. In the early days of banding, these "game birds" were the species of primary interest, and it was natural to recognize the flyways most used by them. Many passerine birds ("songbirds") and hawks also use the flyways, but there are notable exceptions in the species that migrate along the Appalachian Mountains (Hawk Mountain in Pennsylvania is a famous locality for bird watchers). There are also notable exceptions among ducks and geese. For example, the east–west route taken by redhead ducks, *Aythya americana*, between their nesting area in the Bear River marshes of Utah and the estuaries of the Middle Atlantic States is most conspicuous, as is the northeast–southwest route of Ross's goose, *Anser rossi*, from Southampton Island in Hudson Bay to the Central Valley of California.

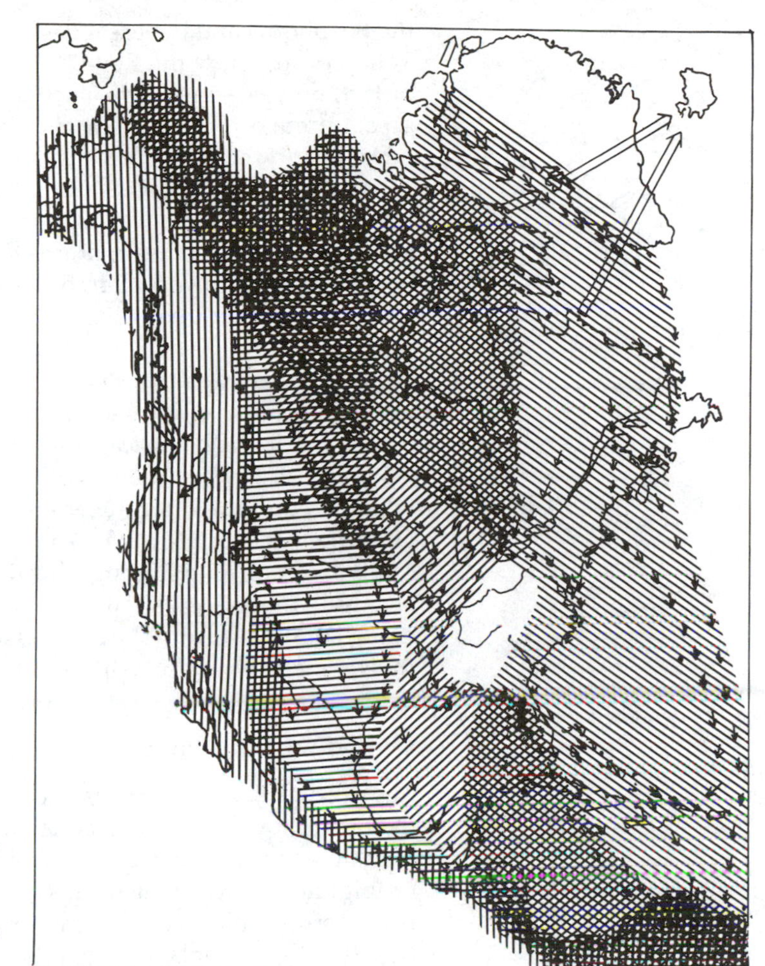

Figure 14.1. *Four major flyways of North American birds. When they are placed on the same map, it is clear that the designations are more or less arbitrary.*

With birds nesting over the whole of Canada, a region that is unsuitable habitat for most species in winter, the flyway designation cannot apply well north of the fiftieth parallel, as birds converge into the flyways only gradually as they make their way south.

European flyways

Birds nesting in Europe and spending the winter in Africa face important barriers: the Mediterranean Sea and the Sahara Desert, in addition to the largely east–west orientation of the Pyrenees, Alps, and Caucasus mountains. If one were advising birds on routes to take, three routes would surely be recommended: to go by way of Gibraltar and the Atlantic coast of Africa, or to go via

Figure 14.2. Routes taken by a few European migrants (e.g., storks) to avoid crossing either the Mediterranean Sea or the Sahara Desert.

the Bosphorus and the east coast of the Mediterranean to the Nile Valley, or to follow the Red Sea to the Ethiopian Plateau. In point of fact, only soaring birds, such as storks, follow those routes in great numbers (Figure 14.2). There is little, if any, concentration of migrants on the other obvious route: that via Italy and Sicily to North Africa. It is obvious that for most migrants the barriers are less formidable than we suppose. Some species fly diagonally across the Mediterranean; others flying via the Iberian Peninsula do not proceed to Gibraltar, but arrive at Cape St. Vincent at the southwest corner, requiring a flight of 200–400 km to Morocco, rather than the 100 km from southern Spain. The Nile Valley and the various oases attract a large variety of species of migrants, but not the numbers of individuals that are known to be moving between Europe and Africa, estimated as 4.3 billion (U.S. billion) (Moreau 1972). We are left with the conclusion that many birds take the entire Mediterranean–Sahara passage in a single flight. How they accomplish that feat is the subject of a later chapter. Many birds do avoid flying directly over high mountains. Extensive use of radar has shown that heavy flights avoid the Alps by going southwest from the Swiss–German border to pass through the valley between the Alps and the Jura Mountains.

Central Eurasian routes

The northern part of Eurasia is covered by tundra near the Arctic Ocean and by conifer forests (taiga) south of the tundra. In summer, this vast area provides productive habitats for great numbers of migratory birds, which must depart before winter. The direct route south is blocked by the inhospitable Tibetan highlands and then by the Himalaya and other formidable mountains. Bar-headed geese, *Anser indicus*, do fly directly over the Himalaya from India to nest beside the Tibetan lakes, but few, if any, other birds take such a route. These geese have been observed flying at 9,000 m, higher than Mt. Everest.

Migration in the desert area east of the Caspian Sea, including the Hindu Kush, Pamir, and Tien Shan mountains, has been followed by Dolnik (1990), using moon-watch and recapture methods. He has estimated that in autumn 1.5 billion and in spring 750 million birds cross the area. Some of the migrants move on the northwest–southeast route between Siberia and the Indian subcontinent; the remainder move on the northeast–southwest route between Siberia and Africa (Figure 14.3). Swallows banded in South Africa have been recovered east of the 90th meridian and north of the 50th parallel. It is likely that birds nesting farther east than the 110th meridian migrate southeast through China. An exception is the Amur falcon, *Falco amurensis*, which nests from Lake Baikal (at 110° E) to the Pacific Ocean north of Korea and spends the winter south of the equator from Kenya to South Africa.

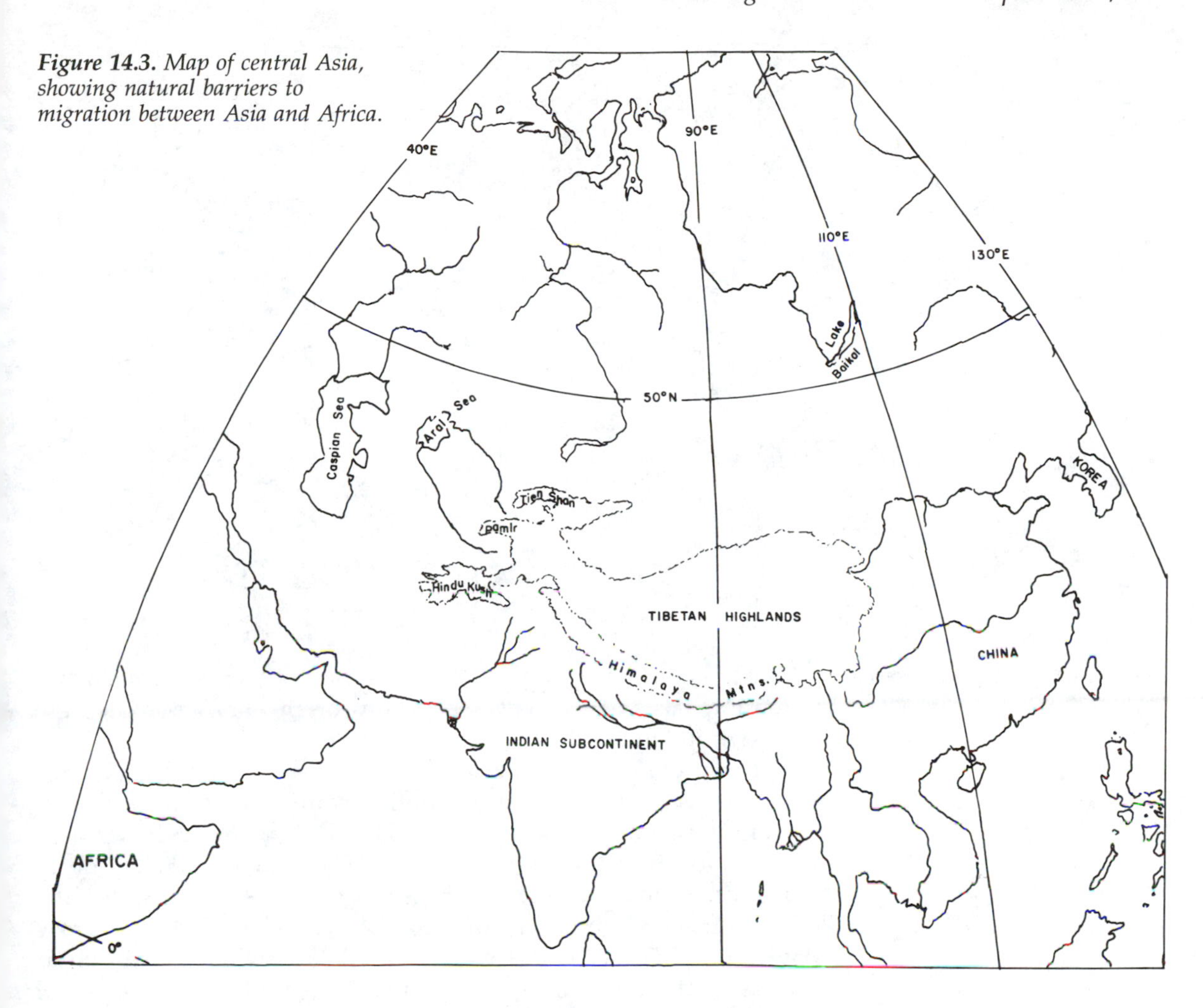

Figure 14.3. *Map of central Asia, showing natural barriers to migration between Asia and Africa.*

Spring and autumn migrants follow somewhat different routes, depending on the relative value of deserts and mountains as stopover sites. The desert is favorable in spring, when the mountains are still very cold; the mountains are favorable in autumn, when they are still warm and the desert is barren. Irrigated valleys, oases, and subalpine areas provide enough food for replenishing fat, so that prolonged flights, such as those crossing the Sahara, are unnecessary.

Oceanic routes

It has long been understood that seabirds migrate over oceans and that there are certain special examples of shorebirds that cross oceans, such as golden plovers, *Pluvialis dominica,* flying from Nova Scotia to Venezuela, or from Alaska to Hawaii and thence to Tahiti. It is only recently, however, that the extent to which land

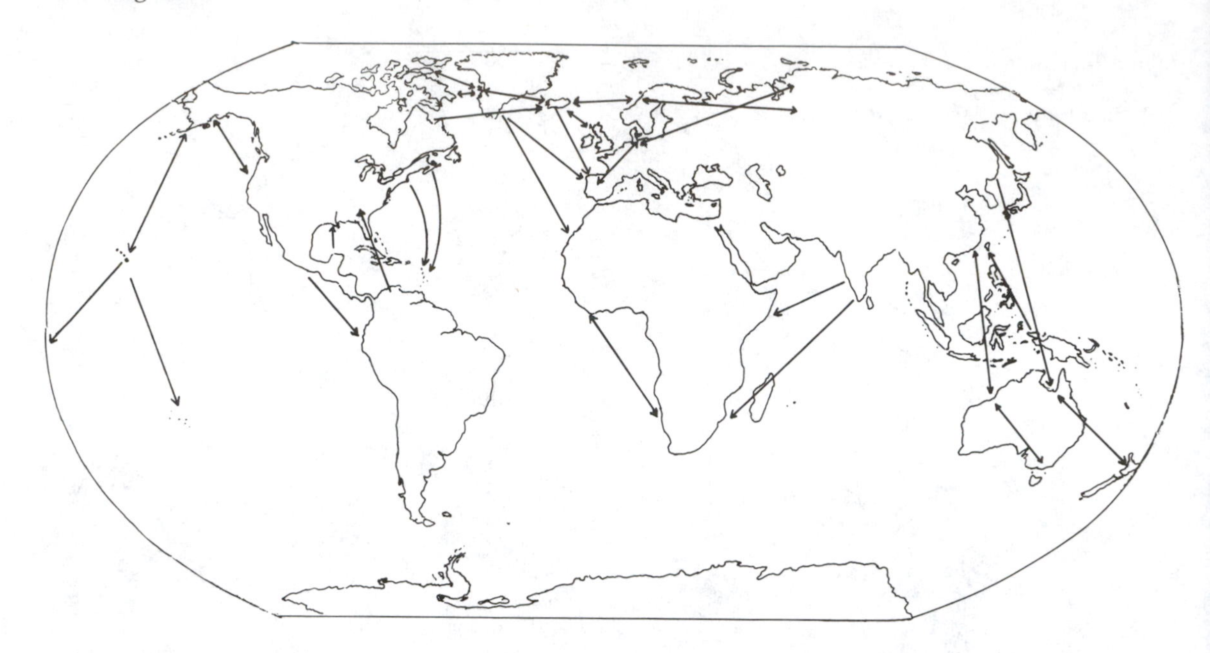

Figure 14.4. *World map showing major transoceanic migration routes.*

birds, including small passerines, regularly fly over oceans in migration has been discovered. Many of the known routes are shown in Figures 14.4 and 14.5.

The route documented best is that over the western Atlantic (Williams et al. 1977). Radar studies from shore, from Bermuda, from ships, and from several islands in the West Indies have established that in autumn large numbers of both shorebirds ("waders" in Britain and Europe) and passerines can be seen leaving the coast from Virginia to Nova Scotia, headed in a southeasterly direction (Figure 14.6). That occurs regularly when a cold front is passing through, thus giving them a tailwind for the first part of the flight. Such fronts normally dissipate before reaching Bermuda, but sightings from that island and from ships between there and the coast have shown that most of the birds fly through the disintegrating front to continue on over the Atlantic. As they go farther, they soon come to the trade winds blowing from the northeast. These constant winds force them toward the southwest and eventually bring them to the Antilles. It is claimed that because of the combination of winds, they do not need to change the heading of their flight. Some individuals have landed on ships, and others are identifiable from ships or from Bermuda. It is significant that the species of passerines seen on the route to Bermuda have been a mixture of wood warblers (Parulidae) and sparrows (Fringillidae), whereas of those seen at and beyond Bermuda have been, with a single exception, warblers. Most of the warblers seen have their wintering distributions in the West Indies and northern South America, though two remain in extreme southern Florida, and two have winter ranges extending north of

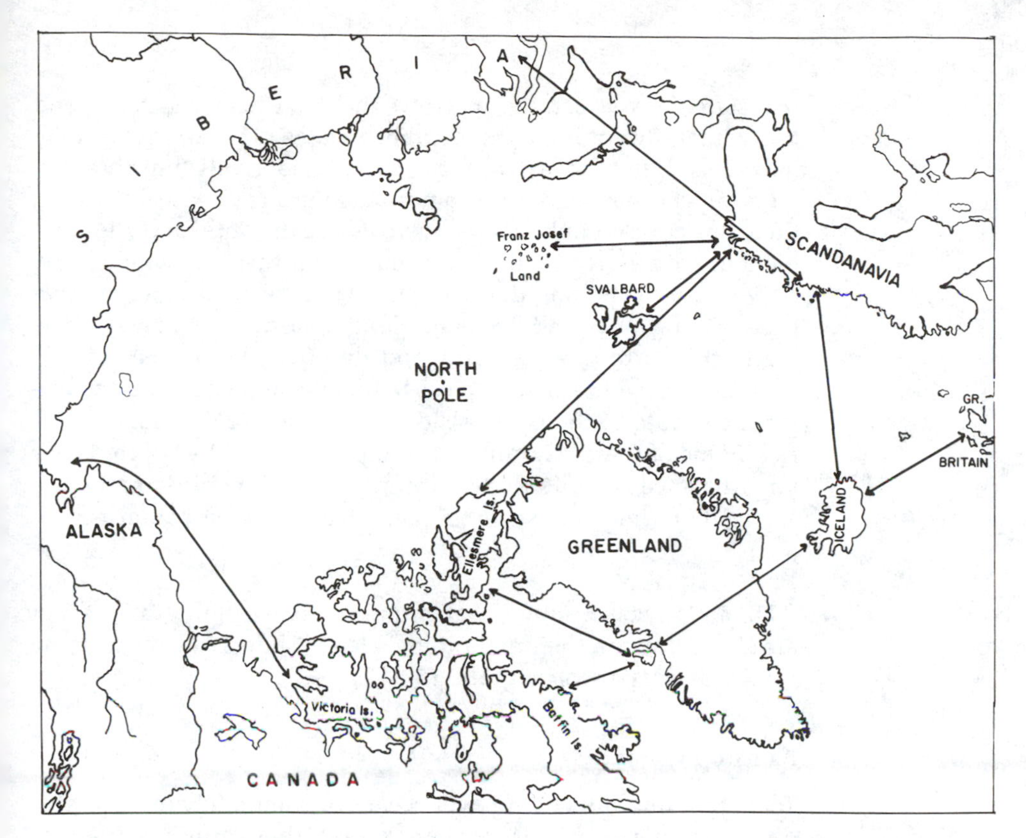

Figure 14.5. Map of the Arctic, showing known migration routes.

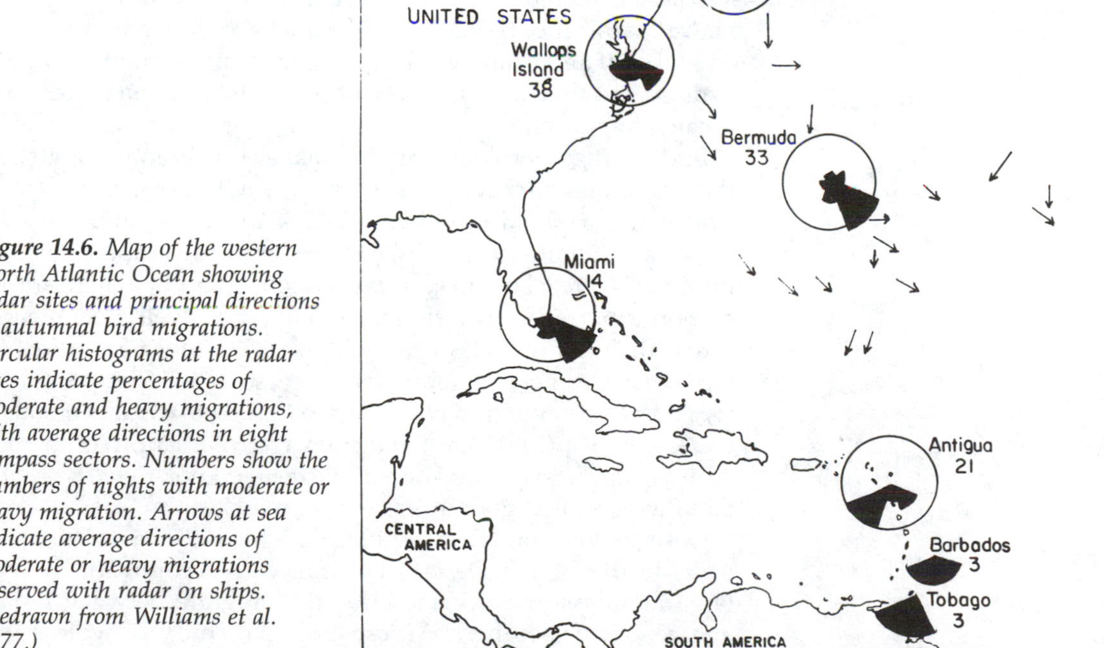

Figure 14.6. Map of the western North Atlantic Ocean showing radar sites and principal directions of autumnal bird migrations. Circular histograms at the radar sites indicate percentages of moderate and heavy migrations, with average directions in eight compass sectors. Numbers show the numbers of nights with moderate or heavy migration. Arrows at sea indicate average directions of moderate or heavy migrations observed with radar on ships. (Redrawn from Williams et al. 1977.)

that state. In contrast, the sparrows that have been sighted spend the winter throughout the southern states. The sparrows may have been blown away from the coast accidentally, but the warblers appear to have been on a normal migratory route. In spring, there are no favorable winds over the ocean route, and the warblers migrate north by the West Indies and then overland.

Two significant long-distance flights over water and ice are that from nesting areas on Ellesmere Island to northern Norway, passing to the north of Greenland, and that from Devon Island and southern Ellesmere Island across Baffin Bay and Greenland to Iceland, en route to the British Isles, the Iberian Peninsula, and Africa. Some of these migrants are also passerines. Most of the other long-distance oceanic flights are made by shorebirds or waterfowl, all of which are stronger fliers than are most passerines.

Achievements of migrating birds

It has already been shown that birds are capable of migrations that are astonishing in their magnitude. Here, we describe some of the feats that are part of migration.

Nonstop distances

Terrestrial migrants flying over oceans or completely inhospitable deserts cannot afford to land; if they do so, they cannot resume the flight. The migrants over the western Atlantic must cover a total distance of 3,000 to 4,000 km, and ordinarily they do not stop at the first land they reach, but continue on to South America. A few species of shorebirds and ducks regularly fly from Alaska to Hawaii, a distance of 3,200 km. The ducks could land on the water and take off again, and golden plovers are also reputed to be able to do so, but they have not been observed to do so on either transoceanic migration.

Birds flying over Arctic regions make equally long flights, and the passerines involved, including the wheatear, *Oenanthe oenanthe*, the Lapland bunting, *Calcarius lapponicus*, and the snow bunting, *Plectrophenax nivalis*, must cover the distance of 2,000 km from Ellesmere Island to Norway without stopping. There are other nonstop transocean flights, though less well documented: across the Gulf of Alaska from Alaska to California, up to 4,000 km, by passerines, shorebirds, and geese; from China to Guam (1,500 km) or beyond to New Guinea (3,000 km); from Greenland to West Africa (3,000 km); and from West Africa direct to South Africa (6,000 km) by two shorebirds, the red knot, *Calidris canutus*, and the bar-tailed godwit, *Limosa lapponica*.

Observations in Libya, Egypt, and Morocco, and at several oases in the Sahara Desert, have failed to account for the very large numbers of passerine birds that migrate between Europe and central or southern Africa. A special study of willow warblers, *Phyloscopus trochilus,* including measurements of body fat at several locations (Greek islands, five locations in Egypt, and three locations in Sudan), showed that it is physiologically possible for the birds to make the 1,800-km flight across the Mediterranean and the Sahara with help from a tailwind of 8 m/s, which can be found at a height of about 1,000 m 95% of the time in autumn.

Aquatic birds cover even greater distances than those just described, notably Wilson's phalarope, *Phalaropus tricolor,* flying from the southwestern United States to Peru (4,000 km), or the theoretical abilities of some shorebirds to make nonstop flights of 7,000 km or more. Those are not reported here in detail either because they can land on the sea (the phalarope) or because such flights have not been documented.

Time consumed

Some species spend a significant part of the year in migration. Estimating from the average times of arrival in spring at different locations in North America, most warblers take 50–80 days to reach their northernmost nesting areas. If they require a similar time to return to their winter ranges, they spend 27–44% of the year migrating. Individual birds may make the trip in much less time. Swallows, *Hirudo rustica,* average 150 km per day between northern Europe and southern Africa, but some have been recorded by banding records to make 260–310 km per day. Other small passerine birds make more leisurely migrations, although the distances covered are much less. The Arctic tern, already mentioned as migrating the greatest distance, must spend over half of the year in migration. A specimen banded in Labrador July 23, 1928, was recovered in South Africa November 14, 1928. As they winter well south of there, the one-way migration probably required 4–5 months, in agreement with an independent estimate of 127 days. Shorebirds on both continents, with migrations of 13,000–15,000 km, require 65–100 days to complete their journey, and they could spend more than half their lives migrating. In contrast to these time-consuming migrations, lesser snow geese, *Anser caerulescens,* and tundra swans, *Olor columbianus,* apparently make only one or two stops between their nesting areas on the Canadian tundra or Arctic islands and their wintering areas on the Gulf Coast or the coast of the Middle Atlantic states. Each long flight requires only 2–3 days, as described later, and the whole process, including stops, may not take more than 2 weeks.

Flight speeds

The figures given for the durations of the migrations of different species include stopover times, not simply the time spent in flight. The most detailed account is that of the blue phase of the lesser snow goose: October 16, 1952, in late afternoon, large numbers were seen leaving James Bay at the southern end of Hudson Bay. October 17, a "blue goose" collided with a plane at an elevation of 6,000–8,000 ft (1,800–2,400 m) just north of Lake Huron. October 18, large flocks were seen over southern Illinois at an elevation of about 3,000 ft (900 m). October 19 (morning), large numbers arrived in southern Louisiana. It appears very likely that the same flight was observed over the 2.5 days. The distance was 1,700 miles (2,735 km), apparently nonstop, and thus the average speed was 45.6 kph. Recoveries of banded birds have been used to estimate flight speeds, with the obvious reservation that resting periods during or at an end of the interval are unknown. The calculations for small to medium shorebirds are 28.6 and 29 mph, or 46–47 kph.

More exact measurements of flight speeds have been made using radiotelemetry or radar, both examples representing short durations. Cochran, Montgomery, and Graber (1967) followed three species of thrushes with radios attached during their spring migrations. Ten individuals of Swainson's thrush, *Catharus ustulatus,* six gray-cheeked, *C. minimus,* and one veery, *C. fuscescens,* were clocked. The range of air speeds was from 24 to 69 kph, with a mean of 47.1. Ground speeds averaged 58.3 kph, reflecting the birds' use of favoring winds, with twelve of the seventeen having faster ground speeds than air speeds, and one showing no difference.

Radar studies of migration over the western Atlantic have permitted measurements of flight speeds of large numbers of birds. Figure 14.7 shows the frequency distributions of air speeds at six locations along the route. Air speeds were calculated by subtracting wind speed from the observed speeds relative to the radar site and varied from 10 to 100 kph for birds leaving Cape Cod, with most birds flying 35–60 kph. A major shortcoming of radar observations is the inability to identify the species or even the major group, such as passerines or shorebirds. The histograms thus probably represent mixtures of those two, with the slower being passerines and the shorebirds being the faster. Even small sandpipers have been timed at 60 kph.

Height of migratory flight

It was once supposed that migrating birds flew at an elevation of 4,500 m, because flight was thought to be easier there. The frequent deaths of many birds flying into lighthouses, tall buildings, monuments, and radio towers showed that the supposition was erroneous, as did a better understanding of the mechanics of flight and the characteristics of the atmosphere. The cited occurrences nearly always were under conditions of night, fog, and low clouds and may have represented somewhat unusual circumstances, but

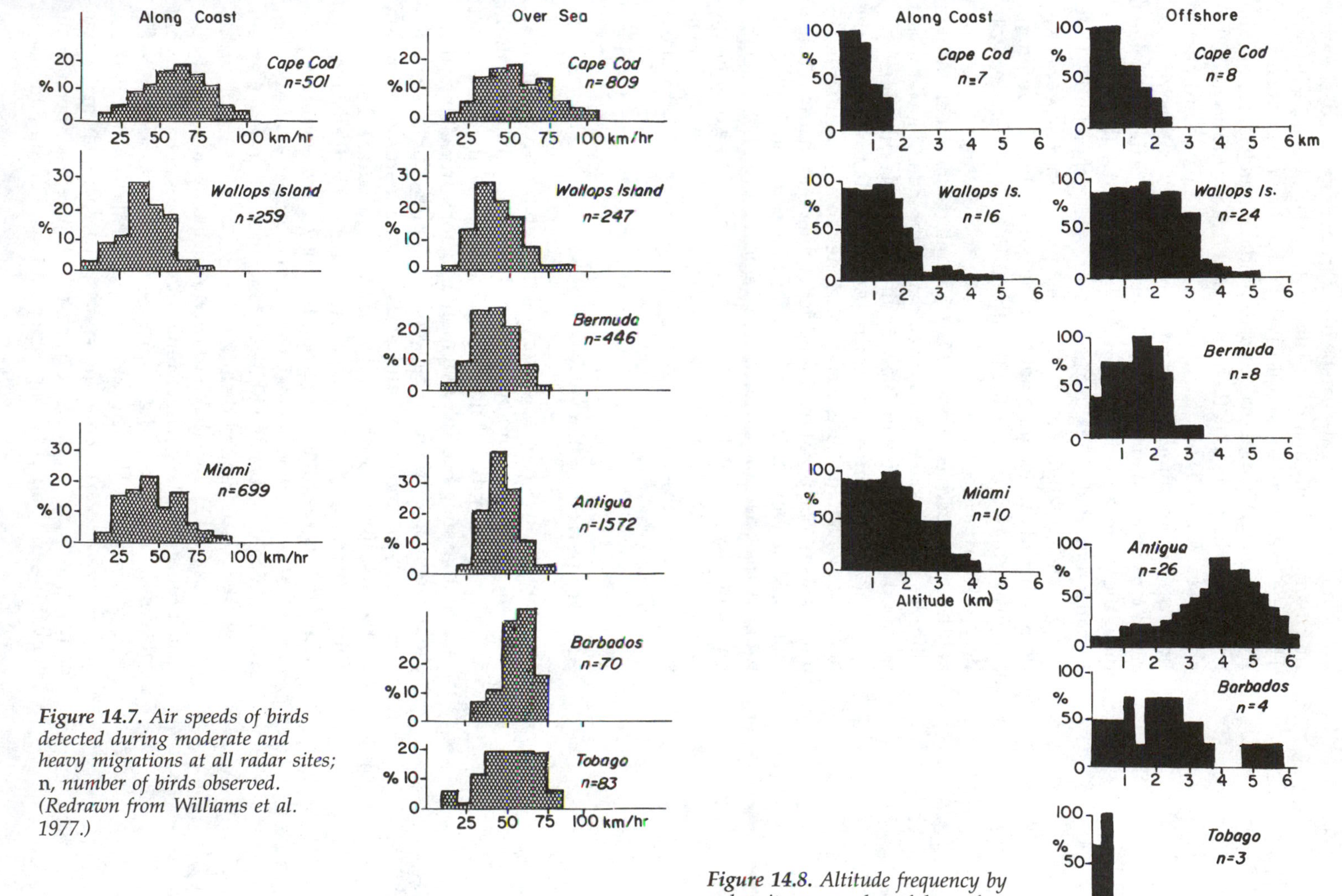

Figure 14.7. Air speeds of birds detected during moderate and heavy migrations at all radar sites; n, number of birds observed. (Redrawn from Williams et al. 1977.)

Figure 14.8. Altitude frequency by radar site; n, number of days of observation. (Redrawn from Williams et al. 1977.)

on clear nights, bird calls can be heard from the ground, showing that many are flying at no more than 100–200 m above the listener. The blue geese whose migratory flight has been described were flying at 900–2,400 m above sea level, and not much less above the ground, as the route took them over land that is only about 150 m above sea level. There have been two studies recording the heights of many hundreds of migrating birds. Analysis of radar reflections from stations around the western Atlantic yielded quantities of information (Figure 14.8). As already stated, the histograms represent a mixture of passerines and shorebirds. They reveal that birds flying along the coast remained at lower elevations than did those setting out over the ocean, and that minimum heights increased during the flight until the birds were ready to descend toward Barbados and thence to Tobago, close to the South American coast (Figure 14.8). The interpretation of the high altitudes flown near Antigua was that the winds at lower levels were too strong, although blowing in a favorable direction.

Dolnik's study in southwestern Asia included details of the heights at which the birds were flying. He reported regular median heights of nocturnal migrants, depending on what was below them (700 m above forest, 1,000 m above desert, and 2,000–3,000 m above the Caspian Sea). His histograms (Figure 14.9) show that most birds fly little higher than necessary when flying over mountains, as they go through passes when possible, but that some species fly as high as 9,700 m above sea level, nearly 2,000 m above the highest peaks.

Accuracy in finding their way

The stamina shown by migratory birds is impressive, but much more mystifying to humans is their ability to migrate over hundreds or thousands of kilometers and arrive at the same exact location that they had left 6–8 months before. In making such a claim, we must distinguish between adult birds that have nested successfully on a territory and birds making a spring migration for the first time. It is the adults that are so accurate. The young birds are able to achieve the general direction and distance, an ability that is remarkable in itself, but they do not ordinarily return to the territory where they were raised. Apparently Audubon was lucky that his banded phoebes came back as close as they did to where he had banded them.

It is less well known that birds become equally attached to specific locations in their winter ranges. An impressive early record concerns a number of indigo buntings, *Passerina cyanea*, banded in a forest clearing at Uaxactun, Peten, Guatemala, in April 1931. Six of them were recaptured at the same spot 1 year later, and a seventh was recaptured there in April 1933. There have been many similar observations. The phenomenon was first brought home to

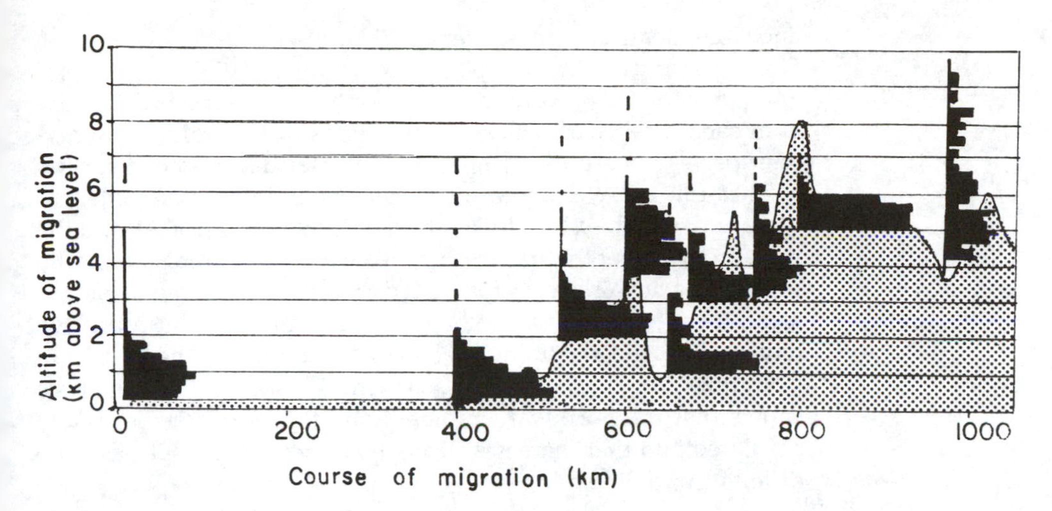

Figure 14.9. *Frequency distributions (filled histograms) of the altitudes of nocturnal migrations above deserts, the Tien Shan, and the Pamirs in a section along longitude 72°$#E. (Redrawn from Dolnik 1990.)*

me when a white-throated sparrow, *Zonotrichia albicollis*, color-banded with plastic rings by a colleague, appeared at a feeding station in our backyard, having flown there from the banding station more than a mile away. It was an adult when banded, meaning that it was more than 1 year old. The easily identified bird returned to the feeding station for four more winters. Besides its accuracy, its survival was unusual, as small passerines have average death rates of about 50% per year after fledging. Thus, that bird was among a group comprising no more than 3% of its original cohort to live as long as it did. That kind of fidelity to a wintering site is characteristic of birds that spend the winter in areas where food resources remain more or less dependable during and between winters. Many European birds wintering in Africa, where the season changes between wet and dry, have been found to shift their locations accordingly. The same kind of behavior has been observed in shorebirds, although winter territories are claimed by some species.

Accuracy of timing of migration

Some of the spring arrival dates of migrating birds are legendary, and probably are exaggerated, but some species do arrive within a few days of the same date each year, as, for example, the swallows at San Juan Capistrano, California, and the Turkey Vultures at Hinckley, Ohio. Most migrants are dependable in their timing, despite the occasional individual that is observed out of phase with its conspecifics. Moreover, in any locality, the sequence of arrival of a series of species is highly predictable. To test the hypothesis that the sequence of arrival of the same species at progressive locations from south to north in the eastern United States remained nearly constant, records were examined for North Carolina (average 35°30′N), the District of Columbia (38°45′N), and Massachusetts (average 42°20′N). Bent's life histories of North American vireos (1950), wood warblers (1953), and orioles and tanagers (1958) were consulted for the earliest recorded arrival in each locality of each of three species of Vireonidae, twenty-one species of Parulidae, one species of Icteridae, and two species of Thraupidae. Average earliest arrivals would be preferable, but were not available. The three sets of dates were compiled, and the rank orderings of the species' arrivals were compared statistically between each pair of locations. All correlations were highly significant, despite the more prolonged season in North Carolina (66 days from earliest-arriving species to latest) than in the District of Columbia (38 days) or Massachusetts (32 days). The analysis shows that the northward progression of each species is closely regulated and that there is a strong tendency for the species to be different in their timing.

Questions about migrating birds

The accomplishments of migrating birds that have been described throughout this chapter raise questions about how they routinely achieve feats that appear to us as little short of miraculous, and the evolution of these abilities has been the subject of much speculation.

Direction finding

Over the past half-century, much attention has been given to answering questions about how birds are able to find their way. Many experiments have been performed, both in laboratories and out-of-doors, and a great deal of ingenuity has been evoked.

Timing

At one time, it was supposed that birds migrated north when it became warm and south when it became cold. Although there is a correlation between northward movement and the progression of spring, southward migrations nearly all begin well before the temperature has begun to fall appreciably. Thus, external stimuli cannot be the sole causes for the initiation of migration. The search for both internal and external causes has been productive and illuminating.

Energy demands

Some species of birds, such as cranes, storks, and many hawks, are able to fly with little expenditure of energy, because they soar using thermal updrafts. Most species, however, progress by flapping flight, which requires large amounts of energy. If they are migrating over a productive countryside, they can land and feed between flights, as is done by many spring migrants in North America. If they are flying over water or unproductive land, they cannot stop; if they did, they would be unable to find food. This involves physiological questions, and much is now known about how they are solved.

The next two chapters will describe many experiments that have been carried out in attempting to answer these questions.

References and suggested further reading

Bent, A. C. 1950. *Life Histories of North American Wagtails, Shrikes, Vireos, and Their Allies.* Bulletin 197, Smithsonian Institution, U.S. National Museum, Washington, D.C.

Bent, A. C. 1953. *Life Histories of North American Wood Warblers.* Bulletin 203, Smithsonian Institution, U.S. National Museum, Washington, D.C.

Bent, A. C. 1958. *Life Histories of North American Blackbirds, Orioles, Tanagers, and Allies.* Bulletin 211, Smithsonian Institution, U.S. National Museum, Washington, D.C.

Cochran, W. W., G. G. Montgomery, & R. R. Graber. 1967. Migratory flights of *Hylocichla* thrushes in spring: A radiotelemetry study. *Living Bird* 6:213–25.

Dolnik, V. R. 1990. Bird migration across arid and mountainous regions of middle Asia and Kasakhstan. Pp. 368–86 in *Bird Migration: Physiology and Ecophysiology*, E. Gwinner (ed.). Berlin, Springer-Verlag.

Gauthreaux, S. A., Jr. 1978. Importance of daytime flights of nocturnal migrants: Redetermined migration following displacement. Pp. 219–27 in *Animal Migration, Navigation, and Homing*, K. Schmidt-Koenig & W. T. Keeton (eds.). Berlin, Springer-Verlag.

Gwinner, E. (ed.). 1990. *Bird Migration: Physiology and Ecophysiology.* Berlin, Springer-Verlag.

Lincoln, F. C. 1939. *The Migration of American Birds.* New York, Doubleday, Doran.

Moreau, R. E. 1972. *The Palearctic–African Bird Migration Systems.* New York, Academic Press.

Schmidt-Koenig, K., & W. T. Keeton (eds.). 1978. *Animal Migration, Navigation, and Homing.* Berlin, Springer-Verlag.

Williams, T. C., J. M. Williams, L. C. Ireland, & J. M. Teal. 1977. Autumnal bird migration over the western North Atlantic Ocean. *American Birds* 31:251–67.

15

Direction finding and navigation by birds

The ability of birds to find their way over long distances, frequently over water or other regions lacking any local cues to the correct direction, has stimulated the curiosity of many workers and has inspired ingenious theories and even more ingenious experiments to test those theories. In some species, especially geese and swans, young birds fly with their parents to the winter habitats and thus can learn the direction and route from them. That is not possible for many species, which apparently travel independently of family groups. The golden plover is a notable example. The adults leave the nesting area in northern Canada and eastern Alaska in a generally easterly direction, flying first to the vicinity of Nova Scotia in eastern Canada. From there, they fly nonstop to northern South America, and after a fairly brief stopover they continue to the pampas of Argentina. The first-year birds take a different route, flying southeast via the central flyway, and thus reach Argentina by a route entirely different from that taken by the adults. Without guidance or experience, they arrive consistently at their destination.

In considering how birds find their way during migration, we should realize that many of the most interesting experiments have been conducted on birds that do not migrate, namely, homing pigeons, *Columba livia*. In this, we are making the assumption that the capabilities shown by pigeons are shared by birds that do migrate but would be more difficult to work with, partly because they ordinarily cannot be recovered if they are released. To the extent that homing and migration involve different challenges, that assumption should be kept in mind.

A number of environmental cues have been proposed as sources of information in direction finding, and new ones are still being offered. The first to be suggested was the use of landmarks, it being thought that birds could recognize major topographic features and use them. Next to be suggested were the celestial cues, the sun and the stars, and aspects of sunlight that are not visible to people, ultraviolet light and polarized light. The magnetic field of the earth has figured prominently in hypotheses of orientation. It has been suggested that characteristic odors can be used in orientation, and most recently it has been shown that infrasound (below 10 Hz) can be heard by some birds. It has long been recognized that true navigation requires more than maintaining a constant heading, as winds and other factors that can cause deviation from the route to the goal must be recognized, and corrections made for the deviation. This chapter describes some experiments, both indoors and out-of-doors, that have been conducted to test the various hypotheses about how birds use the different cues in migration.

Landmarks

In order to locate the exact places where they had either nested or spent the winter 4 months or more previously, it is almost certainly necessary that birds use landmarks near the end of the migration, but it is not known how near to their goals they have to be before landmarks must come into use. They do respond to landmarks, sometimes in disastrous ways, as has been shown by the deaths of many migrants that have flown into tall buildings, lighthouses, and monuments, and they recognize obstacles, as their avoidance of the Alps attests. The most direct evidence is from a study of pigeons that had been released about 70 km from their loft. They were followed in a light airplane. From the release point they followed a compass course until a particular tall building near the loft was in sight. They then corrected their direction and continued to the loft.

Such evidence as is available indicates that geese and other daytime migrants follow rivers, especially if flying at fairly low altitudes. A series of the nocturnally migrating white-throated sparrows, *Zonotrichia albicollis,* released at Wallops Island, on the coast of Virginia, and followed with radar, showed no evidence of using the coastline as they continued their spring migation. They had been captured locally and thus should have been familiar with the terrain (S. T. Emlen, pers. commun.).

Nevertheless, there is other evidence of the use of landmarks by nocturnal migrants. Radar observations have shown that flocks of birds migrating at night under overcast skies still maintain consistent and essentially correct directions. They do require sight of the terrain below them, however. Birds become disoriented when flying between clouds or in fog, and those are the conditions under which they fly into buildings or other lighted obstacles, apparently reacting to them as they might to the moon.

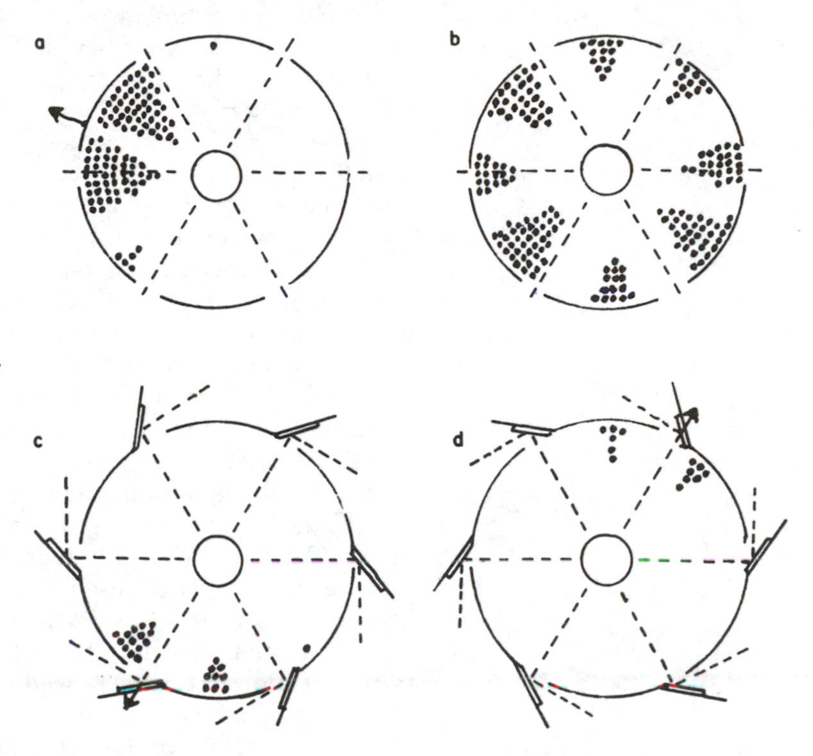

Figure 15.1. *Orientations of spontaneous diurnal migratory activity by a caged European starling under various conditions of sun exposure. The bird was tested outdoors in a pavilion with six windows during the spring migration season. The conditions were (a) clear sky, (b) total overcast, (c) image of sun deflected 90° counterclockwise with mirrors, and (d) image of sun deflected 90° clockwise with mirrors. Each dot represents 10 s of activity. Arrows show mean directions. (Redrawn from Emlen 1975, after Kramer 1952.)*

Sun

For birds that migrate during the day, the sun is the most conspicuous cue for orientation. The first attempt to conduct experiments to test that hypothesis was by Kramer in Germany. He captured European starlings, *Sturnus vulgaris,* kept them in a cage on the roof, and tamed them enough to be used in his experiments. He and others had observed that migratory birds kept in captivity were more active in their cages during the migratory seasons than at other times. They gave that behavior the German name *Zugunruhe* ("migratory restlessness"). With the sun visible to the starlings, the *Zugunruhe* took the form of fluttering against the side of the cage that was appropriate to the direction in which they would have migrated if flying free. In order to quantify the behavior, Kramer placed the cage inside a circular arena with no landmarks in sight, but with six symmetrically placed windows. The arena had a roof, and the birds' only clue was the sunlight coming differentially through the windows. The starlings fluttered in the appropriate direction as long as the sky was clear, but were random in orientation during overcast (Figure 15.1a,b). The next experiment was to mount a mirror beside each window, at an angle that would deflect the sunlight into the arena at a right angle to its actual direction. The mirrors could be reversed. The birds responded by shifting their fluttering direction by 90°, as shown in Figure 15.1c,d. There can be no doubt that their orientation was based on the direction of the sunlight, either actual or apparent.

The perceptive student will find two points to discuss in these figures. The first is that although the birds clearly favored a particular direction whenever there was sunlight, there still was considerable scatter in the direction taken. Thus, a bird orienting solely on the sun would be likely to fly at some small angle away from the optimum direction, and after a few hundred kilometers it would be far from its goal. Sun orientation could not be the sole mechanism of orientation. The second problem is in the distribution of the data points. Whereas the arena had six windows, the data points are grouped into eight sectors. The explanation is that the floor of the arena was transparent plastic, with the observer looking from below, where eight options were available to record each trial. The arrangement provided for an unbiased record, as the observer was prevented from knowing what experiment was being conducted.

But if the sun is used in orientation, and the sun's position in the sky changes during the day, the starlings must have been able to react to the time of day. Otherwise they would have attempted to migrate in a semicircle each day. Two experiments tested that proposition. The first experiment involved putting the cage inside a larger dome in which a strong light bulb was suspended on a track that allowed it to be moved up and down, but not laterally. The starlings then changed the direction of their fluttering gradually over the course of the day, reacting as though the light bulb was moving across the "sky." The second experiment involved the process known as "clock shifting": A bird is kept in a room with only artificial light, which is turned on either 6 h early for one experiment, or 6 h late for another, in each case being left on for 12 h and thereafter kept on the new schedule until the bird has had time to adjust. Then, if the bird is taken out-of-doors, it should now interpret the sun's position incorrectly by 6 h, or 90°, or one-fourth of the circle, as 6 h is one-fourth of the day, as shown in Figure 15.2a. If a starling so trained were then released, it should fly in a direction at a right angle away from the correct one. Unfortunately, the practical result would have been a bird lost to the investigator, and so the test was conducted with homing pigeons, which normally head directly for their loft when released after being removed for 30 km or more. In the experiment, they were first clock-shifted, and then taken in four directions away from their home loft, and the direction they took from each release point was recorded. The results (Figure 15.2b) were as predicted, although there was still a troublesome scatter of data points.

There are two reasons that the ability to use the sun as a compass is inadequate to explain more than a part of the ability of birds to maintain a constant direction. The first of these is that homing pigeons take the correct direction on being released, even if the sky is cloudy. Moreover, if the sky is overcast, clock-shifted

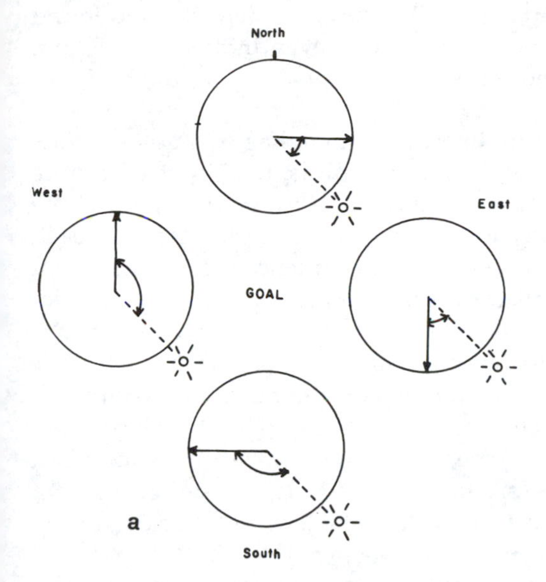

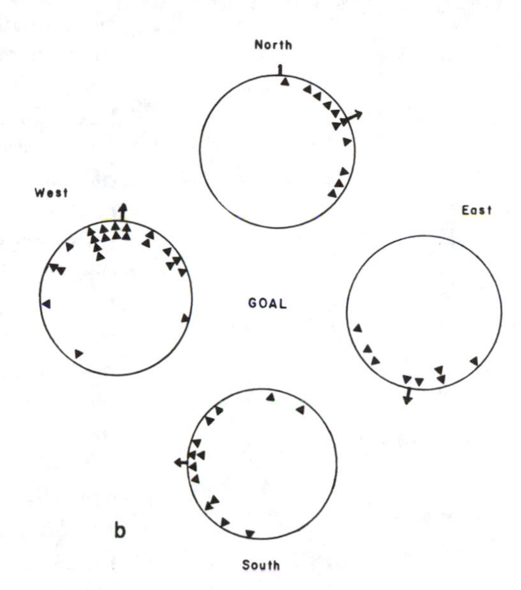

Figure 15.2. *Predicted (a) and observed (b) departure bearings of pigeons returning home. Their clocks had been advanced by 6 h, and they were released 30–80 km from home. (Redrawn from Emlen 1975.)*

pigeons take the correct direction. They must have some additional capability for finding directions. That they were not using landmarks was shown by putting opalescent plastic covers on their eyes. They still came very close to finding their loft. The second reason that the ability to use the sun cannot provide a total explanation is that many birds migrate at night. The position of the sun at sunset has been found to orient them properly, but it seems improbable that that initial orientation could maintain their direction throughout the night.

Stars

Not long after Kramer's experiments, the use of a planetarium was tried for discovering whether or not migratory birds were able to use the stars for orientation. The experiments had technical flaws, but those were eliminated by Stephen Emlen, who used indigo buntings, *Passerina cyanea*. In a planetarium, of course, the projected stars can be placed in any position whether or not that position corresponds to the direction outside. Thus, Emlen could orient the projected stars so that the North Star was on the east, south, or west side of the planetarium, with the rest of the stars correctly oriented to the North Star. He constructed a circular cage with sloping sides and placed a paper funnel inside the cage and an ink pad on the flat bottom of the cage. In their migratory restlessness, the indigo buntings fluttered in a preferred direction, and when they rested, they were re-inking their feet. Thus, the paper recorded the direction of fluttering automatically, eliminating the necessity of direct observation, which would have required a light. Emlen compared the direction taken in the planetarium with that taken under the clear night sky and found that the results were very similar for seven individuals (Figure 15.3). When the planetarium sky was rotated 180°, the birds rotated their orientation similarly.

The planetarium has an interesting advantage over other kinds of tests. Instead of clock-shifting the birds, the sky itself can be clock-shifted to any desired amount, because the stars in the northern sky apparently rotate around the North Star once each day. The indigo buntings showed no evidence that the apparent time of night affected their directions (Figure 15.4). The entire sky was not necessary for orientation, but some parts were more important than others. Most important was the part within 35° of the North Star, although there were major differences between birds in their abilities to orient under different parts of the sky.

When the stars were turned off in the planetarium, and only a dim diffuse light was present, some weak orientation was observed, and that was correct for the outside world, demonstrating that some cue other than the sky was being used.

It has been shown that contrary to generally accepted statements that vertebrates cannot see ultraviolet light, some birds do react to those wavelengths. There is no evidence that they use ultraviolet light in orientation during migration. Polarized light could be used during the day, and especially around sunset, when it could serve to fix the position of the setting sun, even under partial cloud when the sun itself is not visible.

Magnetism

Next to visual cues, the earth's magnetic field has attracted the most attention from students of orientation. The stimulus to investigate this possibility was the discovery that homing pigeons take the correct direction to their home loft even when there is heavy overcast, and from locations to which they have never been taken before (Figure 15.5). The first experiments were attempts to upset their orientation by the use of small magnets attached to their backs. Control birds carried brass bars of the same weight. Because pigeons use the sun as their primary cue, the experiments were conducted under complete overcast. The results (Figure 15.6) show that the magnets had a disorienting effect, although the birds did not disperse from the release site completely at random.

Many other experiments have been carried out, both on pigeons and on migratory birds. Such tests first took the form of exposing caged birds to magnetic fields of tripled strength. The response was an increased activity, but the experiments revealed nothing about the use of the normal field in orientation. More meaningful experiments on European robins, *Erithacus rubecula*, showed that they could orient in a visually cueless room, but that the ability was lost when the cage was placed in a large steel container, which

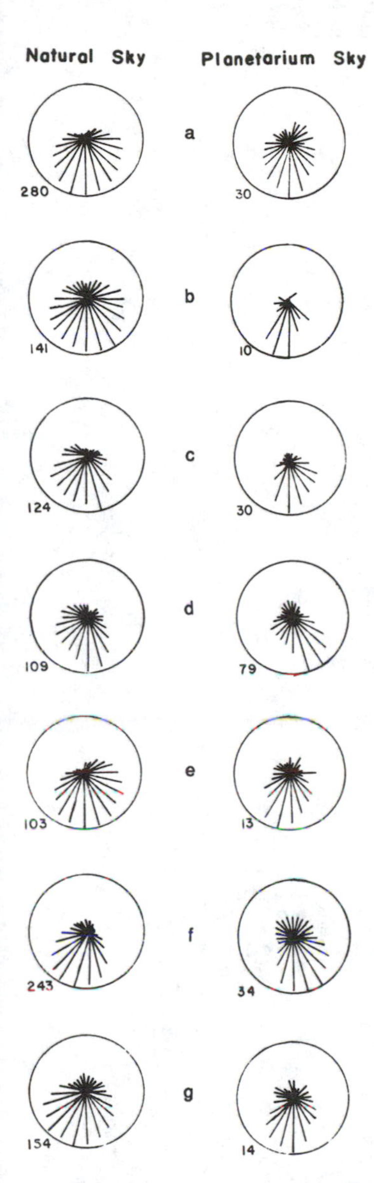

Figure 15.3. *Migratory restlessness orientations of caged indigo buntings* (Passerina cyanea). *Left: outdoors under clear night sky. Right: in a large planetarium under artificial sky matching that outdoors. The length of each vector is proportional to the greatest amount of activity in that direction. Numbers are numbers of observations. (Redrawn from Emlen 1975.)*

greatly reduced the magnetic environment. Subsequent experiments, using large Helmholtz coils to control the direction of the field, showed that the robins responded to the altered direction of the magnetic lines of force in ways appropriate to the direction of migration in the season, whether spring or autumn. Indigo buntings were also shown to orient without visual cues, but the data are widely scattered, showing that the response is weak. They also shifted appropriately when an artificially rotated field was applied. Most interestingly, the orientation was strongest early in the migration season and became progressively weaker, with random orientation late in the season, both for the normal field and for the artificially rotated one.

The anatomical and physiological bases for the detection of magnetic fields have received relatively little attention, with one exception. Pigeons have been found to have magnetite-rich innervated structures next to the brain, and it is possible that these function in detecting magnetic fields. Other birds are not known to have them, and it is not known how they function.

As a final note on orientation according to magnetic fields, there has been some work on humans, although they are birds only metaphorically. It was originally reported that students at the University of Manchester, when blindfolded and driven over different elaborately circuitous routes, could estimate the direction back to the university while the blindfolds were still in place. In a separate set of experiments, one group of students wore bar magnets under the straps of their blindfolds, and the other group wore brass bars of the same weight. The students with the brass bars performed significantly better than those with magnets (Baker 1980). On learning of those results, two groups, one at Princeton University and the other at the State University of New York at Albany, attempted to repeat the experiment, with even more elaborate precautions against visual cues being used. In two cases out of fourteen, significant clusterings of estimates were obtained, one pointing 175° away from "home," and the other 98° away. Magnets had no effect. Later, Baker and a distinguished group of outside observers joined the Princeton group in an attempt to repeat the Manchester experiments. Again, there was no positive result (Gould and Able 1981). Clearly, either Manchester students were sensitive to the magnetic field and Princeton students were not or some visual or proprioceptive cues were available at Manchester. Princeton and the surrounding area are on relatively level ground, whereas the Manchester area is hilly, and it is possible that despite the precautions the Manchester students sensed at least parts of the directions they had been taken. At Cornell University, which lies between flat land in one direction and steep slopes in the other, students are able to repeat the Manchester experience if they are taken in the latter direction, but not if they leave over the flat land (K. Adler, pers. commun.).

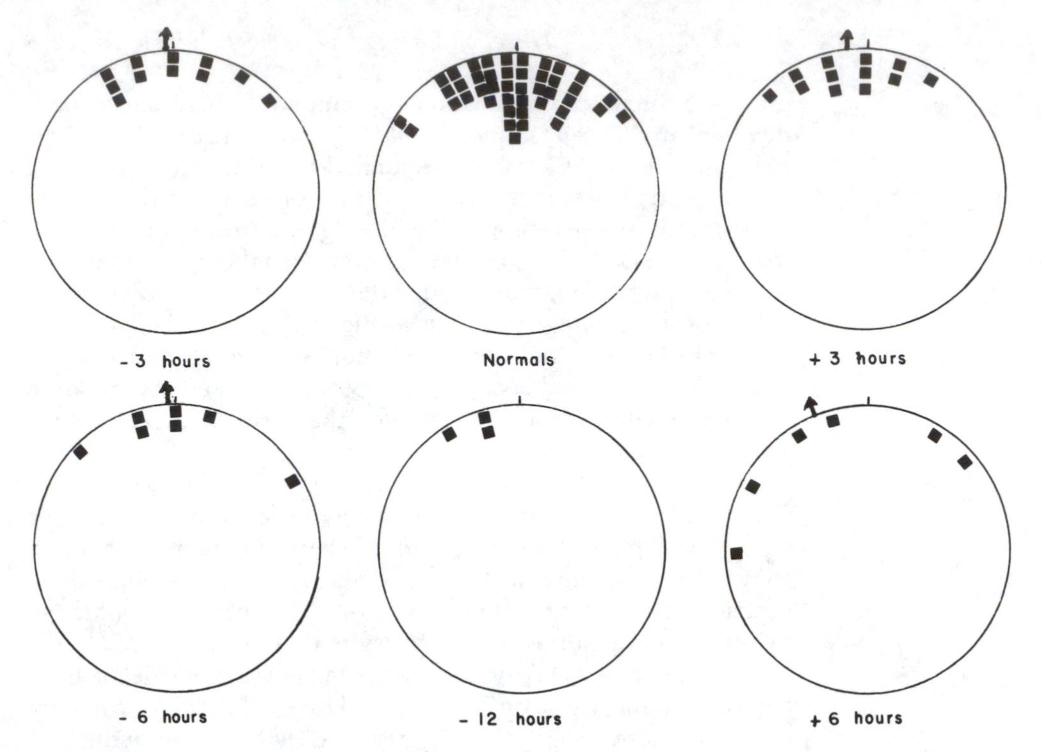

Figure 15.4. Results of experiments designed to test the effect of clock shifting on the orientation of indigo buntings in a planetarium. There was no effect. Contrast with Figure 15.2. (Redrawn from Emlen 1975.)

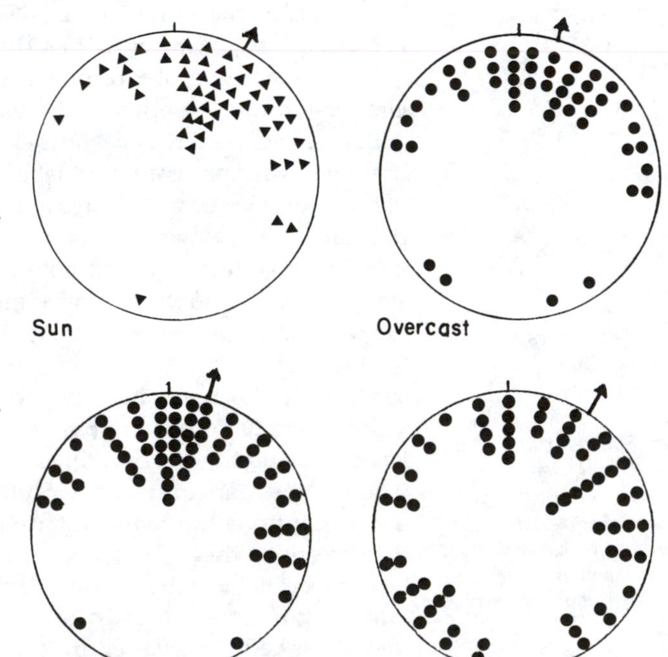

Figure 15.5. Departure bearings for homing pigeons released under clear and cloudy skies at distances of 33–160 km from the home loft. (Redrawn from Emlen 1975.)

Figure 15.6. Departure bearings for experienced homing pigeons released under total overcast at distances of 27–50 km from the home loft. Although the pigeons with magnets attached were not random in orientation, they clearly did not perform as well as those carrying brass bars of the same weight. (Redrawn from Emlen 1975.)

Smell

It is widely thought that among birds only turkey vultures, *Cathartes aura,* and some members of the Procellariformes (tube-nosed swimmers) have more than a rudimentary sense of smell. It came as a surprise, therefore, to learn of experiments in Italy that reported evidence that homing pigeons use odors smelled on their outward journeys to guide their return. The experiments involved moving two groups of pigeons from the same loft, selected to be alike in age and experience, by opposite semicircular routes to the same release point. The hypothesis was that if stimuli received en route were used, those that had been taken to the release point from one direction would depart from the release point in the opposite direction, and vice versa. The birds were first transported in baskets inside a van to prevent the use of visual cues. The departure directions from the release point were significantly different for pigeons taken by the two routes in three of four tests, and in two of them the mean directions taken were both as predicted. In the other two experiments, those brought by a clockwise route still took a slightly clockwise bearing on being released, but the mean bearing in one experiment was significantly different from that of the birds brought by a counterclockwise route. Pooling all of the data gave statistically significant results confirming the original hypothesis that stimuli received on the outward trip influenced the departure bearings. Odors were not the only stimuli that might have been received, even though visual cues were eliminated. The birds might have used magnetic lines of force. Further experiments were indicated. In those experiments, the pigeons were transported in iron containers that virtually eliminated the magnetic field, and controls were transported in aluminum boxes. Air could be supplied either from bottled supplies or sucked in from outside. In the first two experiments, only the birds carried in aluminum containers and provided with outside air departed from the release site in the homeward direction. Subsequent experiments used only outside air and either iron or aluminum containers. In nine tests, only the birds transported to the release site in aluminum containers oriented significantly toward home. Those transported in iron containers either left the release site in random directions (three experiments) or were oriented in directions significantly different from home (six experiments). The overall results of these experiments indicated that the birds could be influenced on the outward trip by either odors or the magnetic field.

Naturally, investigators in other countries tried to duplicate those results, and in neither the United States nor Germany could they do so at first. Repeated trials in both countries finally gave some confirming results, but they were in the minority among tests. Collaboration between Italian and German workers gave some confirmatory results, but again they were a minority of the experiments, even in Italy. As a result, after all that effort, we are not sure that pigeons use odors received on the outward journey to orient toward home.

Sound

Many migratory birds fly low enough at night for their calls to be heard from the ground. The possibility that they are using echoes to identify the kind of terrain over which they are flying, such as cleared land, forest, or water, has been investigated, and it has been found that theoretically they could receive some information in that way, but it remains a hypothetical proposition. They may simply be communicating with each other.

Nocturnal migrants being followed by radar were shown to alter their flight paths when certain sounds were broadcast. They reacted most strongly to the call note of a sharp-shinned hawk, *Accipiter striatus*, to the distress call of a mockingbird, *Mimus polyglottus*, and to the sound of thunder, but there were many inconsistencies on different nights, perhaps because different species of birds were being followed.

There is a world of loud sounds that are inaudible to humans because they are at frequencies below our threshold. Pigeons have been shown to react to such sounds at frequencies below 0.1 Hz and levels below those that occur naturally. Such sounds theoretically could be used in navigation, as they carry over thousands of kilometers with little attenuation, but at present it is not known how birds could use the information.

Navigation

As has been pointed out, direction finding is necessary to migrating birds, but it is not sufficient to lead them to their goals, for which some form of true navigation is required if they are to correct for being blown off course while maintaining a constant compass heading. Because winds frequently are from a quartering direction, the birds must be able to recognize errors arising from such a source and be able to correct for them. This was first demonstrated for European starlings. They nest in the region from southern Denmark to the Baltic Sea, and in autumn migrate southwest to spend the winter in southern Britain and northwestern France. Perdeck (1958) captured and banded a large number of migrating starlings in the Netherlands and transported them southeast to Switzerland, where they do not normally go. After being released, many of the birds were recaptured, the majority in the southern half of France and northern Spain, showing that most of them continued in the direction in which they had been going when captured. A minority of the starlings, all but one of them adults, were recaptured in northern France and even in England. None of the adults reached Spain. The adults somehow sensed that they were in the wrong place and made corrections; the immature birds were unable to do so. The adults could have recognized that familiar landmarks, such as the English Channel, were missing, but how they corrected for the displacement remains unknown.

Some observations by Gauthreaux (1978) in South Carolina indicate how some birds sense and correct for displacement during migration. By observing nocturnal migrants visually and tracking them with radar on calm nights (wind speed less than 2 m per second), he established "preferred" directions for both spring and autumn. Those directions were compared with directions taken by migrants on nights when the wind speed was greater. By knowing the wind speed and direction, he could calculate the extent to which the birds could correct for the wind while in flight. He showed that they could not do so completely.

His next step was to watch some normally nocturnal migrants that had landed in his vicinity on the morning after they apparently had arrived. They flew in daylight, and he recorded the direction in which they vanished. That was significantly correlated with his calculation of their displacement by the winds during the previous night. The birds clearly sensed that they had been blown off course and were correcting for the deviation. That apparently was accomplished by feeling the wind while in flight.

Summary

Many of the different kinds of stimuli that can be used by migratory birds in orienting to a constant direction and even in true navigation have been investigated intensively, and a great deal has been learned about how the birds find their way. For birds to locate the exact spot where they had been months before, they must use remembered landmarks, and a few observations have shown that they do so. The extent of such use is unknown. The sun obviously is a cue to direction, and experiments on many birds have shown that they use it, commonly in preference to other available cues. Its use requires awareness of the time of day, because the sun moves across the sky. Two kinds of experiments have shown that birds respond to the sun according to the time of day. Nocturnal migrants use the stars in orienting, but apparently do not react to the time of night, probably because the North Star and nearby constellations are continually in view, even though the constellations rotate around the North Star. Many experiments have shown that birds are sensitive to the magnetic field of the earth and orient accordingly, but there is evidence that they are unable to use that source of information by itself. The use of local odors and sounds, including infrasound at extra-low frequencies, has been postulated, but the experiments on those sources of information have been inconsistent.

Direction finding is not sufficient for migrating birds; they must know whether or not they are at the correct location, a knowledge that requires true navigation. One of their sources of information about this is feeling the force of wind while flying, and it is likely that landmarks also play a role.

References and suggested further reading

Baker, R. R. 1980. Goal orientation by blindfolded humans after long-distance displacement: Possible involvement of a magnetic sense. *Science* 210:355–7.

Dolnik, V. R. 1990. Bird migration across arid and mountainous regions of middle Asia and Kasakhstan. Pp. 368–86 in *Bird Migration. Physiology and Ecophysiology.* E. Gwinner (ed.). Berlin, Springer-Verlag.

Emlen, S. T. 1975. Migration: Orientation and navigation. Pp. 129–219 in *Avian Biology*, Vol. 5, D. S. Farner & J. R. King (eds.). New York, Academic Press.

Gauthreaux, S. A. 1978. Importance of daytime flights of nocturnal migrants: Redetermined migration following displacement. Pp. 219–27 in *Animal Migration, Navigation, and Homing*, K. Schmidt-Koenig & W. T. Keeton (eds.). Berlin, Springer-Verlag.

Gould, J. L., & K. P. Able. 1981. Human homing: An elusive phenomenon. *Science* 212:1061–3.

Gwinner, E. (ed.). 1990. *Bird Migration. Physiology and Ecophysiology.* Berlin, Springer-Verlag.

Kramer, G. 1952. Experiments on bird orientation. *Ibis* 94:265–85.

Perdeck, A. C. 1958. Two types of orientation in migrating starlings *Sturnus vulgaris* L. and chaffinches *Fringilla coelebs* L., as revealed by displacement experiments. *Ardea* 46:1–37.

Schmidt-Koenig, K., & W. T. Keeton (eds.). 1978. *Animal Migration, Navigation, and Homing.* Berlin, Springer-Verlag.

16

Physiological solutions to problems of migration

Timing of migration

When asked why birds migrate when they do, many people respond that they migrate north when it begins to get warm, and south when it begins to get cool. The spring correlation is fairly good, the best example being the Canada goose (Figure 16.1). The northward migration of these large birds closely follows the 35°F isotherm (the zone across eastern North America where the mean temperature is 35°F at the same time). The mean arrival times of the geese coincide quite well, suggesting that their timing is correlated with the waters becoming ice-free.

Such an explanation cannot be valid for most species, however, because, as already described, many of them tend to arrive within a few days of the same date each year, and the weather is much less regular than that. (As I write this, we have just had several feet of snow in the mountains of North Carolina in May.) Moreover, many species begin their southward migration before the weather begins to become cool. Some sandpipers and plovers arrive on the southestern coast of the United States at the end of July, having migrated there from the tundra. They are still in their summer plumage.

Such phenomena led to the hypothesis that some environmental cue more regular than weather must be responsible. The cue suggested was the changing length of daylight, and experiments by the Canadian Rowan in the 1920s confirmed the hypothesis. He exposed juncos, *Junco hyemalis,* to experimentally lengthened days in winter and found that they became restless and tried to fly north.

Becoming restless is not the only change that birds undergo around the time of migration. They molt first, and at the same time that they become restless, they gain weight, up to 40% of their normal body weight, and their gonads become enlarged. An early experiment demonstrating these changes in response to an abnormal increase in daylength was conducted by Emlen (1969) on indi-

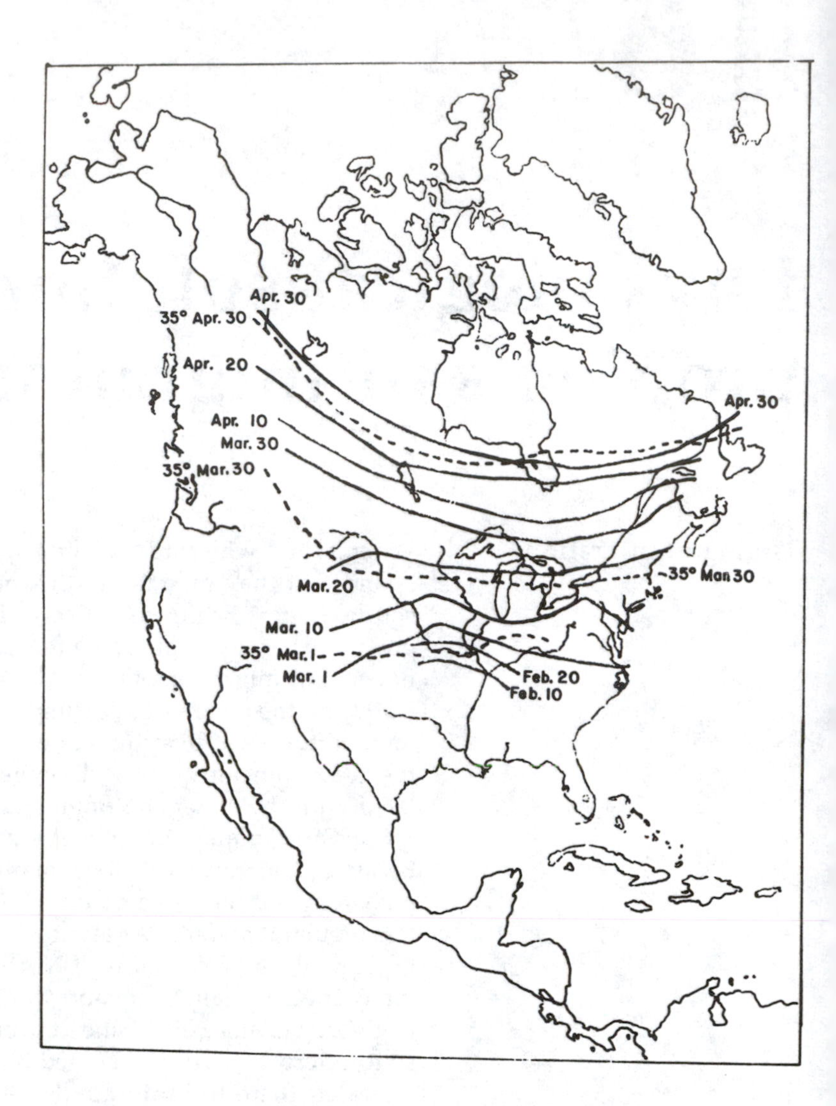

Figure 16.1. *Migration of the Canada goose,* Branta canadensis. *The northward movement keeps pace with the advance of spring, in this case with the isotherm of 35°F. (Redrawn from Lincoln 1939.)*

go buntings, *Passerina cyanea* (Figure 16.2). He allowed one group of birds to experience normal daylengths for 1 year, starting in August; a second group received the same treatment until December, when the daylength was suddenly changed to that of June. It was kept at that level until the first of March, when the daylength was gradually reduced to the December level by the end of June. The two groups were molting (the "postnuptial" molt) at the start of the experiment. That was followed by a period of restlessness and weight gain. The group exposed to normal daylengths had a "prenuptial" molt in late February and March, followed by restlessness and weight gain in April and May. The experimental group had a prenuptial molt in January, followed by a period of restlessness from mid-January to March, but without any clear

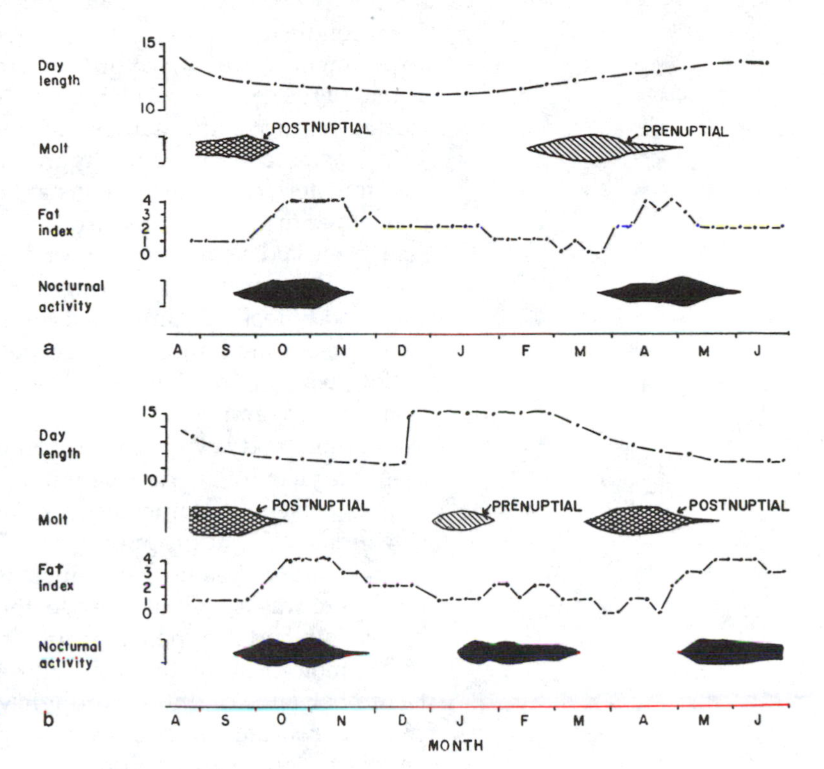

Figure 16.2. *Occurrences of molt, fat deposition, and nocturnal activity in* Passerina cyanea: *(a) control birds on a natural photoperiod, (b) experimental birds exposed to an accelerated photoperiod. (Redrawn from Emlen 1969.)*

weight gain. In April, they had a postnuptial molt, with restlessness and weight gain in May and June.

It appeared that the changes in daylength were the cues to which "calendar" migrants responded by migrating either toward or away from their nesting areas, depending on the physiological state induced. Thus, when daylengths are increasing, gonads become increasingly active, and the birds migrate north. When daylengths begin to decrease, breeding has ended, and the birds respond by migrating south. All of that seems nicely plausible until we remember the golden plover and the bobolink, *Dolichonyx oryzivorus*, both of which spend the winter far south of the equator. Such birds must begin their spring migration when the days are shortening, contrary to the simple form of the daylength theory as stated. Many similar situations are known, especially in Africa, and even more species spend the winter at or near the equator, where the amount of change in daylength is too little to stimulate a spring migration.

In Chapter 15, we saw that birds use the time of day in correctly orienting to the position of the sun, indicating that they have an internal "clock." The migration of the bobolink suggests that some birds, at least, must also have an internal "calendar," which starts them on their spring migration at the correct time, without the stimulus of an increasing daylength. Many experiments have now been conducted showing that suggestion to be true. The first such experiments were carried out in Europe, and an excellent example is that of the garden warbler, *Sylvia borin,* a small bird that nests in northern Europe and winters in central and southern Africa. The bird was hatched in captivity in late May and was hand-reared and transferred in July to a constant regime of 10 h light 14 h dark. The experiment continued until the end of April, 2 years after it had been laid as an egg. The following were monitored: testis length, *Zugunruhe,* body weight, and molting. The postjuvenal molt (body feathers only) had already started when the experiment began, and migratory restlessness and increase in body weight began immediately. By November, migratory restlessness had stopped, and body weight had dropped to its original value. A full molt (body, wing, and tail feathers) began in December and was complete by February. By March, the testes had become enlarged, weight had increased rapidly, and there was marked *Zugunruhe.* Weight dropped in April, and testis length decreased in June, but restlessness continued at a low level through September. There was a molt of body feathers in June and July. By October, weight had dropped, and restlessness had stopped, and there was a molt of all feathers. Weight and restlessness increased in December, and continued through March. Testis size began to increase in February, and to decrease in April. Restlessness declined greatly in January, and body weight was back down in March. There was a full molt beginning in April.

There are two important features of these results. The first is that the sequence of physiological events was the same throughout. That is, molts occurred when the body weight had declined, and the beginning of restlessness coincided with the beginning of the increase in body weight. The second important feature was that the sequence was accelerated in the second year. That is, events that started out being spaced over 11 months became spaced over a shorter interval. In another long-term experiment under the same conditions, both the "winter" and "summer" molts of two species occurred at intervals of 10 months instead of 1 year, but nevertheless were repeated at that interval for 8 years. Thus, a constant sequence is maintained under unchanging conditions, but is speeded up in comparison with natural events. Their rhythm is called a *circannual rhythm,* as the daily activity cycle is called a *circadian rhythm,* the two terms meaning "about a year" and "about a day." Under natural conditions, both the calendar and the clock are reset by natural events once each cycle, thus precluding the speedup that is observed when the event that resets the cycle is prevented from occurring: changing daylength for the circannual cycle, or sunrise for the circadian cycle.

From these experiments we can see that a bird's internal physiological state provides the correct signal for it to fly north or south at the appropriate time of the year. Emlen confirmed that by

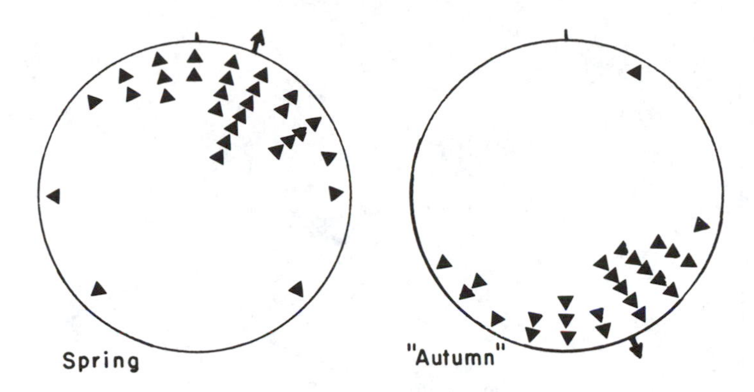

Figure 16.3. Zugunruhe *orientations of indigo buntings artificially brought into physiological states appropriate for spring and autumn migrations. Birds from both groups were tested simultaneously in May. (Redrawn from Emlen 1969.)*

bringing two sets of indigo buntings into opposite physiological states by exposing them to long and short daylengths. One group was thus in a spring physiological condition, and the other in an autumn condition. He placed both groups in a planetarium under a spring sky. The "spring birds," exposed to long daylengths, fluttered in a direction slightly east of north (22°) (Figure 16.3); the "autumn birds," exposed to short daylengths, fluttered in a direction slightly east of south (157°). The fact that they did not orient in exactly opposite directions may be attributed to the common exposure to a spring sky.

All species of migratory birds do not respond identically to artificial exposure to different daylength cycles. The European collared flycatcher, *Ficedula albicollis,* nests largely in central Europe and winters near the equator (10° N to 20° S) in Africa. When exposed to experimental daylengths that simulated conditions 20° south of the equator, at the equator, and 10° north of the equator from October to June, testis length and restlessness responded in the same way for all three conditions, as did body molt; only molting of flight feathers did not occur for the southernmost simulation. Its congener, the pied flycatcher, *F. hypoleuca,* which nests in northern Europe and winters north of the equator in Africa, responded normally to conditions simulating 10° N, but there was no testis enlargement for conditions at the equator or at 20° S; for equatorial conditions, there was a molt of body feathers, but not of flight feathers; for conditions at 20° south of the equator, there was no response to the light regime.

Similar differences have been found between members of the same species. The blackcap, *Sylvia atricapilla,* is found throughout Europe, and there is also a population on the Canary Islands, off the west coast of Africa. Among members of a population from southern Germany, after *Zugunruhe* begins, it rises quickly to a peak of more than 8 h per night, and the level of activity gradually declines through 150 days. Among members of the Canary Islands population, the restlessness rises only to 2 h per night and disappears by the end of 50 days. Hybrids between members of

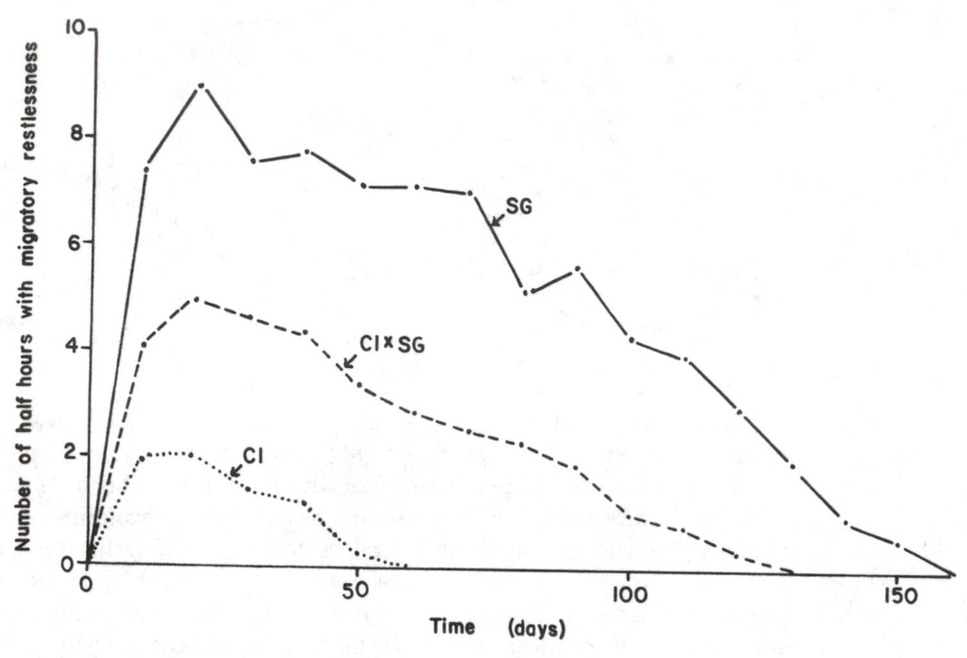

Figure 16.4. Time courses of nocturnal migratory activity in groups of hand-raised blackcaps from two populations (SG, southern Germany, CI, Canary Islands, and their hybrids). (Redrawn from Berthold 1990.)

the two populations are almost exactly intermediate in the level and duration of *Zugunruhe* (Figure 16.4).

Energetics of migration

The responses by birds to increases in daylength function through the endocrine system. The extra light stimulates either the eye or other light sensors deep in the head and causes an increase in the function of the pituitary. This seems to be controlled by the nearby hypothalamus, which is involved in sleep and wakefulness, and the extra light keeps the bird awake longer and longer. The result is the secretion of a prolactin-releasing substance. The prolactin induces migratory restlessness and fat deposition. Many experiments have also shown that the gonads increase in size with an increase of light.

The need for large reserves of energy is obvious, as many species must fly 40–60 h nonstop on their migrations. The fat deposition that has been mentioned obviously serves that function. The present question concerns how the fat is produced, as it must require a large amount of food or a considerable increase in the conversion of food into fat. Both an increase in food intake, called hyperphagia ("overeating") in the jargon of migration literature, and an increase in the efficiency of assimilation have been demonstrated experimentally among captive birds. The extra feeding is better documented, food intake having been shown to increase by up to 40% above normal shortly before migration. It has also been observed that many insectivorous birds change to eating fruits at the same time. In the southeastern United States, both tree swallows, *Iridoprocne bicolor,* and yellow-rumped warblers, *Dendroica coronata,* feed extensively on berries during the winter. In fact, the consumption of myrtle berries by the warbler was responsible for its original common name, myrtle warbler. Captive garden warblers, *Sylvia borin,* shifted their diets under conditions where both insect larvae and berries were provided in amounts of as much as they could eat. The berries were consumed in greatest quantity in December and January, the birds consuming the least amount of insect larvae in December. The extra carbohydrates in fruits are thought to be more readily convertible to fat than is the normal diet. It is also true that for fall migration, at least, fruits are available in greatly increased quantities. Some species have been shown to be selective among kinds of fruit. That is true of the barnacle goose, *Branta bernicola.* There is, however, some opportunism involved. Part of the population of golden plovers, *Pluvialis dominica,* flies east to Nova Scotia at the beginning of fall migration. There they fatten on blueberries before their nonstop flight to South America. The western population flies first to the Pribilof Islands, where they fatten on the maggots of flies that breed on the carcasses of fur seals left to rot each year after the fur harvest. From there they fly to Hawaii, another long nonstop flight over water.

Little appears to be known about the sources of food for premigratory fattening in preparation for spring migration. It is unlikely that fruits would be so conveniently available in quantity. Here, it is necessary to point out a discrepancy between the facts concerning migratory phenomena and the widely held ecological theory that bird populations are limited by their food supply. If the latter is true, it remains a mystery how the birds find the extra food to increase their intake by 40% within a few days. It is possible that it is only during the nesting period that food supplies limit bird populations. The food for fall migration could be available on that hypothesis, but finding the extra food in the tropics before departure on spring migration remains a problem. We have seen that there is some evidence of competition with the tropical residents. No ecologist has attempted to reconcile these conflicting facts.

Despite the problem for some ornithologists, food intake is increased and fat does accumulate, often within 4–10 days, at a rate of 0.1–1.5 g/day. Fat makes up a surprising proportion of the live weight of the bird: 30–47% for small birds that migrate long distances, 13–25% for small birds that migrate shorter distances, and 20% for large shorebirds. As the last are among the species that have the longest nonstop flights, the smaller proportion of fat is puzzling, until we realize that there is a complex interaction among body weight, fat supply, and flight velocity. The efficiency with which a long nonstop flight can be made depends on both endurance and flight speed, and the speed depends on the power/weight ratio, which favors larger birds, as it decreases slowly with weight from small to large birds.

The energy required for migratory flights has been calculated: 0.1 kcal g^{-1} h^{-1} in flight, where g is the mass of the bird in grams. Knowing the sizes of the birds and the speeds at which they fly, the following flight ranges can be calculated:

Small birds, 2,500 km (1,500 miles)
Shorebirds, 10,000 km (6,000 miles)
Hummingbirds, 1,000 km (610 miles)

Presumably, the small birds that take the Atlantic route to the West Indies have sufficient tailwinds over most of their route to hold the elapsed time within the energetically set limit. Thus, under normal conditions, the birds have an ample supply of energy. The ruby-throated hummingbird, *Archilochus colubris,* flies nonstop from the Yucatan Peninsula to Louisiana, a distance of 900 km, or less than 600 miles. That does not leave a large margin for error, unless there is a favoring wind. Many birds of all types arrive at their destinations with the entire fat supply exhausted, indicating that they have consumed all of their reserves.

It has been suggested that evaporative water loss through respiration (a common problem for marathon runners, for example) would be the limiting factor for long flights. Examination of birds at the end of migratory flights, however, has failed to demonstrate dehydration, despite near exhaustion of fat.

References and suggested further reading

Berthold, P. 1975. Migration: Control and metabolic physiology. Pp. 77–128 in *Avian Biology,* Vol. 5, D. Farner & J. King (eds.). New York, Academic Press.

Berthold, P. 1990. Genetics of migration. Pp. 269–80 in *Bird Migration: Physiology and Ecophysiology,* E. Gwinner (ed.). Berlin, Springer-Verlag.

Emlen, S. T. 1969. Bird migration: Influence of physiological state upon celestial orientation. *Science* 165:716–18.

Gwinner, E. 1986. Internal rhythms in bird migration. *Scientific American* 254(4):84–92.

Gwinner, E. (ed.). 1990. *Bird Migration: Physiology and Ecophysiology.* Berlin, Springer-Verlag.

Lincoln, F. C. 1939. *The Migration of American Birds.* New York, Doubleday, Doran.

PART

VI

Vertebrate behavior

Watching animals in the field can be fascinating. The same is true of listening to them. If our sense of smell were as acute as that of most mammals, we could add another dimension to our observations of vertebrate behavior.

The reasons for many different forms of behavior seem to be so obvious that it is surprising to find that we are mistaken in some instances. For example, the romantic opinion that male birds sing to attract and hold mates is only half of the truth. Another important function of the song is to establish and maintain possession of a territory that is in the proper habitat for a nest site and is large enough to support foraging for a brood of young.

If we are mistaken about some forms of behavior that seem obvious to us, those that are far outside our own experience can be truly mystifying. The function of courtship is to secure a mate and thus assure the courting partners of descendants. How, then, can we account for males that court in groups, when only a small fraction of them actually copulate?

The three chapters in this Part VI describe increasingly complex forms of behavior. Chapter 17 concerns behavior that is carried out by individual animals acting independently of others. This involves the apparently simple functions of feeding and avoiding predation. Chapter 18 covers behavioral interactions between two individuals. As these are rarely reciprocal, we need to understand the interactions in terms of the different behaviors of both individuals. Courtship and mating, parental care, territoriality, and predator–prey behaviors are covered. Chapter 19 describes the behavior of vertebrates in groups – in most cases the most complex forms of behavior known.

At all three levels of complexity we seek the immediate reason for the behavior, frequently termed the *proximate* or *physiological cause,* but that can be frustrating without consideration of the benefit of the behavior to the animal or animals involved in terms of the number of descendants they leave, termed the *ultimate* or *evolutionary cause.* Without attempting to understand the evolution of behavior, it makes no more sense than any other part of biology does without a consideration of evolution. Some biochemists and molecular biologists consider their fields to be isolated from evolutionary phenomena, an exceptionally shortsighted view. Discoveries in those fields have had major influences in helping other biologists understand evolutionary phenomena, and reciprocal impacts have been numerous. For example, without evolution, unraveling the almost contradictory facts of immunity would have been virtually impossible. The evolutionary advantage to animals of being able to distinguish between their own proteins and those of an invading organism provided the selective pressure favoring proteins that could neutralize the invaders without harming their own bodies. That, in turn, favored a genetic mechanism that could produce proteins capable of recognizing millions of different chemicals, without requiring millions of different coexisting genes. This phenomenon challenged the dogmas of molecular biology, and the necessary mechanism was eventually found.

As with other parts of science, understanding behavior requires a mixture of observation, experiment, and theory. None of them can stand alone, and each has advantages and limitations. Direct observations, preferably under field conditions, provide the basis of the questions that we raise and of the hypotheses that are proposed in our attempts to answer these questions. In order to formulate acceptable hypotheses, we need a foundation in theory, and to determine whether our hypotheses are correct or false, we must test them by using them to predict something that is not already known. That can be hazardous, and frequently, in both behavior and ecology, previously available knowledge has been used to claim an accurate prediction, which thus involves circular reasoning. Of course, if the prediction were made in ignorance of the existing knowledge, that would provide a valid test, but the surest test of a prediction is of the outcome of an experiment that has not been conducted. Unfortunately, experiments on behavior carry the risk that the result may be determined or at least influenced by the artificial conditions of the experiment itself.

These difficulties, plus a natural facility with mathematics, have led some biologists to concentrate on theories of what animals should be expected to do. Expressed in mathematical terms, these theories are appealingly precise. The approach has achieved a high prestige among behaviorists and ecologists alike, many of whom stand in awe of mathematics, in which many of them have not been adequately trained. The validity of the conclusions reached when applying mathematics to biological questions depends not only on the correctness of the mathematical manipulations themselves but also on the *accuracy and the completeness of the initial assumptions* that were made in formulating the questions. Often assumptions are left unstated, and many assumptions are made in the interest of keeping the mathematics tractable for an analytical solution to the problem. Analytical solutions, in contrast to simulations carried out with computers, are generally considered to be elegant and carry the highest prestige for the mathematical theorist. They tend to be general in application, but frequently lack realism or precision. It is not possible to maximize all three desirable properties of a theory simultaneously, as emphasized by Levins (1968).

As examples of the kinds of questionable assumptions just discussed, we consider two independent studies of a favorite topic in behavioral ecology. As described in Chapter 17, choice of food is an important behavioral trait, and the reasons given for certain choices include the argument that food is chosen as an evolutionary result of competition, leading to less overlap with the diet of a competitor. The catchphrase for this process is *niche partitioning*. Only rarely is the alternative hypothesis suggested: that restricting the diet to reduce competition can be disadvantageous if the different food items fluctuate in abundance and the species is left specializing on a declining resource. On the contrary, specialization could be an advantage if all items remain abundant. If they remain abundant, however, there will be no competition for them.

Two papers, both published in 1976, are most frequently cited as proving mathematically that niche partitioning is the natural result of competition between two species for two food resources. In one of them, Roughgarden (1976) states the assumptions clearly enough for the reader to realize that he has assumed that niche partitioning has already taken place, and the mathematical manipulations start there and proceed to "prove" a conclusion that is circular as far as the present discussion is concerned. The second paper (Lawlor and Maynard Smith 1976) avoids such circularity, but assumes that no other food is used and that no other factors determine the abundances of the two competing species. Therefore, a most unnatural initial situation is proposed. No experiment has demonstrated competition between two species for only two resource species. As a matter of fact, another theory, optimal foraging theory, had already proposed that specialization for such a restricted diet could evolve only when the food species was superabundant. A second important assumption that is critical for the conclusion reached by Lawlor and Maynard Smith is that the changes in abundance of the two resource species are determined solely by their densities, the number of consumers, and the rates of consumption. No influences external to the assumed system are allowed to affect the abundances of the resource species. It is further assumed that the two resource species do not compete, nor do they evolve. Finally, for the mathematical theory to work, there must be an initial difference between the two competing species in the relative rates at which they capture the resource species. The net result of these assumptions is that whatever the result of the mathematical manipulations, the situation postulated is too unrealistic to be taken seriously, but such is the prestige of mathematical theory that the conclusions of that study have been accepted unquestioningly by many behavioral ecologists. This should not be taken as completely invalidating the principle of niche partitioning, but is meant to show the influence of theories based on incomplete or invalid assumptions. Nor is it a plea to abandon mathematically based theory, but an exhortation to mathematicians and biologists alike to examine carefully the assumptions made in the construction of a theory.

From this discussion it can be seen that understanding animal behavior is difficult and requires knowledge of what is happening under natural conditions, carefully considered explanations of those happenings, and rigorous testing of the explanations with predictions, including, especially, well-designed experiments.

References and suggested further reading

Lawlor, L. R., & J. Maynard Smith. 1976. The coevolution and stability of competing species. *American Naturalist* 110:79–99.

Levins, R. 1968. *Evolution in Changing Environments*. Princeton University Press.

Roughgarden, J. 1976. Resource partitioning among competing species: A coevolutionary approach. *Theoretical Population Biology* 9:388–424.

17

Independent behavior of individuals

From the standpoint of behavior, many vertebrates act independently of other organisms in feeding and in acts promoting their survival, even though such acts clearly are intimately involved as part of the ecological interactions of their communities and, from the ecological standpoint, are of concern to many other species and individuals.

Feeding behavior

Feeding behavior of animals when several kinds of food are superabundant involves one major requirement: the ability to select the most appropriate food, namely, that for which each is specialized. Two examples were described in Chapter 11: the giant panda and the koala. The giant panda, *Ailuropoda melanoleuca*, feeds only on bamboo, and the koala, *Phascolarctus cinereus*, feeds only on one of the hundreds of species of eucalyptus, except when that species is temporarily unsuitable, when a few other species are acceptable. A very delicate sense of smell or taste, or both, must be the principal mechanism involved. Neither species can be kept alive on any species of plant other than the one or the few for which it is specialized.

A species equally specialized for a particular kind of feeding is the red crossbill, *Loxia curvirostra* (see Figure 1.19). Its peculiar beak is specialized for extracting seeds from the cones of spruce or white pine. As these trees set seeds at irregular intervals, the birds must be able to locate those stands where cones are available, and therefore they nest in different areas in successive years. Apparently they do this by haphazard search among different stands of appropriate trees. This is related to problems of optimal foraging that are described later, in that leaving a satisfactory area for feeding occurs when its profitability decreases.

For a great many species of vertebrates, including nearly all predators, granivores, nectarivores, and omnivores, no kind of suitable food is superabundant, and the abundances of the different kinds fluctuate both in time and in space. These species must be able to switch between kinds of food as the different kinds fluctuate, and also in many cases they must move from one locality ("patch") to another as these become depleted. The means by which they solve these problems is the subject of the optimal foraging theory (OFT). Long before this theory was worked out mathematically in the late 1960s and early 1970s, the whole problem was questioned as being irrelevant, because it was claimed that such species took food items in proportion to their abundance, and made no choices. The data on which that claim was based came from a study, published in 1932, in which were listed the contents of the stomachs of 80,000 North American birds. The tabulation took the form of the number of food items belonging to a particular family of prey, and that family's proportional representation among all identifications of members of the order or suborder to which that family belongs, compared with the proportion of all species in that family among the species of its suborder or order. That study was taken seriously as late as 1949, despite the fact that there were some major discrepancies between the proportions compared; for example, less than 1% of the 997 amphibians found in the stomachs belonged to the Family Plethodontidae, which comprises nearly 32% of the species of North American Amphibia. Thus, the original claim lacked validity for a number of families of animals, and the comparison itself was only faintly relevant to the question asked, as the correct comparison should have been between the proportions of individuals of a species found in the stomachs and the relative abundance of that species in the habitat searched. The taxonomic representation of the family is only remotely related to its relative abundance. Moreover, as different birds eat different kinds of prey, the food consumed by each species of bird should be tallied separately. We can therefore reject the claim that there was no selection by the birds among the different kinds of prey.

That they do in fact make such selections has been the finding in numerous studies, one of the best-known being an analysis of the different kinds of insects brought by great tits, *Parus major,* to their nestlings in a wood in the Netherlands. It was observed that when a given species was rare or uncommon, that species of insect was represented in the diet of the nestlings less frequently than expected from its abundance in the habitat. As its abundance increased during a normal fluctuation, it began to appear more commonly, and when it reached intermediate densities, it became more frequent in the diet than expected. As the abundance continued to increase, the rate of delivery to the young again fell behind the expected value. The original investigator reasoned that the birds were learning to find each kind of insect separately, a difficult task because of the similarity between the insects and their background. Therefore, they did not encounter rare kinds often enough to learn to find them, but as the rare kinds became more abundant, the birds encountered them frequently enough to learn their appearance. Having learned it, they acquired a "search image" for that particular prey and found it more frequently than would have been expected from its abundance. He supposed that the failure to keep up with the further increase was due to the advantage of a varied diet. The basic observations have been confirmed frequently, and a better theory with fewer assumptions was propounded. The failure to keep up with the increase was explained by the time required to "handle" each individual captured (e.g., killing, returning to the nest, and feeding it to the nestlings). That time is constant, and as the numbers of that kind of insect increased, a larger and larger proportion of the total time was spent in handling, rather than in searching for and capturing the prey. As most forest insects are not distributed randomly, but occur in relatively dense patches, and as different kinds tend to be in different patches, the relative profitability of searching in one patch as compared with another dictates that a moderately favorable patch should not be abandoned until another one becomes more profitable, thus accounting for the lag in taking a kind that is early in its increase in abundance.

This is but one specific example, and the reasons for different forms of feeding behavior are numerous, as detailed in OFT. The theory, as expressed mathematically in terms of choices that an individual animal should make to maximize its energetic intake, makes four predictions:

1. A species should reject items that would lower its overall energetic intake.
2. The greater the overall abundance of food, the more specialized the diet should be.
3. Prey types that individually would lower energetic intake should be ignored, regardless of their abundance.
4. Any given kind of food either should always be taken when found or should never be taken.

The theory on which these predictions are based has the basic premise that energy intake should always be maximized. The premise itself has been challenged on the grounds that not all populations are limited by their food supplies. Predator avoidance, for example, may be much more important than energy intake. In such cases, one should not expect OFT to apply. A major assumption of OFT up to the mid-1980s, including the formulation of the four predictions, was that the foraging animal had perfect knowledge about the energy content of each species of potential prey, its relative abundance, and its distribution in the habitat. As no forager has such perfect knowledge, it must acquire knowledge by sampling different prey items throughout the habitat, and because the abundances of the prey change continually, it must continue sampling. That means a continuing source of error in its knowledge. In recent years, further developments of the theory have made corrections that have helped observers make more accurate predictions, but the following analyses do not incorporate them.

How well does the theory model nature? A great many experiments have been carried out since the mid-1970s to test the predictions, but there is controversy over the interpretation of the results. Not surprisingly, the behavioral ecologists who contributed to the development of the theory find that the outcomes of a large majority of the experiments confirm it, at least "qualitatively." Others are less sanguine. Part of the disagreement stems from the failure of some experiments to meet all of the requirements of proper experimental procedure, in particular the requirement that the experiment's purpose be stated before its outcome is known. Any study conducted before the theory was first presented in complete form in 1974 was therefore automatically an a posteriori test of the theory. Among eighty-six papers analyzed for the period 1974–84, nineteen studies were properly designated a priori tests, and sixty-seven were put in the less exacting category of tests that were less than fully adequate. Among the correctly conducted experiments, seven were in the field, and twelve were in the laboratory. Many of the papers described tests of more than one of the predictions of the original form of OFT. Therefore, the total number of tests exceeds the number of papers cited. All eight proper tests of the first prediction in the field confirmed it; three of the four proper field tests of the second prediction confirmed it, and the fourth gave inconclusive data. None of the eight proper field tests of the third and fourth predictions confirmed them, one test giving an inconclusive result for prediction 3. Among twelve properly conducted laboratory studies, seven of eleven experiments confirmed prediction 1, five of seven confirmed prediction 2, and none of fourteen confirmed either prediction 3 or prediction 4; three of five gave inconclusive answers to prediction 3. Thus, as expected, predictions 1 and 2 fared well, and predictions 3 and 4 failed confirmation. The larger group of papers, those lacking a priori procedures, gave similar but less conclusive results, as follows: For field tests, twenty of thirty-six tests confirmed prediction 1, eight of twenty-two confirmed prediction 2, and none of twenty-seven confirmed either prediction 3 or prediction 4. Among tests conducted in the laboratory, nineteen of thirty-six confirmed prediction 1, six of twenty confirmed prediction 2, and none of thirty confirmed either prediction 3 or prediction 4.

Pooling all tests, regardless of quality or where conducted, 58.7% confirmed prediction 1, and 41.5% confirmed prediction 2. For all tests of all four predictions, 34% confirmed one prediction.

In fairness to OFT, it must be stated that the data just presented were tabulated by a skeptic. Turning to data compiled by a favorable reviewer, we find that no papers published after 1982 were included, but the later results were not importantly different from the earlier ones. As the skeptic had not considered different optimal foraging models, only part of the analysis can be used for comparison. Of thirty-three suitable tests, complete or partial agreement with optimal diet models was found for twenty-four (73%). The skeptic, interestingly, found confirmation with at least the first or second prediction of the models by twenty-three of thirty tests (77%). He did not regard the remaining three tests as suitable. Agreement between the assessments of the individual studies was good, but not perfect: The enthusiast claimed "qualitative agreement" or "consistent" results for three tests for which the skeptic claimed "indecisive" or "non-supporting" results; one paper was rated as "supporting" by the skeptic, but it was rated as not in agreement by the enthusiast.

Thus, the differences of opinion came not from different opinions of individual studies but from judging the success of any of the four predictions versus the success of each prediction. The choice of papers to be included was also an important factor.

A representative study: The European bee-eater

As an example of the tests of OFT and of the way in which progress is made in this field, I have chosen a study of the European bee-eater, *Merops apiaster.* These birds nest in southern Europe and migrate to Africa for the winter. They excavate nest holes 2 m deep in banks or level ground and feed by catching insects on the wing.

Early in the nesting period, and continuing through egg laying, the male brings insects to the female, and later both sexes bring insects to the nestlings and continue to feed the fledglings while they are still learning to forage for themselves. Returning to the female or nestlings with food increases the "handling" time for prey, which in this case must include both the time spent flying to capture insects and the time spent transferring the catch to the female (or to the young). To keep the profitability of prey (energy content per unit handling time) constant, the increased handling time for food brought to the female or young must be compensated by a higher energy content of prey than would otherwise be the case. In this study, an estimate of energy content was based on the size of the insects brought to the female. Those were larger than the insects eaten by the male himself. Fortunately, the difference between large prey (large dragonflies) and small prey (small dragonflies, horseflies, and honey bees) was easily detected by the observers. Depending on the colony being observed, all or a large proportion of the birds were marked individually. Thus, each foraging flight by a known male could be timed, as could the size of prey captured, and either the time required to transfer the prey to the female or the time in handling it for his own feeding. The difference between transfer time and handling time for each of two sizes of prey made the situation more complex than could be predicted by a simple extension of OFT, called central place foraging theory, which originally involved only the time to return to the nest in addition to time handling the prey.

In this example, the theory predicts that if handling time is greater than transfer time for a given size of prey, that size should always be given to the female; if transfer time is greater, the male should always eat that size of prey himself. Such behavior would maximize the energetic intake of food by both sexes, assuming that the two sizes of prey were handled and transferred, respectively, at different rates. Both handling and transfer times were quite different for the two sizes of prey.

From the statements of the theory and the information available, all large prey should have been given to the female, because the time required to transfer one large prey was only 60% as great as the time spent handling it for the male's own consumption. The original data (Table 17.1) show that the prediction was only 71% correct. For small prey, the difference between handling and transfer times was very small, and the prediction that the male should eat all small prey was only 41% correct. Thus, for the initial form of the theory, the predictions were fair for large prey but very poor for small prey.

Next, a refinement of the mathematical model was tested. The initial form had omitted consideration of the energetic maintenance cost for the male. Physiological experiments had already determined his energy requirements while resting and in flight. Measurements of the energy content of the two sizes of prey completed the necessary data. The new model allowed a sharing of the small prey, although a female monopoly of the large prey was still predicted. The theory was therefore improved (Table 17.2), but not completely satisfactory. It was suggested either that the physiological measurements were not sufficiently accurate or that the male had to gain weight, rather than simply meet his maintenance requirements, in anticipation of the demands of feeding the nestlings, and therefore consumed more than expected of both large and small prey.

Two other parts of this study tested different aspects of OFT. The first was foraging for nestlings. In that case, theory stated that the choice of large or small prey should depend on the relative abundances of the two kinds and the distance from the nest at which captures are attempted. The choice was between selecting large prey only and taking both sizes of prey in proportion to their abundance. Travel time to the areas where captures were made and back to the nest was the important variable, as pursuit, capture, handling, and transfer times were constant for each kind of prey. The longer the travel time, the relatively less important the

Table 17.1. *Courtship feeding of European bee-eaters: Numbers of large and small prey given to the female or eaten by the male*

	Large prey	Small prey
Given to female	112	662
Eaten by male	45	451
Percentage given to female	71	59

Source: Kamil, Krebs, & Pulliam (1987).

Table 17.2. *Mean handling time and transfer time (seconds) for European bee-eaters during courtship feeding*

	Handling	Transfer	Handling – Transfer
Large prey	16.6	10.1	+6.5
No. of observations	44	74	
Small prey	4.9	5.04	–0.1
No. of observations	444	551	

Source: Kamil, Krebs, & Pulliam (1987).

other time costs became. If the search area was distant, only large prey should be captured, because of the extra cost in travel time. The theory stated, however, that below a predictable threshold travel time, both prey sizes should be captured as they are encountered. As small prey were five to ten times as abundant as large ones, small prey should dominate the catches up to the threshold travel time, beyond which large prey should be selected preferentially. The small prey were five times as abundant as large prey in one year and ten times as abundant in the next year, and these were the predicted relative abundances of the two sizes of prey at short flight distances from the nest in the two years. Beyond 35 s of flight time in the first year and 40 s of flight time in the second year, there was a marked drop in the percentage of small prey brought to the nest, from 65–85% to 20%. Thus, although large prey were not taken 100% of the time beyond the predicted threshold, the birds foraged essentially as predicted by the theory.

The final test was of the theoretical allocation of large and small prey by both parents to fully fledged young. The differences between handling and transfer times for the two sizes of prey led to a prediction that all small prey should be fed to the young, and that large prey should be divided between parents and young. The model failed: 81% of large prey and 67% of the small prey were fed to the fledglings. It was suggested that the model failed because the mathematics did not take into account the necessity to add premigration fat, although it is difficult to understand why that should have altered the allocation of prey sizes, as both adults and young migrate.

Thus, in two of the three parts of this carefully planned and well-conducted study, the mathematical models gave poor predictions, although a modification of the model for courtship feeding did improve the predictions. In defense of the approach, the authors (Krebs et al. 1987, p. 188) commented as follows:

> While the idea of successful predictive models is clearly an ultimate goal to aim for, the optimality approach is recognized as playing an equally valuable role as a guide to the kinds of questions that might be asked of field data and suggesting the kinds of data that might be collected in the future. In this role, failures of a specific model's predictions are potentially just as interesting and creative as success.

Notice that no mention is made of possible experiments that could test the mathematical predictions in a truly a priori manner. There is no assurance in the description of the work that the observations were made after the predictions. This raises the question of circularity, which has been an ongoing problem for theoretical behavioral ecology ever since the preliminary work of the late 1960s. It remains to be seen whether or not modifications of the models can improve predictions generally. If they can, we should be alert to the probability that the modifications can restrict their applicability, as it is a well-accepted principle that models cannot simultaneously maximize realism, precision, and generality. Thus, the closer the model approximates the actual situation being used to test it, the less likely it will be applicable to other situations.

Predator avoidance

The survival of many animals in nature depends to a large extent on their ability to escape from predators. Many of the means by which they accomplish that are obvious, but some are not.

Running away

The most obvious escape is by running, flying, or swimming faster than the predator can. In practice, this method is normal only for large animals in the open, such as antelope on the African plains. As nearly all of them are faster than their predators over more than a minimal distance (20–30 m for lions, 100 m for cheetahs or wild dogs), they can be caught only when surprised at close range. Mountain hares (*Lepus timidus*) in Scotland browse on pioneer heather even though it is less nutritious than taller, more mature heather nearby, apparently thus making it easier to watch for predators, especially red foxes (*Vulpes vulpes*). Rabbits, however, use a combination of speed and subsequent concealment. Few birds or fishes are able to move fast enough to be successful with such a simple avoidance method. They use different methods, as described later, some in Chapter 19. Smaller mammals, such as squirrels, use rapid dodging rather than simple speed. Among flying vertebrates, such erratic flight has been called protean behavior. Its special advantage is said to lie in the difficulty the predator has in anticipating the next direction.

Taking cover

Many species of small birds can escape by diving into dense bushes, where the greater speed of an attacking hawk gives it no advantage, and brush piles are used by small mammals, lizards, and snakes in the same way. Many frogs escape terrestrial predators by jumping into the nearest body of water. Rooted vegetation in shallow water provides many species of fishes with cover from predatory species. Indeed, predatory bass, *Micropterus dolomieui* and *M. salmoides*, regularly patrol the open water at the edge of such vegetation, searching for prey that have left the cover.

Freezing

When predators are detected at such close range that the prey species do not have sufficient time or space to escape by other means, remaining perfectly still is a tactic that is used by many birds and mammals. In most such cases, they are specially colored or marked to match their background. A classic example is the coat of a spotted fawn, which is similar to the sunlight-dappled forest floor. Because most predators on vertebrates depend on movement as a primary means of locating prey, the tactic of freezing can be very effective. Bitterns, members of the heron family, point their beaks upward, and their streaked brown plumage helps to conceal them among the cattails and marsh grasses where they live. Pipefishes, relatives of the familiar sea horse, are well concealed among the eelgrass, owing to their elongate bodies and small size.

Using special defenses

Animals having protective shields, such as turtles, armadillos, pangolins (anteaterlike mammals covered with overlapping horny plates), or dangerous spines, such as porcupines and hedgehogs, have defensive behaviors that make full use of the protection. Turtles withdraw their heads, legs, and tails into their shells, whereas the mammals roll into balls in order to cover vulnerable parts and present the maximum amount of protective structures to the predator. Even so, such protection is not always effective. Raccoons are able to kill some turtles, and the fisher, a large weasellike mammal (*Martes pennanti*), kills porcupines by attacking the nose. The penalty for armored protection is some loss of mobility, but turtles are among the longest-lived vertebrates, and they have been in existence since early in the Mesozoic era, far longer than other reptiles, birds, or mammals, and probably longer than the living groups of Amphibia.

Skunks have a quite different form of protection: special glands under the tail that secrete noxious spray. This is so effective in repelling most potential predators that skunks are boldly marked in black and white, presumably a warning that most other species respect. When provoked, the spotted skunk, *Spilogale putorius*, stands on its forelegs and ejects the spray over its back in the direction it is facing. But even skunks are not completely immune from predation, as great horned owls, *Bubo virginianus*, feed on them regularly (Forbush 1927).

Confronting the predator

When cornered and unable to flee, some prey species will turn and fight. These are nearly all very large species, such as the African elephant, rhinoceroses, and buffalo. Many such species are also gregarious, and their group behaviors are described in Chapter 19.

Two special cases

From among many fascinating special behavioral adaptations used in avoiding predation, two examples are chosen.

Cowbird victims

The brown-headed cowbird, *Molothrus ater*, is a nest parasite of small birds: warblers, vireos, sparrows, and a number of other species. The female makes no nest, but like the European cuckoo, *Cuculus canorus*, lays its eggs in the nests of other species, always those smaller than itself. The timing of egg laying is close to the start of egg laying by the host, and the cowbird incubation period is normally shorter than that of the host, giving the early hatching young cowbird an advantage. Many birds give the food item that they have brought to the nest to the young that raises its head highest (ordinarily the hungriest), and the cowbird gets most of the food, frequently starving its nestmates. Most species of hosts appear not to notice the foreign egg, and incubate it along with their own. A few species, such as robins, catbirds, orioles, and yellow-breasted chats, will eject the cowbird egg from their nests. Some other species will build a new platform over the cowbird egg, sometimes covering one or more of their own eggs beneath the new floor, where, of course, they cannot hatch. Fourteen species have been recorded as having this behavior, some more regularly than others. The champion is the yellow warbler, *Dendroica petechia*, which will continue to cover cowbird eggs if they are laid

Figure 17.1. Outlines of cardboard models used to test the reactions of birds to birds of prey. Only the models with short necks (+) evoked escape reactions. (Redrawn from Tinbergen 1948.)

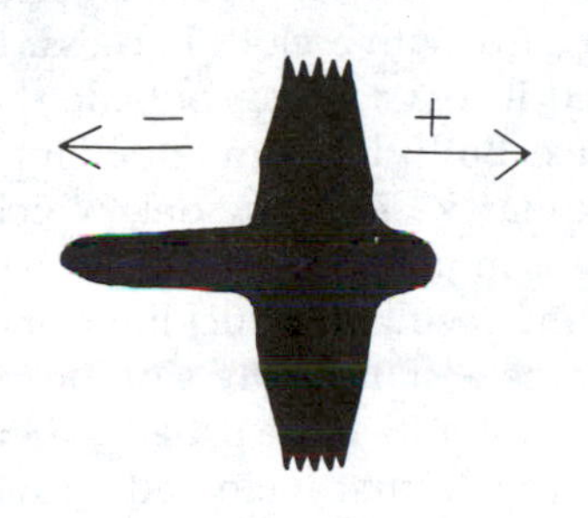

Figure 17.2. Outline of a cardboard dummy that evoked escape reactions when sailed to the right ("hawk"), but was ineffective when sailed to the left ("goose"). (Redrawn from Tinbergen 1948.)

after earlier ones have already been covered with an upper nest. The record number of stories for a yellow warbler nest is five.

The notorious hawk–goose model

Many birds that nest on the ground, including gallinaceous birds (grouse, quail, turkeys, etc.), most ducks, and shorebirds, have *precocial* young, which hatch in an advanced state of development and within minutes are able to follow the parent and forage for themselves. The alternate kind of young, *altricial*, are characteristic of nearly all tree-nesting birds. They hatch in a helpless state and must be fed by their parents. Precocial young also have the ability to identify and react to hawks flying overhead. The first experiments were inspired by casual observations of the behavior of domestic chickens to the sight of a hawk. Those experiments were conducted on chicks of capercaillie (*Tetrao urogallus*), a large European grouselike bird, ptarmigan (*Lagopus lagopus*), an Arctic grouse, and turkey (*Meleagris gallopavo*). Naive birds of those species were exposed to cardboard silhouettes of various objects, including geometric shapes and those of various birds in flight (Figure 17.1). "Escape reactions" occurred when the shapes of hawks were shown overhead, but not for other shapes. The escape reactions were either to run away or to crouch motionless when the hawk silhouette appeared. Further experiments narrowed the necessary stimulus to a short neck, as shown in Figure 17.1, and that was further refined. One such test was to "fly" the same silhouette overhead, both forward and backward. In the forward direction, the neck is short and the tail long, like those of a hawk (Figure 17.2); in the other direction, the neck is long and the tail short, like those of a goose. The chicks were reported to react to the "hawk," but not to the "goose." As the chicks had never encountered predators, it was claimed that the behavior was innate, not learned from experience with parents' alarm calls in the presence of predators.

Attempts to repeat those experiments were not always successful, and the experiments themselves were criticized as lacking quantitative measures of the "escape response," as well as not controlling for prior experience of the chicks with flying birds.

Clearly, the interpretation of the birds' behavior depended on the observer's subjective impression. Subjectivity is difficult to avoid in science. Even when we are aware of the problem, we cannot always anticipate our own perceptions, and that is especially difficult in making observations of behavior. It was suggested that those who expected the experiment to give a positive result "saw" the turkeys crouch, whereas the skeptics did not. In most accounts of the controversy, the importance of innateness is not stressed sufficiently. Only a little experience is needed for chicks to become highly selective in their alarm responses. The issue is important because this form of innate recognition of a stimulus may be the only documented case for a configurational stimulus, one that depends on a *pattern* in the responses of sensory cells.

Eventually, the subjectivity was eliminated by the use of an electronic device attached to the young bird with a cloth harness. It monitored the heartbeat and radioed it to a receiver outside the arena where the experiment took place. Both chicks and ducklings thus equipped were exposed to the hawk–goose model, which was towed overhead in both directions in random order. Analysis of the *rate* of heartbeat gave conflicting results for ducklings and chicks, but for each of them there was a statistically significant increase in the amount of *variation* in the rate of heartbeat when the "hawk" was presented. The heart rate first increased above normal, and then decreased below normal, which is what happens with other cases of alerting or alarm. The criticisms of the earlier experiments were avoided, but in eliminating the subjectivity of the observations, the possible survival value of the observed effect was somewhat obscured. Behavior reflecting the survival value might be detected by monitoring the experiment with video equipment and analyzing the tape in some prearranged manner.

Summary

Two aspects of the behaviors of individual vertebrates that are most important in their lives have to do with obtaining food and avoiding being preyed upon. Foraging is described with relevance to optimal foraging theory (OFT), the dominant theoretical approach to the subject. In a careful study of the European bee-eater, OFT made one quite accurate prediction, failed for a second, and was only partly accurate for a third, which was the most detailed part. The various forms of behavior that vertebrates use to avoid predation are running away, taking cover, freezing in place, confronting the predator, and taking measures that are appropriate to special defense mechanisms, such as shells, spines, and offensive odors. Two special behavioral adaptations have been described: frustrating the cowbird nest parasitism by ejecting the foreign egg or covering it with a new floor to the nest, and the innate reaction of some precocial birds to the appearance of a hawk overhead.

References and suggested further reading

Forbush, E. H. 1927. *Birds of Massachusetts and Other New England States,* Vol. 2. Commonwealth of Massachusetts.

Grey, R. D. 1987. Faith and foraging: A critique of the "paradigm argument from design." Pp. 69–142 in *Foraging Behavior,* A. C. Kamil, J. R. Krebs, & H. R. Pulliam (eds.). New York, Plenum Press.

Kamil, A. C., J. R. Krebs, & H. R. Pulliam (eds.). 1987. *Foraging Behavior.* New York, Plenum Press.

Krebs, J. R., M. I. Avery, & A. I. Houston. 1987. Delivering food to a central place: Three studies of Bee-eaters *Merops apiaster.* Pp. 173–91 in *Foraging Behavior,* A. C. Kamil, J. R. Krebs, & H. R. Pulliam (eds.). New York, Plenum Press.

Morse, D. H. 1980. *Behavioral Mechanisms in Ecology.* Cambridge, Mass., Harvard University Press.

Mueller, H. C., & P. G. Parker. 1980. Naive ducklings show different cardiac response to hawk than goose models. *Behaviour* 74:101–12.

Schoener, T. W. 1987. A brief history of foraging ecology. Pp. 5–67 in *Foraging Behavior,* A. C. Kamil, J. R. Krebs, & H. R. Pulliam (eds.). New York, Plenum Press.

Tinbergen, N. 1948. Social releasers and the experimental method required for their study. *Wilson Bulletin* 60:6–51.

18

Behaviors of two individuals interacting

Much behavior involves interactions between two individuals. In some instances the two individuals behave in much the same way, as at the boundaries of territories; in others, as in mating behavior, parental feeding of young, or parental defense of young, the two individuals behave very differently.

Territorial behavior

Many animals remain within a relatively small area throughout most of their lives. Some of the possible advantages are obvious: Knowing the best places to find food and knowing where the nearest escape route is at all times are the best examples. Such an area is a *home range*. To be a *territory*, the area must be defended against other individuals, usually of the same species, and almost always of the same sex. The territory is defended because it contains one or more resources that are essential to the territory holder. The behavior involved in defending territories is rarely actual fighting, although some initial fighting to establish a territory may be involved. Much more usual is some form of advertisement that the area is occupied. The advertisement takes the form of a display – visual, auditory, or olfactory – and may include the use of combined visual and auditory displays.

Auditory displays

Most familiar are bird songs. These often call attention to the presence of the territory holder long before it can be seen. The individual characteristics of the song serve both the singer and the hearer. In some species of woodland songbirds there are many small differences among the songs sung by different individuals. These differences permit the birds to recognize each other, and after the territories have been set, the familiar song of a neighbor elicits little response from a territory holder, a phenomenon called dear-enemy recognition. An unfamiliar song from a member of the same species stimulates an immediate trip to the vicinity of the boundary of the territory. The recognition of a neighbor's song is not simply a matter of "familiar versus unfamiliar." In a series of experiments with hooded warblers, *Wilsonia citrina*, tape recordings of the songs of different individuals were played near boundaries with neighboring territories. When played from the familiar direction, they caused no special reaction. In contrast, the same familiar song, played from a direction different from that of the neighbor's territory, elicited a strong response, much the same response as was evoked by the song of an unfamiliar hooded warbler. Such territorial confrontations involve visual displays of the brightly colored and boldly marked males. Chasing and fighting can follow as a final resort. That nearly always results in the territory holder winning, probably because it is familiar with the area, and also because it has more to lose than the intruder.

Birds are not alone in advertising their territories by vocal means. The tokay, *Gekko gekko* (see Figure 1.8), a lizard of the East Indies, utters loud cries at night and is physically aggressive to back up its voice. The name is undoubtedly onomatopoetic, as anyone who has heard it can attest.

That song deters intrusion into a territory has been demonstrated experimentally. When great tits, *Parus major*, were removed from their territories, the vacated space was promptly occupied by newcomers. If the owner's song was recorded and played in the territory after his removal, the territory remained unoccupied for 20–30 daylight hours. The same experiment with thrush nightingales, *Luscinia luscinia*, resulted in empty territories for the remainder of the nesting season.

Visual displays

Fishes and birds are most prominent among vertebrates in using visual displays to defend their territories. One of the best-studied is the three-spined stickleback fish, *Gasterosteus aculeatus*. These small fishes, which inhabit freshwater ditches in Europe and North America, are drab gray during the winter, but as mating season approaches, the males acquire bright red coloration, first on the throat and then on the belly. At the same time, they become strongly territorial and will display at and then attack any other individual or even a crude dummy that has a red underside, but will ignore even a realistic model of a male stickleback that does not have a red belly.

Male birds are famous for being brightly colored or boldly marked. Some of these colors are used in attracting mates, as described later, but they are also prominent in territorial display. The scarlet tanager, *Piranga olivacea*, probably the most brilliantly marked bird in North America, with its flaming red head and body, and black wings and tail, perches in the canopies of the tallest trees in its territory, where it sings loudly. Even birds that forage and nest near or on the ground, such as the bright red cardinal, *Cardinalis cardinalis,* and the boldly marked rufous-sided towhee, *Pipilo erythrophthalmus,* display from high perches.

Some birds use bright markings to indicate aggressiveness toward intruders. Red-winged blackbirds, *Agelaius phoeniceus,* erect their red shoulder patches and face any intruder. Experiments indicate that the red color itself contributes to keeping intruders out of the territory. Snow buntings, *Plectrophenax nivalis,* which are boldly marked in white and black, use their markings in a similar manner.

The lapwing, *Vanellus vanellus,* a conspicuously marked shorebird of Europe, displays in flight over its territory, as do a number of other open-country birds.

In addition to colors and patterns, size is used in visual displays, as size commonly goes with superior ability to eject intruders. Many vertebrates erect hair, feathers, or special appendages in display, thus giving the appearance of extra size. Flying lizards, *Draco bimaculatus,* of the Philippines erect crests and extend dewlaps in territorial confrontations (Figure 18.1).

The male of *Agama agama,* a large African lizard, is brightly colored, with a yellow head, blue body, and banded tail. They display continually at the borders of their territories. These large males are dominant over all other individuals of whatever age or sex in their territories, to the extent that none of the others have bright marks, being dull brownish gray. In fact, an observer has difficulty in finding them unless the dominant male is chased away, at which point there will suddenly be many small lizards moving about.

Chemical advertisement

Some vertebrates that are not as mobile as birds, but still require large areas, mark their territories with scents. Odors have the advantage of duration, which no other displays achieve, and thus compensate for the inability of the territory holder to move quickly to repel intruders. These scents are applied in different ways. The commonest is to mark the boundaries with feces or urine, as is practiced by wolves, other canids, some rodents, and at least one marsupial, the hairy-nosed wombat, *Lasiorhinus latifrons.* These large herbivores live in small colonies in open country, where they inhabit extensive burrow systems. The territory boundary is marked by defecating on top of large rocks. Like the feces of some other grazing mammals, they are dry and firm. Each scat or dropping has several flat surfaces, and thus remains in place, instead of rolling off of the rock.

Salamanders of the genus *Plethodon* are also territorial. They use

Figure 18.1. *Territorial behavior of the flying lizard,* Draco bimaculatus: *(a) lizard in normal posture on the trunk of a coconut palm; (b) the same lizard (note white mark on trunk) and a territorial invader in mutual display.*

a

b

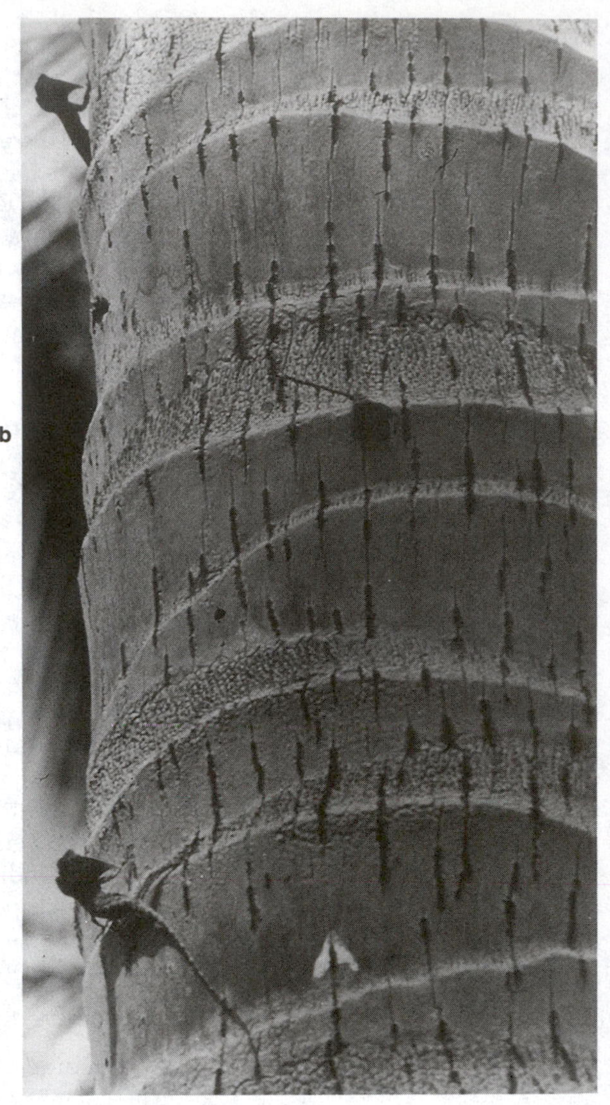

pheromones, which are secretions that carry messages, in several ways, one of which is to mark their territories. The secretions come from skin glands or from cloacal glands and are applied by pressing the chin, body, or cloacal opening to the substrate. It has been shown experimentally that *Plethodon cinereus*, the common red-backed salamander of eastern North America, is able to recognize neighboring individuals and to react to them differently from unfamiliar ones, in a way analogous to birds recognizing neighbors' songs.

Kinds of territories

Territories vary greatly according to the resource that is being defended. The type held by most land birds, and by many lizards and mammals, covers an area large enough to supply the food resources required to support a brood of young, in addition to a nest site. Small woodland birds commonly have territories ranging in size from 0.1 hectare to 4 or 5 hectares. (One hectare equals 2.47 acres.) Open-country birds have larger territories, ranging from the western meadowlark, *Sturnella neglecta,* 9 hectares, to the golden eagle, *Aquila chrysaetos,* 9,000 hectares. Although not as well documented for vertebrates as for desert spiders, it appears that each territory is large enough to provide adequate resources even in poor years. The sizes are not completely inflexible, as they vary with habitat for most species, but they are not completely dependent on population density, as in many species there are "floaters" – individuals that have no territory, but can occupy one if the owner is removed.

In the case of colonial seabirds, territories are small, as the defended resource is only the nest site itself. The value of the site comes from the fact that many seabirds nest on islands or cliffs because they are free of terrestrial predators. Examples include the territories of ring-billed gulls, *Larus delawarensis* (Figure 18.2). Other species may have even smaller ones, such that the area controlled is exactly that which the sitting bird can reach by pecking. Food for the birds comes from the surrounding sea, and it is not possible for one bird or a pair to hold a territory there, especially as the food moves from one part of the sea to another. The actual nest site is the only resource that is worth defending.

A third kind of territory is that held by male fur seals, *Callorhinus ursinus,* during the breeding season, when they arrive early on the beaches. The few that are large enough and strong enough to do so take possession of short stretches of beach. When the females arrive later, each territorial male takes control of and mates with as many as he can protect from neighboring males. The resource in this case is the females. The territory is the means by which he maintains control of them. Other large mammals with harems, such as the wapiti or American elk, *Cervus elaphus,* do not have territories, but keep the harem by chasing potential rivals away.

Some male animals have mating territories of a different type. They gather in groups, each one holding a small arena in which he displays. The group of territories is a *lek.* Females come to the lek, and most of them mate with one of a small fraction of the males holding mating territories. Leks are formed by a variety of vertebrates (and a few insects). Examples include such open-country birds as prairie grouse, plus some African antelopes, birds of paradise, the ruff (a European sandpiper), and the tropical American manakins. One of the best-studied is the sage grouse, *Centrocercus urophasianus*. These birds of the western plains of North America assemble in early spring at the same locations year after year. The males occupy small territories in the lek, where they display by inflating air sacs under conspicuous yellow pouches and white ruffs and spread their tails into impressive fans. They are able to mate with any female without interruption, unless they are too close to their boundary, in which case the male in the adjacent territory will attack and prevent the completion of copulation. It was discovered that there are individual differences among males in the pattern of white spots beneath the fanned-out tail (Figure 18.3), and that allowed the observer to follow the mating success of as many as twenty individual males in the same lek. Each lek has a "mating center," where virtually all of the copulations take place. The males holding territories at this center are thus so strongly favored that it seemed a problem how any individual not there could leave descendants. The investigator, R. H. Wiley, found that individuals progressed toward the mating center by occupying territories nearer the center than their own whenever one became vacant (survival among males is only about 50% per year). The youngest birds are those on territories farthest from the mating center. Thus, mating success depends on both survival and good fortune in having a male disappear whose territory is closer to the mating center.

As forest-inhabiting grouse species do not form leks, it appears that the advantage of lek formation lies in having many potential lookouts for predators while a large fraction of those gathered are distracted by displaying and mating. Perhaps those birds, both males and females, in the center of the lek are better warned of the approach of a predator, which would favor mating there.

Benefits and costs of territoriality

Most discussions of the advantages and disadvantages of territoriality have centered on the large multipurpose territories. The advantages that have been claimed are both direct and indirect. The direct advantages are as follows:

1. The owner has the exclusive use of an area for attracting a female and mating without interference from competitors.
2. The owner has the exclusive use of food resources near the nest site, sufficient to raise a brood of young.

The postulated indirect benefits, which are less easy to document, are as follows:

1. The uniform dispersion of territory holders makes searching by predators difficult, because most foraging predators continue to search in the immediate vicinity after having located one prey, thus remaining away from other territories.

Figure 18.2. *Three nests of ring-billed gulls,* Larus delawarensis, *to show the small territory of a seabird.*

Figure 18.3. *Displaying males of the sage grouse,* Centrocercus urophasianus, *from behind, to show distinctive patterns of the white spots. (Photographs by R. Haven Wiley.)*

2. The use of space, which does not fluctuate, rather than food, makes for population stability.

There appear to be three kinds of potential costs of maintaining a territory:

1. Advertising by the territory holder to keep competitors away can also attract the attention of predators.
2. There is some energetic cost to advertising and defending a territory, when the time could be spent feeding or resting.
3. The territory holder may have to fight a persistent intruder, running the risk of injury or death.

Mating behavior

Mating behavior consists of two parts for many vertebrates: attracting a mate, and performing the series of acts that must precede copulation. The first frequently is identical with territorial display, which thus serves a dual function. The second is the subject of this section.

In most species, the male is the more active partner, and the sometimes elaborate series of acts can have several functions. The first is to ensure that both individuals belong to the same species. This may seem trivial, but if a mistake is made, the result usually is a waste of the entire act, because if any offspring are produced, they are unlikely to be well adapted to the requirements of either parent, and thus unlikely to pass on the genes of either parent. The second function is to ensure that the two individuals are of opposite sexes. As will be shown, sometimes the male must begin his behavioral sequence, which then may elicit either an aggressive male behavior, or a different female response. The third function is to test the physiological readiness of the female to mate.

Being near another individual of the same or even larger size involves risk of attack. Therefore, the preliminary acts of the male are normally performed at some distance, although that is not possible where chemical communication is through the skin, as is the case with salamanders of the Family Plethodontidae, described in Chapter 1. The details must be different for different species, even when they are closely related. These can be minor differences in the behavioral acts, or they can be differences between the chemical signals that are used, as must be the case for *Plethodon jordani* and *P. glutinosus*, which have nearly identical behavior, so much so that in a few areas they interbreed successfully.

Lizards have complex and interesting behaviors, and as mentioned earlier, territorial and mating displays grade into one another. Male flying lizards of the genus *Draco*, as shown in Figure 18.1, extend their dewlaps and raise their nuchal crests when they encounter another individual. They also bob up and down on their forelegs, both actions being characteristic of lizards gener-

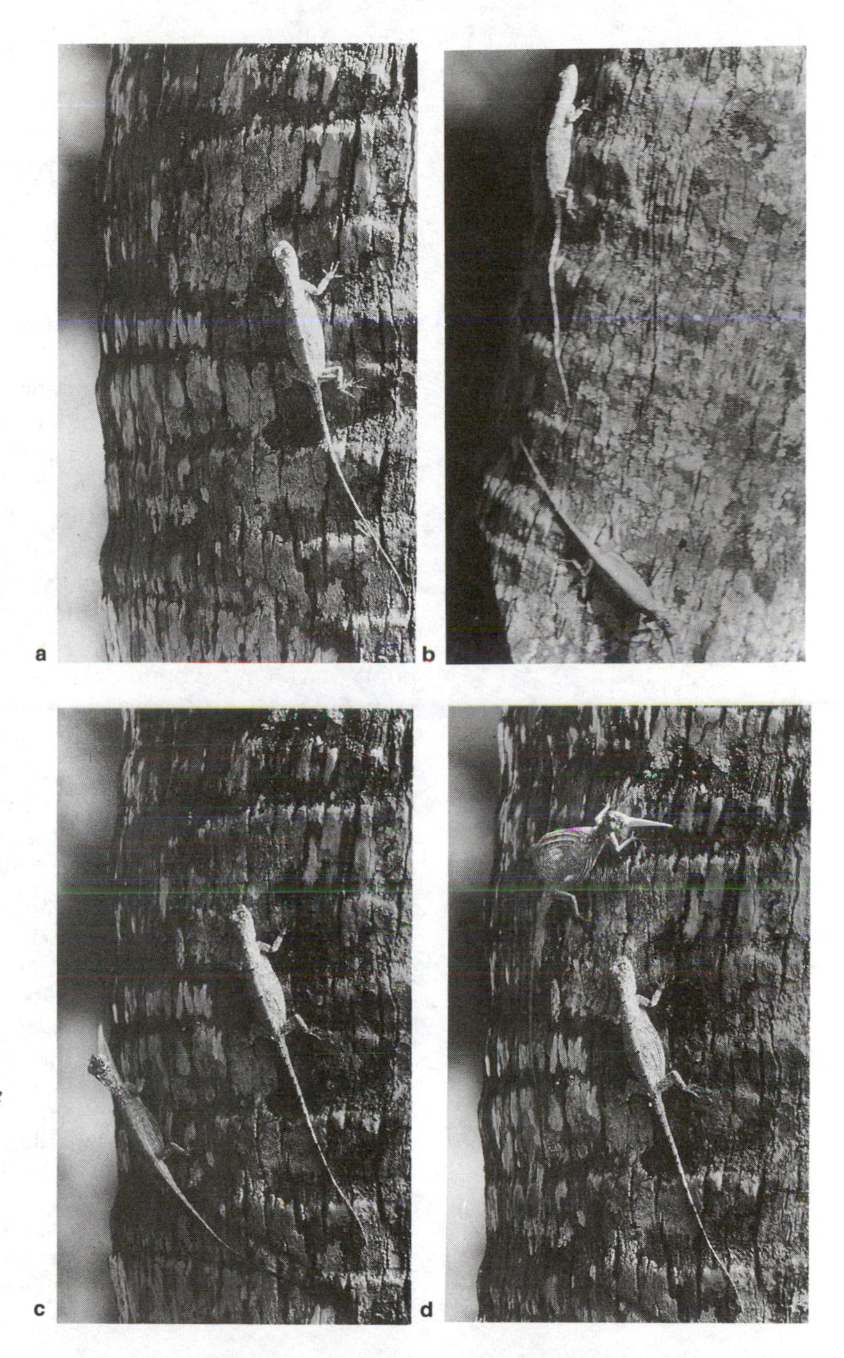

***Figure 18.4.** Mating behavior of the flying lizard,* Draco volans: *(a) a female alone on the trunk of a coconut palm; (b) the female in the same position being circled by a male; (c) the female with the male circling in the opposite direction, dewlap extended; (d) the male with partly opened wings.*

ally. *Draco volans* apparently is able to extend its dewlap directly forward, unlike *D. bimaculatus* (Figure 18.4). If the second individual does not respond, the male circles around it several times with his dewlap extended so that it looks like a beak. During this process, the male opens his wings, which are lateral flaps supported by the ribs. The photograph shows the wings only partly open.

When fully open, the leading edge is perpendicular to the side. During these observations, the pair was frightened, and actual mating was not observed. That the passive individual was a female was deduced by its failure to react to the approach of the other lizard. A male would have reacted as in Figure 18.1. In another lizard species, *Anolis carolinensis,* the female simply permits the male to mount and then bends her neck in a characteristic motion. Among birds and mammals, the female is a more active partner and shows behavior that is called soliciting or presenting. The female lion, for example, moves in front of the male, with her body held close to the ground.

The most elaborate mating behavior is found among the bowerbirds, Family Ptilonorhynchidae, of New Guinea and Australia. The name comes from the display "bowers," built by the male. These vary among species from simple platforms of twigs and moss to huge hutlike structures, 1 m high and 2 m across, constructed by a bird 26 cm long (about 11 inches). In all cases, the bower is decorated by the male with conspicuous objects, such as brightly colored flowers, yellow leaves, snail shells, and, when near humans, manmade objects, such as plastic tops from ballpoint pens. One experimenter tested the color preference of the male by making different plastic poker chips available. The male displays in front of and around the bower. The male satin bowerbird, *Ptilonorhynchus violaceus,* of Australia is a dark shiny blue when fully adult. His bower consists of two parallel walls of sticks, with a runway between them, in which he places an assortment of decorations, preferably blue. When a female approaches, the male adopts a rigid stance, with the neck outstretched and stiff and the tail fanned and cocked up. In that position, he hops stiff-legged around the bower with an ornament in his bill, uttering harsh cries. An even more elaborate display is by Archbold's bowerbird, *Archboldia papuensis,* of the mountains of New Guinea. According to one observer, at the approach of the female, the male

> flopped prostrate on the bed of ferns, began a low churring and started crawling toward the female now sitting on a low branch at the edge of the bower. While crawling, the male held his wings and tail partly open and pressed down against the ground, but the golden crest was folded flat along the crown and nape. His head was raised up and the bill, in which was often held a piece of vine, was pointed toward the female. The bill was opened and closed while the wings were fluttered. (Cooper 1977, p. 228)

Parental care

Incubation

At least some representatives of all major groups of vertebrates protect their developing eggs from various dangers, most frequently from predation, but also from such adverse factors as low temperatures and undernourishment. The simplest and commonest type of protection consists in guarding the eggs until they hatch. Although most egg-laying vertebrates simply deposit their eggs in a favorable location and depart, nest guarding is known for many fishes, most species of salamanders, a number of frogs, caecilians, alligators, a few snakes, nearly all birds, and the egg-laying mammal, the duck-billed platypus, *Ornithorhynchus*, but no turtles and only a few lizards. More elaborate care of the developing eggs involves placing them in a body cavity: the mouth, the vocal sacs of some frogs, folds of skin of other frogs, or simply entwining the strings of eggs around the legs of the male of the midwife toad *Alytes obstetricans*. Most remarkable is the gastric-breeding frog *Rheobatrachus* of southeastern Australia. The female swallows the eggs, which develop in her stomach, in which digestion is suspended until the larvae hatch and escape. Unfortunately, this fascinating species is near extinction. The egg of the other egg-laying mammal, the spiny anteater *Tachyglossus*, is laid directly into a pouch similar to those of marsupials.

In addition to nest guarding, nearly all birds add the incubation of the eggs by means of the parents' body heat. The exception is the Australian "brush turkeys" and their relatives, Family Megapodiidae. These birds build mounds of earth and decaying vegetable matter into which the eggs are laid. Incubation is by the heat generated by the decay process, and the young are completely independent of the adults as soon as they hatch. The American alligator, *Alligator mississippiensis*, builds a similar nest, and it may serve a similar function.

Retaining the eggs within the body of the mother can be considered a form of parental care. Some fishes, a few amphibians, a few lizards and snakes, no birds, and nearly all mammals are ovoviviparous (the eggs hatch before leaving the mother's body) or viviparous (the developing embryos are nourished by the mother as they develop).

Feeding the young

Young mammals receive milk, and some of them, especially Canidae (dogs), have food brought to their den by the parents, who regurgitate it. The most interesting and complex feeding behaviors, however, are found among birds.

Altricial young respond first to the vibration of the parent landing on the edge of the nest, and then to the outline of the parent. Their response is to raise their heads with their mouths open. The color and markings of the mouth provide the stimulus for the parent to put food into it. American robins, *Turdus migratorius*, for example, have red mouths with yellow borders. In some African finches, the young have red mouths with a specific pattern of black spots, and that pattern is necessary to stimulate the parent to

Figure 18.5. *Two nestlings of the ring-billed gull,* Larus delawarensis, *to show that the black markings on the heads are distinctively different.*

feed the young. A related species of finch is a nest parasite, like the cowbird, and it has evolved a similar pattern of black spots, mimicking the host and getting fed.

Among precocial birds, the feeding of gull chicks has been most thoroughly studied. The chicks have a set of black marks on the tops of their heads, and each chick is uniquely marked (Figure 18.5). For the first 5 days after hatching, the chicks of ring-billed gulls, *Larus delawarensis,* can be exchanged between nests without eliciting any negative behavior by the adults. After that, they will be rejected and attacked. An experiment was performed using a black felt-tipped pen to alter the marks of some chicks that were 12–20 days old. The parents also rejected them. A more radical change on herring gulls, *L. argentatus,* making the whole head black, did not elicit a negative reaction by the parents. Herring gulls recognize their chicks by their calls, unlike ring-billed gulls.

Herring gulls have yellow beaks, with a red spot near the tip, and it has been shown that the chicks react to the beak by pecking at the red spot. The pecking stimulates the parent to regurgitate the food that it has brought. Experiments showed that both yellow and red colors are necessary to elicit the pecking, and one might suppose that a chick's response would be greatest to an imitation beak that was as much like the original as it could be made; however, it was not the most realistic model that was most effective, but a longer and slenderer model with three red bands across it. Evolution, as might be expected on the basis of a little thought, cannot predict. Natural selection can act only on existing natural differences, and the different beak acted as a "supernormal stimulus."

Protecting the young

When vertebrates are with their young, many will react to predators in ways that are quite different from their reactions under other circumstances; they will risk their lives in attacks. I have been charged by a ruffed grouse, *Bonasa umbellus*, and birds of prey will attack humans who come too close to their nests. Owls are well known for that behavior, and sometimes inflict injury with their talons.

Other forms of defensive behavior are more specific to the protection of the young. Among birds that have precocial young, the adult feigns injury to itself, keeping just out of reach of the predator in the famous "broken wing act," as it leads the predator away from its young, which at the appropriate call note have remained motionless. The wing is dragged along the ground in a most realistic manner as the bird makes alarm calls. Even when the observer knows what is happening, it is difficult to resist the temptation to follow and find out if perhaps this time the wing really is broken.

Summary

There are three situations in which there are behavioral interactions between two individuals: territorial defense, mating displays, and parental care of young. Territorial displays, in which one animal shows ownership of a specific area, can be auditory, visual, or olfactory, and frequently there are combinations of visual and auditory displays. For some birds, there are variations in individual songs that permit the listener to identify the singer, and it has been shown that the listener's reaction depends on the direction from which the song comes. If it is from a familiar direction, there is little reaction; if not, the territory holder rushes to the border of his territory. Visual displays include showing bright colors or bold patterns, as well as specific behaviors. The display may include extending appendages, making the animal look larger, and presumably more dangerous. Chemical advertisement commonly takes the form of scent marking by urinating or defecating along the territorial boundary, as is common among members of the dog family. Some salamanders mark their territories with secretions from skin and cloacal glands. Besides the large all-purpose territories, there are smaller ones, including only the area immediately around the nest, as among island-nesting seabirds, those held by harem-holding fur seals, and the small areas for display and mating used by animals that group in leks during the mating season. Mating behavior commonly involves attracting the female and then carrying out some reassuring behavior to allay the fear that many animals exhibit when near another individual of similar size. These displays often begin in much the same way as territorial displays, proceeding to more elaborate acts if the partner does not react like an intruding male.

Male displays reach their pinnacle of elaborate behavior in the construction of bowers, some of them very large, for the sole function of serving as a place to carry out their very complex series of acts to induce a female to mate. Parental care involves both feeding and protecting the young. In addition to protecting their eggs during incubation, in common with many different groups of vertebrates, they must feed their young. There is mutual stimulation, the behavior of the young stimulating feeding by the adult, or the adult first stimulating the young, which in turn stimulate the adult. The signals back and forth include vibration of the nest by the adult landing on it, and color marks in the mouths of the young. Among species with precocial young, spot patterns on the heads of the young and color marks on the beaks of the parents are the stimuli. Adult birds frequently lure predators away from their young by feigning injury to themselves, or by feigning attack on the predator.

References and suggested further reading

Cooper, W. T. 1977. *The Birds of Paradise and Bower Birds* (text by J. M. Forshaw & W. T. Cooper). Sydney, William Collins.

Drickamer, L. C., & S. H. Vessey. 1992. *Animal Behavior. Mechanisms, Ecology, and Evolution.* (3rd ed.). New York, Wm. C. Brown.

Gould, J. L. 1982. *Ethology. The Mechanisms and Evolution of Behavior.* New York, Norton.

Hairston, N. G. 1957. Observations on the behavior of *Draco volans* in the Philippines. *Copeia* 1957:262–5.

Miller, D. E., & J. T. Emlen. 1975. Individual chick recognition and family integrity in the ring-billed gull. *Behaviour* 52:124–44.

Morse, D. L. 1980. *Behavioral Mechanisms in Ecology.* Cambridge, Mass., Harvard University Press.

Wiley, R. H. 1978. The lek mating system of the sage grouse. *Scientific American* 238(5):114–25.

19

Behavioral interactions in groups of vertebrates

The interactions between individuals usually are understandable in relation to reproduction. When animals interact in groups larger than the reproductive unit (family), their reason must be something other than the direct advantages of successful reproduction or of favoring the survival of direct descendants. Yet a great many animals do belong to groups of mutually interacting individuals, including a large variety of fishes, birds, and mammals. Assembling into groups involves risks, such as attracting the attention of predators more than would single animals, or risking aggression from being too near individuals not the mate. It is therefore necessary to examine group behaviors to discover their offsetting advantages. Several such potential advantages have been proposed.

Predator avoidance

The commonest proposal is that forming a group reduces the danger of being preyed upon. This can take three different forms. As mentioned in the description of lek behavior (Chapter 18), having many potential prey near each other means that many individuals are available to detect the approach of a predator. If the group is foraging, a small proportion of the animals can alert the whole group. As suggested for sage grouse in leks (Chapter 18), birds on the periphery of flocks in the open are more vulnerable and hence are more likely to be watching for predators than are those in the center. In a study of starlings, *Sturnus vulgaris,* in England, that expectation was confirmed (Figure 19.1). In small flocks of less than 50 individuals, the birds in the center spent between one-sixth and one-third of their time "vigilant," standing still with the head up and the beak above the horizontal. In flocks larger than 150, centrally placed birds spent only about 2% of their time vigilant. Peripheral birds spent between one-third and one-half of their time vigilant, regardless of the size of the flock.

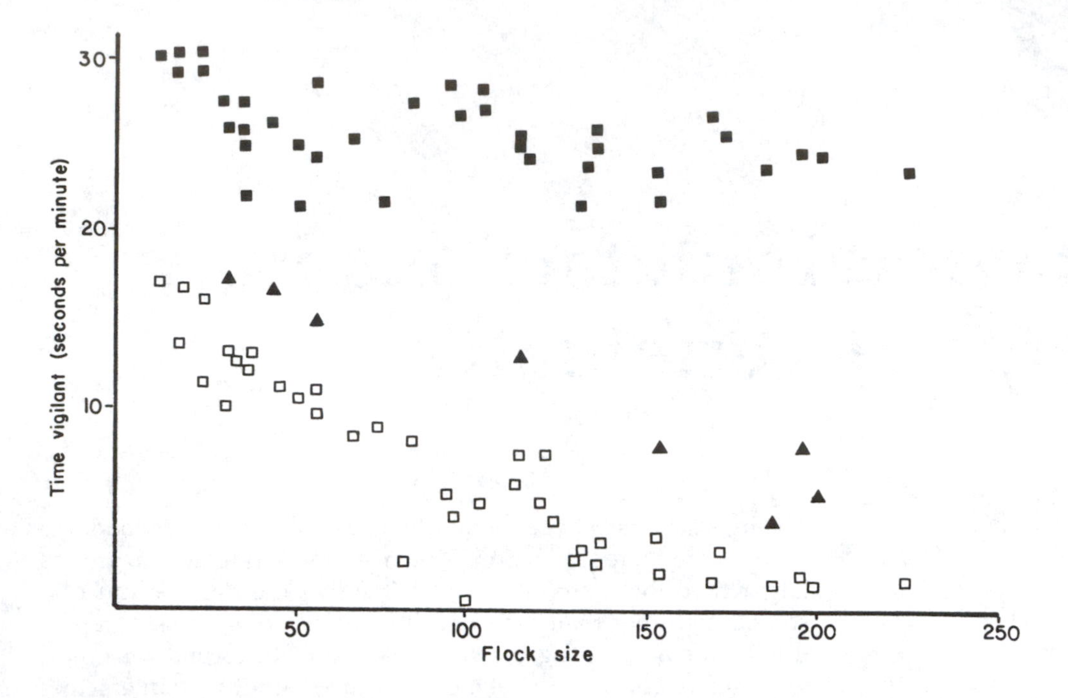

Figure 19.1. Relationships between flock size and time spent vigilant for central (open squares) and peripheral (filled squares) birds and some in midway positions (filled triangles). (Adapted from Jennings & Evans 1980.)

A second advantage of flocking or schooling to avoid predation is the safety of being in the group instead of being separate. It has been shown that hawks selectively attack single birds that get separated from the flock, and the same phenomenon has been observed among fish. Therefore, it appears that the flock or school itself is a place to hide. Flocks of shorebirds are especially interesting in their ability to keep together, even when turning quickly. Avocets, *Recurvirostra americana*, make a spectacular sight as they turn together. The stimulus to make the turn is obscure, whether from one individual or from something external to the flock. The evident advantage of being in the center of the flock has given rise to the term "You first." An interesting theoretical paper on the subject was Hamilton's "Geometry for the selfish herd," published in 1971. He postulated a circular pond around which frogs are sitting at random intervals. At some time during the day, a snake appears and eats the nearest frog. What should the frogs do before the snake appears? It is assumed that they fear terrestrial predators too much to leave the edge of the pond. Obviously, each frog should move into the smallest gap available, but if it does so, that will leave the one over which it has hopped in a vulnerable position. That frog should then take the same kind of action. This goes on all around the pond. A computer program predicted that the frogs would finish in a single pile, or occasionally two piles. Clearly, those in the middle of the pile would be safest.

A quite different approach that a group can take to avoid predation is mobbing. Small birds, commonly of several species, will respond to the first alarm calls by approaching the predator, continuing to call, and even swooping at it, sometimes actually striking it with beaks. Mobbing occurs only if the predator is perched. Frequently, the result of such harassment is that the predator will leave the vicinity. Cases have been recorded, however, in which a mobbing bird has been caught.

Many large herbivorous mammals living in open country form herds that are more organized than simple assemblages. When attacked, they go into coordinated defensive behavior. The most famous are musk-oxen of the tundra. Their only significant predators, other than human hunters, are wolves. When wolves approach a herd of musk-oxen, the latter form a circle facing outward, with the calves inside. Their horns make a formidable barrier, which the wolves can rarely penetrate.

Elephant herds consist of females and their young in various stages of growth. They have a specific behavior when threatened. Although at present only lions and humans can threaten them, during the Pleistocene the saber-toothed cats (now extinct) were specifically adapted to kill elephants. The adult females arrange themselves between the threat and the younger animals and show the preliminaries to aggressive acts: raised trunks and ears extended to the sides. As they retreat, the dominant female will be the last, and will charge, or at least feign a charge, as the group leaves. Males remain solitary or in loose groups elsewhere, except when the females are in estrus.

Baboons living in rich savannah forest form large social troops, dominated by groups of three or more large males. These males join in threatening or driving off potential predators, even leopards.

Probably the most complex form of predator avoidance by a group of vertebrates is that of beavers, *Castor canadensis*. These rodents live in family groups consisting of the parents and their 1-year and 2-year classes of young. The predators are deterred by the beavers' pond, which they create by damming a stream. Trees are cut down with the large incisor teeth and cut into pieces. Branches and small pieces of trunk are brought to the dam site and placed into position, with the butt ends downstream. Mud is then carried to the dam and forced in among the branches until the stream is stopped. The dam is then extended beyond the streambed to create a large pond. The dam can reach several hundred feet in length and 6 ft deep. On the bank of the original streambed, the beavers begin building a lodge by piling mud, branches, and small trunks in a solid mass to a height of 5 ft or more. They then dig entrances from below the water line toward the center of the pile, cutting off any branches encountered. An entrance chamber is dug just above the water level, where they will stop to let the water drain, thus keeping the higher level of the pile dry. They then dig the main chamber at that higher level. The outside of the structure is then sealed with more mud, and more branches and trunks may be added. During winter, when potential predators can approach over the ice, the lodge is frozen and unassailable.

Energetic benefits to vertebrates in groups

There have been several different suggestions that vertebrates in groups may have advantages in increasing their rates of obtaining food. There are two possibilities: improving their rate of finding food, which implies that individuals will be attracted to places where others are feeding, or taking the cue that the presence of others indicates an area that has already been searched and therefore should be avoided. The first suggestion has been supported by some laboratory experiments, in which the rate of finding food by great tits was significantly greater for birds in groups of four than for pairs, and greater for pairs than for single birds. An anecdotal observation also supports the suggestion that birds in groups can find food more readily. We maintain a feeder at our summer home in the mountains of North Carolina. Seeds of various kinds, and a cake made of corn meal, bacon drippings, peanut butter, and small seeds, are the only food provided. That birds are attracted to the location where other birds are feeding is shown by our observation of five different species of purely insectivorous birds perched on the feeder along with seed-eating species, but not attempting to feed: great crested flycatcher, *Myiarchus crinitus*, black-throated blue warbler, *Dendroica caerulescens*, black-throated green warbler, *D. virens*, Kentucky warbler, *Oporornis formosus*, and hooded warbler, *Wilsonia citrina*.

The suggestion that birds in a flock will key on the locations of others to avoid previously searched areas has received relatively little confirmation. The implication is that they should be spaced out as they forage, which would imply that they would forage along a broad front, which does not seem to happen. One example that seems to apply involves the behavior of blackbirds, grackles, and starlings when foraging in open fields. Flocks of hundreds or thousands in compact formations move along in one direction, with birds flying up over the rest to reach the front when they find themselves at the rear of the flock. The result is a continuous "rolling" appearance to the whole flock. It might be argued that they are moving from an exposed position, but avoiding an already searched area seems more plausible.

A third suggestion is that groups of birds flush more insects and other prey per bird than can individuals alone. There is little confirming evidence from observations of mixed-species flocks in the temperate zone in winter, but some species of tropical birds do seem to benefit, although only a minority of the species seem to gain.

An entirely different kind of energetic benefit comes from slipstreaming. It has been shown that the aerodynamics of the V-shaped flocks of geese and other large birds allow the birds that are following the leader to expend less energy in moving through the air. On a long migratory flight, the saving could make the difference between risking an extra stopover, with the danger of being preyed upon, and making the long flight nonstop. Fishes in schools likewise derive an energetic benefit from the hydrodynamics of the school. When moving rapidly, the schools tend to have an optimal teardrop shape, which is lost when they slow down to feed.

Like the organized herds of large herbivores in their defense against predators, organized groups of predators depend on interactions among the individuals. These make the predators' group behaviors both more complex and more interesting than those flocks, schools, and herds that derive benefits simply from being in an aggregation. Pack hunting, as practiced by wolves, African wild dogs, and hyenas, consists in running down prey, after close examination of the herd to select an individual that appears to be weaker than the rest, whether because of its youth, old age, or sickness. In wolf and dog packs, the leading animal is the first to attack, frequently seizing the victim by the nose or throat, after which the other pack members attack from other angles. The rump is another favored point of attack, being away from the hooves, which in the case of moose are formidable weapons of defense. Indeed, a moose standing at bay instead of running can nearly always discourage a pack of wolves from attacking. Only running animals are actually attacked. Among hyenas, the females dominate, and there appears to be no single member that is the first to attack. Lions practice group hunting, but in a different manner, one that is less organized than those of canids and hyenas. Rather than chasing their prey until it is exhausted, several female lions will stalk the prey until they are within a short distance. Moving slowly, and stopping whenever one of the prey herd raises its head, they will attempt to encircle a herd. Then, one or two will make a rush, scattering the herd. Should one of the herd flee in the direction of one of the other lions, a kill often ensues. Males do not participate in the hunt, but then they move in to claim the kill for themselves and their offspring, after which the females are permitted to feed. Lions also are strong enough to take a kill away from other predators. For that reason, leopards, which hunt singly, carry their kills up into trees.

Some aquatic birds also engage in group hunting. White pelicans, *Pelecanus erythrorhynchos,* form a line and with flapping wings move toward the shore. That drives the fishes into shallow water, where the pelicans are able to catch them. Double-crested cormorants, *Phalacrocorax auritus,* sometimes show similar behavior.

Origins of group behavior

There appear to be two quite different origins of groups that act in some form of concert. The first is simple aggregation, which turns out to be advantageous for one or more of the reasons already described. The second, which is more likely to give rise to complex forms of group behavior, arises through prolongation of parental care, without immediate dispersal of the grown young. A familiar example is the beavers, *Castor canadensis*. The young remain with their parents for 2 years before being forced out of the lodge. The social coordination of effort by the adults and two age classes of young has an impact on the environment that is far greater than could be exerted by an equivalent number of solitary mammals.

In the more elaborately organized groups of animals, individuals have distinct behavioral roles. The simplest form of this is the *dominance hierarchy* or "pecking order." In groups showing a pecking order, there is a rigidly maintained pattern of social dominance. This was first observed among the hens of domestic chickens: One hen is dominant to all of the others and can peck them without retaliation. A second hen can peck any but the dominant one; a third can peck any but the top two, and so on down to the hen that can be pecked by all of the others. The dominance quickly becomes recognized by all individuals, and actual pecking becomes rare, each bird yielding to all higher-ranking ones at feeding time, for example. Dominance hierarchies frequently develop in groups of normally territorial animals when they are confined in a space that is too small for territories to be held.

Group selection

The forms of behavior in groups that have already been described are beneficial to all members of the group. There are, however, a number of examples of behavioral and other traits that have been claimed to be harmful to the individual showing such a trait, but beneficial to the group to which it belongs in relation to other groups that lack any members with the trait. The proposal that genes associated with such traits can spread in a population is called *group selection*, and among evolutionists this has been a source of great contention. Group selection has been claimed to work because it is "for the good of the species," "for the good of the population," and other similar phrases. Not all of these traits involve behavior, but for the sake of completeness, those not involving behavior will be treated briefly here. Before taking them up, we must realize a few things about natural selection. The first has already been mentioned: *Natural selection is never predictive*. There is no evidence for such a phenomenon, nor is there any evidence that mutants or recombinants appear in response to an individual's needs or a population's needs. Thus, *the present environment selects among organisms that are present*. Those with favorable characteristics will leave more descendants, so that any genes associated with favorable characteristics tend to spread in a population. This process is called evolution by natural selection. The question we must consider is this: Can characteristics favorable for the population or the species spread even when they are not favorable for individuals belonging to the population or species?

The following traits have been claimed to fit this condition:

1. Many birds and mammals give warning signals at the approach of predators or other danger. Whitetail deer, *Odocoileus virginianus,* "flash" the undersides of their tails – a conspicuous act. Many birds have special calls that they give under similar circumstances. Invariably, these warning signals are conspicuous to us, and it seems reasonable that they would attract the attention of predators also. If that is the case, why should the individuals give signals that would seem to put them at risk, if not for the exclusive benefit of the group?

2. Territoriality often seems to involve defense of more area than is needed to raise the young. Thus, it is claimed, reproduction is restricted by individuals to avoid overpopulation by the group.

3. Epideictic behavior: It has been claimed that birds flying in flocks can be assessing their numbers, and regulate their reproduction accordingly, even though an individual that "cheated" (i.e., failed to do so) would be at a huge advantage in leaving descendants.

4. Helpers at the nest: In more than sixty species of birds it has been observed that whereas only one female in a group produces all of the eggs, there are one or more other adult birds that help in raising the young. These helpers do not contribute any genes to the next generation, and so genes associated with helping should lose out in natural selection.

5. Risking one's life to save that of an unrelated individual is clearly altruistic. The potential to lose all of such a person's genes should therefore be selected against, and yet the behavior is commonly observed. Can this behavior spread by natural selection?

6. Many species contain noxious chemicals that will repel or, if eaten, will sicken predators, which thereby learn not to prey on any animal resembling the one previously eaten. The first one eaten by the predator, therefore, is sacrificed for the good of the rest, although it cannot pass on its genes.

7. Social insects, among which most of the individuals are sterile, are donating their effort for the good of the group and cannot possibly pass on genes responsible for that behavior.

8. Sex: It is obvious arithmetically that species that are all-female have a numerical advantage over sexual species, because it takes only one instead of two parents to produce the same number of offspring. The usual argument is that sex provides for genetic recombination and thus for variability among the offspring, allowing at least some of them to cope with a changing environment, a possibility denied to parthenogenetic species. Thus, sex is a trait that favors the species over the individual, which gives up

half of its reproductive capacity to provide for a *predicted* changing environment.

9. In regard to genetic dominance, the argument is similar to that for sex. By masking true genetic variation under the dominant allele, adaptation to present environmental conditions can be preserved at the same time that some potential variation is retained against the possibility of future environmental change. Again, the advantage of dominance is in *predicting* future environmental change, instead of simply eliminating the less favored allele by natural selection.

These questions led to the development of sociobiology. Its originators pointed out the logical difficulties associated with claiming that any trait could be selected for and be truly altruistic, because any gene for such a trait would decrease in subsequent generations, as its possessor would leave fewer descendants than would other individuals that did not possess the gene. Sociobiologists have shown that the apparently altruistic traits just listed can be explained in ways that do not involve the logical difficulties.

1. Warning signals, while alerting other members of the signaler's group, for example, are given while the predator is sufficiently far away for the signaler to be safe from attack. I have noticed that when deer are surprised at close range, they keep their tails tucked against their bodies. It has been suggested that the deer are signaling to the predator that it has been seen and therefore its chances for a kill are poor. Any benefit to other members of the group would thus be incidental. It has been suggested that small birds' alarm calls are acoustically difficult to locate and therefore do not represent as much risk as might be supposed. That hypothesis has been contested, but experiments suggest that it may be valid.

2. As already described, territoriality is advantageous directly because it gives the owner exclusive use of a feeding and nesting area, thus inceasing the chances of success in raising a brood. Population regulation or stability would be a by-product of the behavior.

3. No evidence has been presented that flocking has any other basis than predator avoidance or confusion.

4. In nearly all cases, helpers at the nest are close relatives of the nestlings they are feeding. Thus, at least large percentages of their genes are held in common, so that raising relatives can spread genes associated with helping. Also, the helpers are learning the complex behavior that makes for successful nesting. Finally, in at least one case, the oldest helper "inherits" the territory if anything happens to the owner.

5. Heroic behavior can be interpreted as advantageous if the risk to the hero is less than the benefit to the person helped by the heroic act. Saving another person from drowning frequently qualifies. If, at some future time, the beneficiary is able to recipro-

cate in any way, both parties gain in the potential to have the "altruist" genes spread. The process of "reciprocal altruism" has been claimed to work widely in human societies, in part because a refusal to reciprocate can carry heavy social penalties.

6. Many species of insects that have noxious chemicals that make them repellent to predators are boldly marked, making learning easy for the predator. Moreover, many species remain together as they grow, so that a predator would learn to shun the siblings of the first one attacked, and they would be the ones most likely to share the noxious trait. In salamanders, the unpleasant secretions are concentrated in the tail; this frequently breaks off and continues to wiggle, attracting the attention of the predator, while the animal crawls away and lives to grow another tail.

7. The nonreproductive workers are sisters of the younger brood members whose feeding and other care they carry out. Thus, in their efforts to promote the welfare of the hive, nest, or other organization, they are effectively helping their own genes to spread to future generations.

8. Genetic dominance represents selection for dominance of the better-adapted allele through modifier genes – those that affect the expression of the locus in question. The recessive allele is thus prevented from being expressed in most individuals, and that is an immediate advantage that has nothing to do with future environmental change.

9. The process of meiosis, which is necessary for the formation of gametes in nearly all cases, involves the pairing of chromosomes and hence DNA. DNA is constantly under bombardment by radiation and thus is subject to deleterious mutational changes. When pairing occurs in meiosis, the damaged part of the DNA can be repaired by an enzyme that copies the undamaged partner. As the damaged part has a vanishingly small chance of being the same on both strands of DNA, the system obviously preserves the undamaged parts.

Although these arguments make group selection less necessary in explaining evolution, they do not exclude it altogether. It is necessary to discover the extent to which such a phenomenon is possible.

Requirements for group selection

The argument for group selection is always that the advantage to other members of the group by having at least one member engaged in altruistic behavior is sufficient to more than offset the loss of the reproductive capacity of the altruistic individual(s). It is clear, though, that if any genetic basis for altruistic behavior is to be passed to future generations, the gene must spread within the group by some means other than natural selection. Such a mechanism, known by the term "genetic drift," was proposed more than 60 years ago, when the mathematics of population genetics was founded. It is the chance loss of an allele when the population is small, even when the allele confers some advantage, or conversely the chance inclusion of the allele in the next generation, even though it may confer some disadvantage.

The smallest possible group of sexually reproducing organisms is one female and one male. If one of them is heterozygous for the allele in question, and the other is homozygous for the opposite allele, the odd allele has a very good chance of not being incorporated into either of the two zygotes that will make up the next generation. Each parent produces large numbers of gametes, and only two from each parent will be included. Chance determines which two, and the odd allele has only a 50% chance of being included in one of them. Conversely, the same chance process could result in its increasing from 25% to 50% of the alleles in the next generation. It is easy to see that a larger population will be less likely to lose an allele by chance, because more zygotes will make up the next generation, and chance will determine the composition of each one, independently of the composition of the others.

It is easy to understand that if some natural event were to reduce a population to very low numbers, the total number of different alleles in that population would also be greatly reduced, except for the impossibly small chance that all individuals would be the same genetically. Such a natural event might be the isolation of a small part of the population from the rest, as when colonizing a small island, or a disaster that would wipe out most of the individuals. The phenomenon is important enough to have acquired a name: a *genetic bottleneck*. Biochemical techniques are now used to determine the amount of genetic variability in a population, and among vertebrates there is an example of the consequences of a genetic bottleneck in the northern elephant seal, *Mirounga angustirostris*, which was hunted nearly to extinction in the nineteenth century, there being barely 100 alive in 1892. Now there are about 100,000, distributed over most of their former range, from islands off Baja California to the Farallon Islands off San Francisco. They have been found to be homozygous at a very high proportion of their genetic loci.

An example incorporating a suggestion that deleterious genetic effects were involved is that of the heath hen, *Tympanuchus cupido*, the eastern representative of the prairie chicken and a relative of the sage grouse (Chapter 18). Early in the nineteenth century, the heath hen was common in the northeastern United States, south to Virginia. By 1840, its numbers were obviously declining because of excess hunting, and by 1870 it could be found only on the islands off the coast of Massachusetts. Quantitative data have been kept since 1890 (Figure 19.2), by which time it was confined to the island of Martha's Vineyard, with an estimated 200 birds present. After a fire on the nesting grounds in 1894, there were fewer than

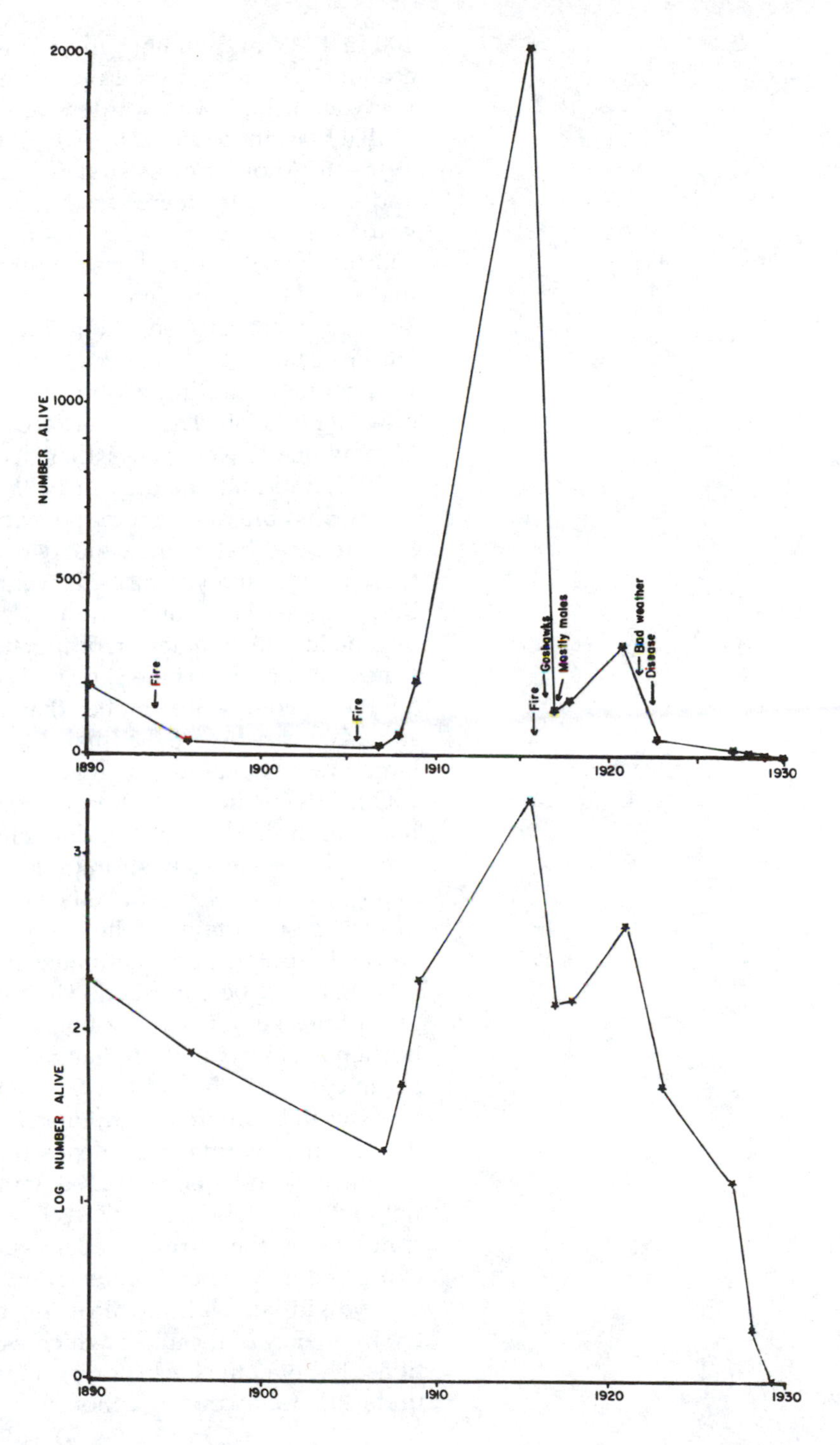

Figure 19.2. *Population history of the heath hen,* Tympanuchus cupido, *after the establishment of a large refuge on the island of Martha's Vineyard, Massachusetts. The upper graph shows the number of birds alive at each recorded time and thus reflects the* amount *of change during each interval; the lower graph shows the logarithm of the number alive and thus reflects the* rate *of change per individual alive at the beginning of each interval.*

100 in 1896, and another fire left only 21 alive in 1907. In 1908, attempts were begun to save the heath hen, and a large reserve was established, with wardens to protect the birds. The response to that was good: 45–60 in 1908, 200 in 1909, and around 2,000 in 1915–16. Another disaster struck in 1916, when there was a fire, and in the winter there was an incursion of goshawks, *Accipiter gentilis,* a large bird-eating hawk. In 1917 there were fewer than 150 heath hens, and it was observed that most of those were males – the first evidence of a possible genetic abnormality. In 1918, 155 were counted, and in 1921 there were 414, and the population seemed to be recovering. Two factors intervened: bad weather in the nesting season, and disease, and by 1923 there were only 50 birds left. They never recovered. By 1927 there were only 13, of which 11 were males; they declined to 2 males in 1928, and 1 in 1929, and that one died in 1930.

It is possible to attribute the extinction to fire, goshawks, rain, and disease, but there were two dramatic increases following those events, and yet the birds were unable to sustain the population. The early records showing mostly males, and the final decline in which males predominated, indicate that some kind of genetic disorder was the principal cause of their inability to make another recovery. Remember that the population dropped to 21 some 20 years before the final stage, and there must have been a large proportional loss of genetic variation at that time.

Our interest in that history centers on the evidence that a small population size permits random changes in genetic composition, even with moderately strong selection against the genes that remain by chance. There have been several mathematical attempts to specify quantitatively the requirements for group selection. For an easily understood treatment, the paper by Levin and Kilmer (1974) should be consulted. They chose to consider a set of ten groups of sizes ten, twenty-five, and fifty, with the following proportions of individuals exchanged ("migration rates") each generation: 0, 0.04, 0.05, and 0.1. Selection against the altruistic allele was varied in different computer simulations: 0, 0.05, 0.1, 0.15, and 0.2. In other words, that means that an individual's chances of leaving descendants, compared with the chances for an individual without the allele, were 100%, 95%, 90%, 85%, and 80%. In most simulations, the altruistic allele was assumed to be recessive – expressed only when homozygous. Selection for groups containing individuals with the altruistic allele was varied according to the frequency of the allele, which was varied for different simulations. The overall conclusion was that group selection *could* work, under the following conditions:

1. The size of each group must be small, no greater than ten.
2. Migration rates must be low, no more than 0.05 of the individuals involved.

3. Selection against the altruistic individuals within groups cannot be larger than 0.1.
4. Selection for groups with altruistic alleles must increase steeply with an increase in the proportion of altruistic individuals.
5. The frequency must rise initially by genetic drift to a level above any reasonably expected mutation rate.

All of their calculations were made on the assumption that the individuals were unreleated, or if they were related, the relationship had no importance for the spread of altruism. It must be obvious, however, that many natural groups consist of families, perhaps of more than two generations, but still of quite closely related individuals. That is important in the effect of altruism and its spread through a population. For the details of the mathematics of this effect of relatedness, see Hamilton's 1964 paper, but it does not require mathematics to understand that each individual will receive half of its chromosomes from each parent and that therefore two siblings will, on the average, share one-half of their genes. A similar calculation shows that an individual will share one-fourth of its genes with an aunt or uncle, and so forth. A famous mathematical evolutionist once said that he would lay down his life to save the lives of any two siblings or any eight first cousins.

The point is that for an altruistic trait to be selected for directly, it must be shared by close relatives to begin with, and then must represent a total gain to relatives that is a little greater than the total relatedness of those relatives to the altruist. The mathematical principle has been used to make remarkably accurate predictions, especially concerning social insects. The original name for the phenomenon was *inclusive fitness*, but it is commonly called by the more easily visualized term *kin selection*. It is the best explanation for many of the "altruistic" traits that were described earlier in this chapter.

Notice that the gene or genes for the trait must be shared initially with at least some of the relatives. It must spread among those relatives in the same way that altruistic genes must spread originally for group selection to work, that is, by genetic drift, as there must, by definition, be selection against the trait within the group.

Summary

Despite possible disadvantages of membership in a group, such as attracting the attention of predators and risking aggression from close neighbors, assembling in groups has major advantages. A number of vertebrates close to each other can depend on part of the group to be watching for predators while the others forage. Being a member of a group is safer than being alone, and both flocks and schools maintain tight spacing. Some groups of birds apparently benefit from mobbing a sitting predator, frequently driving the predator from the area. Groups that are more organized than flocks or schools can mount a coordinated defense, such as those used by musk-oxen and elephants. More elaborate cooperation is seen in beavers in the construction of dams and lodges for defense against predators. There are also energetic benefits to members of groups. Simply watching others and observing their success can lead individuals to food. A different kind of energetic benefit comes from slipstreaming behind a leader, as is observed for flocks of geese and other large birds and for schools of fish. Again, group organization makes it possible to capture large prey, as packs of wolves and African wild dogs demonstrate. White pelicans fish in groups by driving fishes into shallow water.

The most fascinating form of behavior among vertebrates occurs when one or more individuals act in ways that appear to benefit their groups, but to be detrimental to themselves. Alarm signals, for example, though alerting other group members, would also appear to attract the attention of the predator to the signaler. Many such examples can be explained as really benefiting the individual, but the possibility of altruistic behavior remains. The evolution of an altruistic trait because of its advantage to the group can be shown to be theoretically possible, but the conditions are very restrictive. A more likely explanation arises when the members of the group are close relatives, so that altruistic behavior benefits genes in relatives that are identical with the altruist's own genes.

References and suggested further reading

Gould, J. L. 1982. *Ethology. The Mechanisms and Evolution of Behavior.* New York, Norton.

Hamilton, W. D. 1964. The genetical evolution of social behavior. *Journal of Theoretical Biology* 7:1–52.

Hamilton, W. D. 1971. Geometry for the selfish herd. *Journal of Theoretical Biology* 31:295–311.

Jennings, T., & S. M. Evans, 1980. Influence of position in the flock and flock size on vigilance in the starling, *Sturnus vulgaris. Animal Behaviour* 28:634–5.

Levin, B. R., & W. L. Kilmer. 1974. Interdemic selection and the evolution of altruism: A computer simulation study. *Evolution* 28:527–45.

Morse, D. H. 1980. *Behavioral Mechanisms in Ecology.* Cambridge, Mass., Harvard University Press.

Index

Notes: Because an animal may be listed under its common and/or scientific names, as well as under more general terms, the reader may wish to consult all relevant entries (e.g., "Mustelidae" as well as "weasel," "bear" as well as "black bear," "grizzly bear," etc.). Mountains and mountain ranges appear under the entry "mountain"; likewise, lakes are grouped under "lake," geologic epochs, eras, and periods under "epoch," "era," and "period," and zoogeographic realms under "realm." Italic page numbers refer to tables or figures.

aardvark, 54
Acanthodia, 15
Accipiter, 56
  gentilis, 332
  striatus, 278
accommodation, 66
acorn,
  burial, 125
  crop, 209, 210
acorn woodpecker, 126
Actinoptergii, 17–18
actinost, 17
activity, seasonal patterns, 121–31
adaptation
  behavioral, 120
  eutherian, 53–4
  gliding, 53
  to physical environment, 131–43
  physiological, 120
  see also counteradaptation
adaptive radiation, 20
advertisement, 44–5, 307
aestivation, 129
Africa, 5, 58, 216, 219, 232, 224, 255, 256
  plains, 300
  savannah, 38
  southern, 260, 283
  west coast, 201
  see also specific country or region
African antelope, 311
African elephant, 302, 323
African finch, 317
African hunting dog, 225, 325
African mole rat, 53
Agama agama, 309
Agamidae, 84
age at maturity, 145–9, 179
age class, 145
age distribution, stable, 148
Agelaius phoeniceus, 309
aggression, 157, 181, 308
Agnatha, 11–14, 236
Ailuropoda melanoleuca, 212, 293
alarm substance, 86
albatross, 57–8
alewife, 14
allantois, 31, *32*
alligator, 35, 75, 85, 124
Alligator mississippiensis, 317
Alopex lagopus, 132
altricial young, 303, 317
altruism, 326–33
Alytes obstetricans, 317
Amazon Basin, 16
Amboseli Reserve, 212
Ambystoma, 115
  maculatum, 29, 128
  opacum, 128
  talpoideum, *29*
  tigrinum, 169
Ambystomatidae, 28
American Ornithologists Union, 104
American robin, 317
amino acid, 108
Ammospermophilus leucurus, 138
amnion, 32
amniote, 66, 96
  oviparous, 47
Amphibia (amphibian), 23–30, 96–7, 187, *188*, 191, *196*, 221, 229, 231, 237, 242
  community, 169–74
amphioxus, 12
Amphisbenidae, 231
amplexus, 30
amplitude (sound), 72
ampulla(e), 71–2
  of Lorenzini, 87
Amur falcon, 256
Anableps, 70–1
anadromous fish, 236
analysis
  population, 118, 145–9
  statistical, 154
Anapsida, 33
Anaspida, 13
ancestry, common, 93
Andrias japonicus, 28
Aneides, 219
Anguilla anguilla, 242
Anguis fragilis, 38
Ann Arbor, Mich., 135

anole, 70
Anolis carolinensis, 316
Anser
  caerulescens, 103, *105*, 261
  hyperborea, 103
  indicus, 256
  rossi, 254
Anseriformes, 59
Antarctica, 9, 228
  fish of waters, 132
  marsupial fossil in, 228
anteater, 51, *52*
antelope, 51, 83
  pronghorn, 198
anthracosaur, 26
Antigua, 264
Antilles Arc, 228
Anura, 27, 30, *171*, *188*
aortic arch, right, 55
Apodemus sylvaticus, 248
Apteriges, 230
Apteryx australis, 82
Aquila chrysaetos, 311
Arabia, 7, 193
Aranaspida, 12
Archaeopteryx, 54
Archboldia papuensis, courtship, 316
Archilocus colubris, 288
archipelago, 185
Archosauria, 33
Arctic, 53, *134*, 185, *259*
Arctic fox, 44, 132, 133, *134*
Arctic gull, 134
Arctic Ocean, 237, 256
Arctic tern, 237, 261
Argentina, 269
argon, 5
Arizona, 138, 158, 163
armadillo, 91, 216, 301
arousal, from hibernation, 122
Artiodactyla, 34
Ascaphidae, 231
Ascaphus truei, 30
Ascension Island, 226, 237
Asia, 225, 226
  central, 226, 257
  southwestern, 264
asteroid, 43
Atlantic eel, 236
Atlantic Ocean, 190
  western, 259, 264
  see also North Atlantic; South Atlantic
audiogram, 72, *73*, 75, 76, *77*, *81*
auditory ossicle, 41, 78
Austin, Tex., 145
Australia, 7, 210, 213, 222, 224, 227, 228, 316, 317
  west coast, 201
Australopithecus, *46*, 226
autumn, 129–31
Aves, 54–62, 96, 188, 196, 221, 229, 231
avocet, 322
Aythya americana, 254
Azores, 246

baboon, 69, 323
bacteria
  decomposing, 117–19
  in hydrothermal vents, 117
  luminescent, 22
Baffin Bay, 260
Baja California, 237, 247, 330
Bali, 193
Baltic Sea, 278
bamboo, 212, 293
bandicoot, 51
banding (of migrants), 252
Barbados, 264
barbel, 86
bar-headed goose, 256
barnacle goose, 287
barn owl, 77
barracuda, 21
barrier, distributional, 186, 193, 228, 255, *257*
Barro Colorado Island, 82
bar-tailed godwit, 260
Basiliscus, 35
bass, 301
bat, 54, 80, *81*, 82, *188*, *221*
  hibernating, 123
  migrating, 247, 249
Bavaria, 54
Bay of Florida, 216
beak, special adaptations of, 59–61
bear, 53
  hibernating, 50, 122
Bear River Marshes, Utah, 254
beaver, *46*, 53, 54
  building dam and lodge, 323
  food storage, 125
  group organization, 326
Bedfordshire, 208
behavior, 289–334
  group, 321–34
  individual, 293–305
  innate, 303
  interactive (two individuals), 307–20
  see also specific behavior
behavioral ecology, 291–2
Belgium, 162
Bering Sea, 237
Bermuda, 222, 258
bichir, 20
biochemical, 25, 108
biogeography, 185–233
  vicariance, 232
biomass, 118, 171
biome, 201–2
bipolar cell, 68
bird, 54–62, 76, 98–9, 188, *190*, *217*
  beak, 59–61
  classificatory groups, 59, *60*
  desert, heat loss, 134–7
  flightless, 58, *59*, 224; *see also* ratite
  flight range, 288
  forest, competition, 161–3
  ground-nesting, 210
  names, 104
  stomach contents, 294
bird feeder, 324
bird of paradise, 311
bison, 53
bite, 16, 39–41
Bitis gabonica, 42
bittern, 301
black bear, 50
blackbird, 324
blackcap, 285, *286*
black-throated blue warbler, 324
black-throated green warbler, 324
Blarina brevicauda, 124, 126
blubber, as insulation, 133
blueberry, 287
bluegill sunfish, 124, *125*, 167–9
blue goose, 103–5, 261–2
bluegrass, 180
blue jay, food storage, 125
blue tit, 166–7
boa, *40*
  tree-living, 87
bobolink, 283
body weight, 284, 287
  gain, premigratory, 281
  of rodents, *123*, *159*
Bonasa umbellus, 319
bony fish, 17–23
Borneo, 193
Bosphorus, 256
bottle-nosed dolphin, 82
Boulder, Colo., 135
boundary, 205
bowerbird, decoration, 316
bowfin, 21
branchial basket, 12, 14
Branchiostoma lanceolatum, 12
Branta
  bernicola, 287
  canadensis, *282*
brass bar (as control for magnet), 274–6
breeding
  alternate year, 27
  timing, 126–8
British Columbia, 239
British Isles, *see* Great Britain
broken-wing act, 319
brood, number, 163
brown lemming, *178*
browser, 47
brush turkey, 317
Bubulcus ibis, 216
Bubo virginianus, 302
buffalo, 302
Bufo, 127
  terrestris, 169–71, 244

bulla, 80
bulldog, 70
bullfrog, 74
burrow, 51, 53, 74, 158, 160

Caeciliidae (caecilian), 27, 74, 231
Calacrius lapponicus, 260
calendar, internal, 284
Calidris canutus, 260
California, 85, 106, 126, 135–6, 158, 185, 237, 254, 260; *see also specific city or region*
Callorhinus ursinus, 311
Cambridge, Mass., 235
Camelidae (camel), 50, 51, *138*, 139–41, 226
Camelus dromedarius, 119
camouflage, *40*, 41
Canada, *198*, 104, 252
  eastern, 237, 269
  northern, 269
  northwest, 201
Canada goose, 281, *282*
canary, 76, *77*
Canary Islands, 285, *286*
Canidae, 50, 69, 225, 309, 317
canine, *see* tooth
Cape Canaveral, 246
Cape Cod, 262
Cape St. Vincent, 256
Captionidae, 231
capybara, 54
Carassius auratus, 242
carbon-14, 4–5
carcass, as food, 117
Carcharrhinus leucas, 15
Cardinalis cardinalis, 309
Caribbean Arc, 223
caribou, 237, 248
carnivore, 46, 47, 132
Carolina junco, 237
carpal, 25
capercaillie, 303
Carphophis amoenus, 41
Carpodacus mexicanus, 135
carrying capacity, 156
cartilage, 12, 14–17, 20
Casmerodius albus, 129
Caspian Sea, 251, 256, 264
cassowary, 58, 224
Castor canadensis, 125, 323, 326
cat, 51, *52*, *77*, *81*, 83
catadromous fish, 236
catbird, 302
category, classificatory, 91, 93–6
catfish, *73*, 86
Cathartes aura, 82, 277
Catharus spp. 262
cattail, 168
cattle egret, 216–17
Caudata, 27, 96, 187, *188*
cause, 289–90
cave salamander, 244
caviar, 20
Celebes, 193
cement, 47
census
  lizards, 146
  sunfish, 167
Central America, 51
central-place foraging, 298
Central Valley, Calif., 254
Centrocercus urophasianus, 312, *313*
Cephalaspida, 12
Cephalochordata, 12
Cervus elaphus, 311
Cetacea, 54, 82
Ceylon, 222
chacma baboon, *69*
chamber, orobranchial, 21
chameleon, 36, 70
Characidae, 231
character, 108
  shared derived, 111
characteristic, 91
Charadriiformes, 59
Chateau de Menton, 205, *206*
Chelidae, 231
Chelonia, 188
chemoreceptor, 82–6, 239
chickadee, 56
  hypothermia in, 124
  memory, 125
checklist, for bird names, 104
cheek pouch, 158
cheetah, 300
chemical
  noxious, 328
  sense, 82–6
Chile, 131
chimaera, 15
chimpanzee, *81*, 110
China, 7, 193, 219, 256
chinchilla, *81*
chipmunk, food storage, 125
Chiroptera, 54, 122
Chondrichthyes, 14–17, 19, 187
chondrostean, *19*, 20
Chordata, 95
choroid, 66
chromosome, 33, 114–15
Chrysemys
  picta, 132
  scripta, 75
chuckwalla, 143
Ciconiiformes, 59
cilium (hair cell), 12, 72, 83
Cincinnati Zoo, 211
cineol, 213
circadian vs. circannual rhythm, 284
circular reasoning, 227, 290
cladistics, 33, 111–13
Clark's nutcracker, 125
clasper, 16
class, 11, 94, 96
classification, 91–116; *see also specific hierarchical level*
cleaner fish, 21
Clemson College, 101
Clethrionomys, 177, *178*, *179*, *183*, 248
climate, 185, 188
  factors determining, 198–201
  global pattern, 193
cloaca, 85
  lip, 29
cloacal gland, 310
clock, internal, 284
clock shifting, 272
clone, 114
cloudiness, and orientation, 271, 273
clutch (of eggs), number and size, 146
Cnemidophorus, 114
Coastal Plain, 189, 202
cobra, 39–41
cochlea, 78–9
cochlear duct, 74
cod, *73*
cold, adaptations to, 131–7
cold front, 258
collared flycatcher, 285
collared lemming, 134
color vision, 68
Colorado, 4, 38, 135–6
Colubridae, 84
colobus monkey, 44
Columba livia, 269
columella, 73, 76
commensal, 34
communication, 73, 85
community, 167–84
  amphibian, 169–74
  fish, 167–9
Comoros Islands, 18, 223
comparison, biochemical, 91
competition, 15, 24, 205, 207–8
  coefficient (α), 156–7
  evolutionary result of, 291–6
  field experiments, 151–65
  interspecific, 118, 152, *155*, 156
  intraspecific, 155
  migrant–resident, 252
  tadpole, 170
  theory, 154–7
competitor, 185
Compsognathus, 54
cone (in vision), 67
confrontation, territorial, 308
conifer, 205, *206*, 256
Connochaetes taurinus, 248
conodont, 11
consumer, in ecological context, 117
continental drift, 5–7, *8*, *9*, 220, 226–32
continental shelf, islands, 222
control, experimental, 152, 174
convection, 124
convenience, 94
convergence, *52*, 93, 224
cooling, evaporative, 140
copulatory organ, 30
Coraciiformes, 60

coral, 21
coral snake, 40
cormorant, 70
cornea, 66
Corvidae, 125, 162
Costa Rica, *188*
counteradaptation, 82
courtship behavior, 289, 314–16
cowbird, behavior of victims, 320
Craniata, 12
creationism, xiv, 94
creeper, 124
Cricetidae, 122, 158, 177
crocodilian, 34–35, 74, *75*, 76, 93, 188
crossbill, 236
Crossopterygii, 18
  rhipidistrian, 25
Crotalinae, 87
Crotalus
  cerastes, 142
  viridis, 247
crowding, 168, 181
crust (earth), 6, 227
crustacean, 168
Cuba, *188*, 223
Cuculus canorus, 302
cupula, 72, 87
current, electric, 22, 87
cusp, 46
Cyanocitta cristata, 125
cyanogenic glycoside, 213
cycle
  of matter, 118
  microtine population, 177–84
  physiological, 122
Cyon, 225

dab, *73*
Dall sheep, fur, 132
Dasypus novemcinctus, 216
daylength, effect on migration, 281
dear enemy recognition, 307–8
decomposition, 117–18
deep sea, 22, 328
deer, 46, 83
  white-tailed: gestation, 126; warning, 327
deermouse
  adaptive coloration, 44
  seed caches, 125
defense, special, 301
Dendroica, 103
  caerulescens, 324
  coronata, 130, 287
  petechia, 302
  virens, 324
Denmark, 278
density dependence, 184
dentary bone, 41
dentine, 47
Dermoptera, 53
desert, 53, 137, *138*, 193, 199–200, 264; *see also specific site*
desert iguana, 142
desert tortoise, 143
Desmognathus, *29*, 97–8, 172–4
  monticola, 95
Dermogenes pusilus, 242
dewlap, 314
Diapsida, 33
Dicamptodontidae, 99
Dicrostonyx, 177
  groenlandicus, 134, *179*, *183*
Didelphis marsupialis, 216
diet, change in, 34
difference, morphological, 167
dinosaur, 32, 43, 54
Dipnoi, 17, 25
Diposomys, 80, 143, 158, *159*, 160, *161*
Dipsosaurus dorsalis, 142
direction finding, migratory, 269–80
dispersal, 220–2, 229
display
  auditory, 307–8
  chemical, 309
  visual, 308–9, *313*
distribution
  disjunct, 219–33
  ecological, 151, 172, *173*
  extensions, 216–17
  geographic, 185–233
  pattern, 186, 188; within continents, 195–8; worldwide, 193–5
  theories of, 219–33
District of Columbia, 265
DNA, 329
  mitochondrial, 25
dog, *52*, 83, 84, *138*
dogfish shark, 237
Dolichonyx orizivorous, 283
dolphin, 54, 82
dominance hierarchy, 326
donkey, 99, *138*
double-crested cormorant, 325
dowitcher, *61*
Draco, 36
  bimaculatus, 309, *310*, 314
  volans, 314–16
dragonfly, 297
Dryophis acuminatus, *42*
duck
  beaks, 61
  diving, 70
  flightless, 58
duck-billed platypus, 317
dugong, 54
dusky langur, *69*
dwarf mudpuppy, 28

eagle, 56
ear
  in hearing, 71–82
  mammalian, *78*
  in temperature control, 50, 137
eardrum, 73
earth, rotation, 199
earthquake, 6, 88
East Africa, 53
Eastern Deciduous Forest Biome, 202
Eastern red-spotted newt, 244
East Indies, 222
echo, 278
echolocation, 80–1, 278
ecosystem, terrestrial, 117–20
ecotone, 202
Ectopistes migratorius, 210
ectotherm(y), 28, 220
  preferred temperatures, 142
Ecuador, 222
Edentata, 53
eel, 21, 236, 242
eel grass, 301
egg, 28, 30, 31–2, 96, 146
egret, 129
Egretta thula, 129
Egypt, 260
Elasmobranchii, 15
electric catfish, 23
electric current, 22, 87
electric eel, 23
electric field, weak, 88
electric signal, 23
electrocyte, 23
electrophoresis, 108
electrosensitivity, 87
elephant, 46, 302, 323
elephant bird, 58, 224
Ellesmere Island, 260
embryo, 91
emigration, 160
Empidonax, 103
emu, 58, 224
enamel, 47
endolymph, 71, 72
endotherm(y), 49, 55, 122, 220
  evolution, 50, 55
  maintenance, 133
enemy, 185; *see also* escape
energetic benefit, from group behavior, 324–5
energy
  demands on migrants, 266
  flow, 117–18, *119*
  required in flight vs. resting, 298
  stored, 49
England, 208–10, 279, 321; *see also* Great Britain
English Channel, 279
Ensatina escholtzii, subspecies, *106*, 107, 111
environment, physical, 118
  adaptions to, 131–43
epideictic behavior, 327
enzyme, 49, 109
epoch, geologic
  Eocene, 4, 226, 228
  Holocene, 4
  Miocene, 225
  Paleocene, 53
  Pleistocene, 4, 51, 53, 194, 219, 223, 226, 252, 323

Pliocene, 226
see also era, geologic; period, geologic
Eptesicus fuscus, 123
equation
exponential growth, 154, 182
logistic, 155, 182
Equus burchelli, 248
era, geologic
Cenozoic, 2, 252
Mesozoic, 2, 27, 32, 43, 227, 302
Paleozoic, 2, 26
Precambrian, 12
see also epoch, geologic; period, geologic
Eremophila alpestris, 216
Erinaceus europaeus, 122
Erithacus rubecula, 274
Erithizon dorsatum, 122
ermine, 44
escape, 49, 303
Eschrichtius gibbosus, 247
estrus (heat), 84
Ethiopian Plateau, 256
Eucalyptus viminalis, 213
Eurasia, 188, 226, 256
Europe, 7, 23, 38, 226, 232, 256, 258
northern, 201, 261
southern, 297
see also specific country
European bee-eater, 297–300
European cuckoo, 302
European eel, 236, 242
European hedgehog, 122
European rabbit, 83
European red squirrel, 209–10
European robin, 274
European smooth newt, 85
European starling, 271–2, 321, 325
Eurycea lucifuga, 244
Eusthenopteron, 26
Eutheria, 51, 53
evolution, 99, 103, 190, 289
of behavior, 312, 318
of competitive ability, 157–8
components of, 100
convergent, 51–2
direction of (*Desmognathus*), 173
by natural selection, 327
theory, 100
evolutionary cause, 289
experiment, 159, 290
control of conditions, 169
failure, 164, 174
field, 151–65
removal, 153
repetition, 163
transplantation, 157
experimental design/procedure, 290
complexity of, 169
requirements, 296
explanation, necessary vs. sufficient, 152, 179
extinction, 207
of dinosaur, 43
of heath hen, 330, 332
rate, *214*, 215
eye, 65–70
shine, 66
eyeglasses, 66

Falco
amurensis, migration, 256
concolor, predation on migrants, 251
sparverius, winter food, 130
falcon, 56
Falconiformes, 59
family, 94, 194
fang, snake, 39
Farallon Islands, 330
farsightedness, 66
fat
brown, 122
deposition of, 281, 286
nocturnal consumption, 129
premigration, 299
storage, 164
fauna, 188
fawn, 301
feather, 55–6, *57*
feces, 83
fecundity, 47, *48*
age specific, 146–8
feeding behavior, 21, 289, 293–300
adult feeding young, 317–19, 320
feigning injury, 319
femur, 25
fence, 34
interception, 242
fer-de-lance, 40
fertilization, internal, 16, 28, 30
fibula, 25
Ficedula, 285
Fiji, 185
fin, 15
fish, 14–23, 114–16, 124, 322
community, 167–9
freshwater, 190
hearing in, 71–3
types, 236
fisher, 301
flight
displays, 309
kinds, *57*
in migration: height, 262–4; range, 288; speed, 261–2, *263*
flightlessness, evolution of, 58
flipper, 54, 93
floater, 311
flocking, *see* group behavior
Florida, 110, 196, 216, 246
Florida caerulea, 129
flycatcher, 60, 103, 324
flying lemur, 53
flying lizard, 309, *310*, 314–16
flying phalanger, 51
flying squirrel, 51, 137
flyway, 254–5
food, 117, 157
amount, 180
fluctuations of, 294
intake, *125*, 286–7
nutrient quality, 180
species, and distribution, 186, 212–13
storage, 125–6
superabundant, 291
types, for migratory birds, 130, *131*
foot, 36, *37*, 50
webbed, 59
forest
conifer, *206; see also* taiga
deciduous, *119*, 198, 202
pine, 216
fossil
age, 1–5, *229*
amphibian, 7
record, 94
fovea, 68
Fowler's toad, 244
fox, 225
France, 185, 205
northwestern, 278
Fraser River, 239
frequency (sound), 72
French Broad River, 102
Fringillidae, 258
frog, 27, 30, 72–4, 187, *190*, 223, 237, 243, 322; *see also* tadpole
fungi, decomposing, 117
furcula, 55

Gaboon viper, *42*
Galapagos, 185, 222
ganglion cell, 68
gar, 21
garden warbler, 284, 287
gas, transfer of, 190
gas bladder, 18
Gasteriostus aculiatus, 308
gastric-breeding frog, 317
Gaviiformes, 59
Gazella thomsoni, 248
gazelle, 50
gecko, 36, 67, 75, 76
Gekko gekko, *37*, 75, 308
Gekkonidae, 84
gel, 109
gene, 104, 109, 290
locus, 110
generality, in theories, 291
genetic bottleneck, 330
genetic dominance, 328
genetic drift, 329–30
genetic effect (deleterious), 331
genetic identity, 110
genetic recombination, *see* recombination
genome, 110

genotype, 104
genus, 94, 96–9
Georgia, 216
Germany, 252, 256, 271, 277, *286*
gestation period, 126
giant ground sloth, 53
giant panda, 212, 293
Gibraltar, 256
gill, 12
gill arch, 14
gill raker, 20
glacier, 194
gland
  cloacal, 310
  mammary, 43
  salt, 35, 55
  scent, 84
  skin, 310
  sweat, 50
Glaucomys volans, huddling, 137
glide, 35, 53
glycoprotein (antifreeze), 132
Gnathostomata, 11, 14
Gobi desert, 199
golden eagle, 311
golden-mantled ground squirrel, 122
golden mole, 53
golden plover, 257, 269, 283, 287
goldfish, 242
gonad enlargement, 281
Gondwanaland, 7, 228
Gopherus agassizii, 143
goshawk, 332
grackle, 324
grade, structural, 94
Grand Canyon, 1
granivore, *159,* 160, *161,* 293
grassland, 198
gray-cheeked thrush, 262
gray gull, 131
gray jay, 134
gray squirrel, 125, *207,* 208–10
gray tree frog, 132
gray whale, 237, 247
grazer, 47
great auk, 58
great blue heron, *61*
Great Britain, 161, *188,* 189, *207,* 208–10, 222, 260
great crested flycatcher, 324
Greater Antilles, 223
great horned owl, 302
Great Plains, 198, 219
Great Rift Valley, Africa, 6, 211
great tit, *77,* 161, 308, 324
  insects brought by, 294
Greek Islands, 261
Greenland, 5, 6, 188, 260
Green River Shales, 4
green sunfish, 167–9
greyhound, 70
grizzly bear, 132
grosbeak, beak, 60
ground sloth, 211, 223
group behavior
  advantages, 321–26, 333–4
  origins, 326
group selection
  explanation, 328
  requirements 329–33
  traits claimed, 326–9
growth, 162, 164, 168
  rate (tadpole), 171
Gruiformes, 59
Guam, 260
Guatemala, 264
guinea pig, *79*
Gulf of Alaska, 260
Gulf of Mexico, 201, 261
Gulf Stream, 201, 236, 249
gull chick, recognition of, 318
Guyana, 217
Gymnophonia, 27, 30, 223

Hadley cell, 199
hagfish, 11, 14, 71
hair, 41, 43
  types, 44
hair cell, *see* cilium, 73
hairy-nosed wombat, 309
halfbeak, 242
half-life, 5
handling time, 295, 297
harbor seal, 248
harem, 311
Hawaii, 208, 222, 248, 257, 287
hawk, 322
hawk–goose model, 303–4
hearing, 71–82
heart
  rate of beating, 304
  three- vs. four-chambered, 44, 49
heat
  adaptions to, 137–143
  exchange, 50, 55, 124, 137
  sensing, 87
  storage, 137
heath hen, 330, 332
hedgehog, 122, 301
Helmholz coil, 275
helper at nest, 327, 328
herbivore, 46, 117–19
herding, *see* group behavior
heron, 129, 252
herpetologist, 28, 96
herring gull, 318
heterodont tooth, *see* tooth
Heteromyidae, 122, 158
heterozygote, 104
hibernaculum, 122
hibernation, 50, 122–4
Hinkley, Ohio, 265
Hirudo rustica, 261
Hispaniola, 223
Holocephali, 15
Holostei, 21
homeothermy, maintenance, 136
home range, 307
homing pigeon, *see* pigeon, homing
homologous structures, 91
Homo sapiens, 53, 226, 231
homozygote, 104
honeybee, 297
hooded warbler, 308, 324
hornbill, beak, 61
horned lark, 216
horse, 46, 53, 99, 211
horsefly, 297
horse latitudes, 199, *200*
horsetail, 23
house finch, 135, *136*
huanaco, 226
Hudson Bay, 254, 262
Hudson's Bay Company, 177, *178*
human, *138*
  altruistic behavior, 328–9
  ancestor, 226
  distribution, 211–12
  genetic similarity to chimpanzee, 110
  hearing, 79–80, *81*
  heat resistance, 139
  and magnetic field, 275
  and other species' dispersal/survival, 211, 222
humerus, 25
hummingbird, 58
  beak, 60
  flight range, 288
humpback whale, 248
hurricane, 217
hybridization, 105
hybrid sterility, 99
Hydromantes, 185
hyena, *46,* 325
Hyla
  chrysoscelis, 169, *171*
  cinerea, 74
  gratiosa, 169, *171*
  versicolor, 132
hyomandibular, 16, 78
hyoid, 14
hypermetropia, 66
hyperphagia, 280
hypothalamus, anterior, 133
hypothermia, shallow, 124
hypothesis, 152, 160, 290
  alternative, 173
  testing, 174
Hyracoida, 54

Iberian Peninsula, 256, 260
ice, in body fluid, 132
Iceland, 226, 260
Ichthyostega, 24, *26*
Icteridae, 266
Iguanidae, 84, 185, 231
Illinois, 262
immigration, 148, 160
  rate, 213, *214*

immunity, 290
implantation, 84
impregnation, 84
imprinting, 86
  salmon, 240
  snow goose, 104
incisor, *see* tooth
inclusive fitness, 333
incubation, 316
incus, 41, 78
India, 7, 225, 227, 228, 232, 256
Indiana, 101, 110, 135
  north-central, 216
indigo bunting, 264, 273, *275*, *276*, 285
information retrieval, 93
infrasound, as orientational cue, 278
insect, 82, 92, 168, 327
  in diet, 294
Insectivora, 43, 53, 54, 122, 223
interaction
  interspecific, *see* competition; predation
  intraspecific, *see* behavior, interactive
intergradation, 101–2, 105, 110
International Commission on Zoological Nomenclature, 92
interpretation, physiological vs. ecological, 172–3
invertebrate, 168
iris, 66
Iridoprochne bicolor, 287
iron container, experimental, 277
island, 101, 222, 330
  biogeography, 213, 215–16
  -hopping, 53
  oceanic, 58, 207, 222,
  types, 221–2
  see also specific island group
Italy, 185, 256, 277
Ithaca, N.Y., 245

jackass, 99
jackdaw, 162
jackrabbit, 137, *138*
Jamaica, 210
James Bay, 262
Japan, 7, 28
jargon, 111
jaw, 11, 14, 21
  hinge, 41
  lower, 78
  muscle attachment, 32
  snake, 39
jellyfish, 12
Junco hyemalis, 281

kangaroo, 51
kangaroo rat, 51, 80, *138*, 143, 158, 198
Kentucky warbler, 324
Kenya, 256
kestrel, 130
kinglet, 124
kind, 94
kinocilium, 72, 87
kin selection, 333
kiwi, 58, *59*, 82, 224, 230
  beak, 61
koala, 51, 212, 293
Korea, 256
krait, 39

Labrador, 201, 261
Labyrinthodontia, 26
Lacertidae, 84
Lacertilia, 33, 188
lagena, 72
Lagomorpha, 51, 54
Lagopus lagopus, 303
lake
  Baikal, 256
  Chilko, 241
  Erie, 13
  Gatun, 82
  Huron, 262
  Malawi, 227
  Manyara, 211
  Nicaragua, 15
  Superior, 13
  Tanganyika, 227
  Tibetan, 255
  Turkana, 227
  Washington, 241
lake trout, 13
lamellae (on gecko foot), 36
lamprey, 11–14, 66, 71
landmark, as orientational cue, 248, 270
Lapland bunting, 260
lapwing, display, 309
larch, 205
Larus
  argentatus, 318
  delawarensis, 311, *313*, 318
  hyperboreus, 134
  modestus, 131
larval stage, 97, 172
Lasiorhinus latifrons, 309
Latimeria chalumnae, 18, 25
Laurasia, 7, *9*, 227
leaching, 118
leaf-eating monkey, 51
learning, of migration route, 269
leglessness, 39
Leiopelmidae, 231
lek, 311–12, 321
lemming, 177, 181
Lemmus, 177, *178*, *179*, *183*
lemur, 223
lens (eye), 66
leopard, 325
Lepidosauria, 33
Lepidosirenidae, 231
Lepidosiren paradoxus, 17, 25
Lepomis, 167–9
  macrochirus, 124, *125*
Leptodactylidae, 231
Lepus
  alleni, 138
  americanus, 133
  timidus, 300
lesser snow goose, 103–4, 261–2
Leurognathus, 98
life cycle, 152
life history, 27
life table, 146–9
life zone, 198
light, as orientational cue
  polarized, 240
  ultraviolet, 274
Limosa lapponica, 260
lion, 53, 83, 300
  female soliciting, 316
  group hunting, 325
Lissamphibia, 27
little blue heron, 129
lizard, 35–9, 74, 75, 84, 93, 142, 145–9, 188, 309, 314–16
  fecundity, *48*
  legless, 38–9
  mating display, 314–16
  parthenogenetic, 114
llama, 226
locomotion, 21
logic, 111
Lombok, 193
London, 208
longevity, 28
Long Island, 135
loon, 70
Lophodytes, 61
Lord Howe Island, 210, 222
loudness, 72
Louisiana, 262, 288
Loxia curvirostra, 293
luciferase, 22
luciferin, 22
luminescence, 22
lung, 17, 74
lungfish, 17, *19*, 224
Luscinia luscinia, 308
Lutra canadensis, 126
Lybia, 260
lynx, 177, *178*

mackeral, 21
macula, 72
Madagascar, 18, 58, 185, 222, 224, 228, 232
maggot, 287
magnetic field, 88, 274
  disruption by magnet, 274
  experimental shift, 241
  human orientation to, 274–5
  reversal, 6
  vertical component, 245
magnetite, 275
magpie, 162
malleus, 78
Mammalia (mammal), 41–54, 96, 188, *190*, *196*, *221*, *229*, 231
  anatomy, *78*, *83*
  fecundity, *48*

Mesozoic, 43
nonflying, 189
placental, 51–4, 221; primitive, 211
tooth specialization, 45–7
mammoth, 53, 211
manakin, 312
manatee, 54
mangrove, 216
manna gum, 213
mantle (earth), 6, 226
mare, 99
marmot, 51, *52*, 122
Marmota monax, 122
marsh rabbit, 54
Marsupialia (marsupial), 43, 45, 48, 51, *52*, 224, 228, 231, 309
fossil, 51, 228
marsupial "mole," 51
Martes pennanti, 301
Martha's Vineyard, 330
Masai, 212
Massachusetts, 266, 331
mast cropping, 212
mastodon, 211
mathematics, use in theories, 290
mating behavior, 289, 307, 314–16
salamander, 28–9
mating success (individual male), 312
maxilla, 40, 41, 78
mean generation time, 28, *29*, 37–8, 147
Meantes, 96
Meckel's cartilage, 78
Mediterranean Sea, 23, 256
Megapodidae (megapode) 48, 317
Megaptera novaeangliae, 248
meiosis, 329
Melanesia, *217*
melon, 82
Melanerpes formicivorous, 126
Meleagris gallopavo, 303
Melospiza, 77
membrane
extraembryonic, *32*
nictitating, 70
olfactory, *83*
respiratory, 24
tectonic, 74
tympanic, 73
Mephitis mephitis, 45
Mergus, beak, 61
Merops apiaster, 297–300
metabolic rate, standard (SMR), 135
metabolism, of fat, 49; brown, 122–3
metamorphosis, size at, 171
Metatheria, 51
Mexico, 114, 135, 252
Michigan, 124, 135–6, 167
Micronesia, *217*, 222
Micropterus, 301
Microtinae, 177, *178*
microtine cycle, explanations for, 179–84
Microtus, 177, *179*, *183*
pennsylvanicus, 180
microvilli, 83, 87
Mid-Atlantic Ridge, 6, 226
Middle Atlantic States, 121–31, 254, 261
Middle East, 230; *see also specific country*
midwife toad, 317
migrant
achievements, 260–6
calendar, 283
competition from resident, 252
nocturnal, 253
tracking/watching, 253
migration, 235–88
bird: and food types, 130, *131*; patterns & capabilities, 251–67
defined, 236
energetics of, 286
landmarks in, 248, 270
nonbird vertebrates, 239–50
nonstop distances, 260
orientational cues, *see* orientation
patterns, Mid-Atlantic States, 129–31
problems, 269–80
route, 254–60; learning, 269
solutions, physiological, 281–8
timing, 261, 265, 281–6
migratory restlessness (*Zugunruhe*), 27, *275*, 281, 283, *285*
milk, 43, 44, 317
Mimus polyglottus, 278
mink, 54
Minnesota, 217
Mirounga angustirostris, 330
mirror, experimental use, 271
missing link, 226
Mississippi Drainage, 190
Mississippi River, 20, 254
moa, 58, 224
mummified, 231
mobbing, 322–3
mockingbird, 278
molar, *see* tooth
mole, 51, *52*, 53
Molothrus ater, 302
molt, 44, 284
function in birds, 128
postnuptial vs. prenuptial, 282
Moluccas, *217*
mongoose, 40, 210
monkey, 51, 69, *81*
Monotremata (monotreme), 51, 78, 228
monument, 270
moose, 325
Morocco, 256, 260
mortality, 168, 179
moth, 82
mountain, 101, 189
Alps, 188, 205, *206*, 256
Andes, 188, 226
Appalachians, 44; southern, 101, 152, 172
Balsam, 101, 152–8
Black, 151, *152*, 205
Blue Ridge, 101
Cascade, 138
Caucasus, 188, 255
Everest, 256
Grandfather, 101
Great Smoky, 101, 152–8, 205–6
Hawk (Penn.), 254
Himalaya, 188, 193, 256
Hindu Kush, 256
Jura, 256
Kilimanjaro, 212
Nantahala, 101, 105
Pamir, 256, *265*
Pyrenees, 255
Rocky, Front Range, 254
Sierra Nevada, 107
Tien Shan, 256, *265*
mountain hare, 300
mourning dove, 142
mouse, 46, 84
mouth, 108
markings in, 317
mechanism for opening, 97–8
mucus, 83
mule, 99
muscle, flight, 55
musk, 83
musk-ox, 323
muskrat, 54
Mustela erminea, *45*
Mustelidae, 54
mutation, 93, 113, 326
Myiarchus crinitus, 325
myopia, 66
Myxiniformes, 12

Nandidae, 231
nasolabial groove, 85
native cat, 51
natural selection, 100, 167, 173, 185, 326–7
navigation, true, 240, 270, 278–9
Near East, 80; *see also specific country*
nearsightedness, 66
nectarivore, 293
Necturus punctatus, 28
neighbor recognition, 307–8, 310
Neoceratodus forsteri, 18
neoteny, 28, 231
nerve
auditory, 71
optic, 68
nest
guarding, 317
helpers at, 327, 328
parasite, 302, 318
predation at, 146

nesting habit, 163
nestling, 294
Netherlands, 278, 294
net reproductive rate, 146
neuromast, 87
Nevada, 158
New England, 237; *see also specific state*
Newfoundland, 222
New Guinea, 51, 58, 224, 260, 316
newt, 28, 169–72
  training to magnetic direction, 244–5
New York, 110, 216
New Zealand, 33, 58, *188*, 208, 210, 222, 223, 224–8
Niagara Falls, 13
niche partitioning, 158, 167, 172, 291
nighthawk, 60
Nile crocodile, 85
Nile River Valley, 256
nitrogen-14, 4
Noctiluca, 22
nomenclature, 94, 114
North Africa, 193, 256
North America, 7, 20, 21, 53, 193, *200*, 211, 220, 224, 225–6, 230, 232
  eastern, 129, 162, 219, 310
  northern, 188
  western, 125, 198, 219, 222
North Atlantic, 201, 227, *247*, *259*
North Atlantic Gyre, 246
North Atlantic Ridge, 226
North Carolina, 101, 135, 151, 186, 206, 216, 265, 324
  as part of continent, 189
  vertebrate distribution in, 187–91
North Carolina State College, 101
North Carolina State Museum, 101
northern elephant seal, 330
Northern Hemisphere, 20
North Pole, 7, 225
North Star, 273
North Temperate Zone, 190
Northwest Territory, 217
Norway, 260
Notophthalmus viridescens, 169, *170*, 244–5
Nova Scotia, 217, 248, 257, 269
Nucifraga columbiana, 125
numbat, 51
nut burial, 125
nuthatch, 124
nutria, 54

observation, direct, 290
Oceanodromus, 82
occipital condyle, 41
Ockham's razor, 113
Odocoileus virginianus, 126, 327
odor, 45, 83
  as orientational cue, 277–8
Oenanthe oenanthe, 260
Ohio, 135–6, 216, 265
olfactory bulb, 83
olfactory pit, 240
olfactory surface, 83
Olor columbianus, 261
omnivore, 46, 160, *161*, 293
Onchorhynchus nerka, 239
oogenesis, 114
Oporornis formosus, 324
opossum, 48, 51, 224
optic chiasma, 68
optimal foraging theory (OFT), 291, 294
  assumptions and predictions, 295
  tests, 296–300
orbit, 33
order, 94
organ
  of balance, 71
  electric, 23
  Jacobson's, 84
  lateral line, 86
  light-emitting, 22
organic matter, dead, 117
orientation, cues for
  celestial, 240, 271–4; polarized light, 240, 244, 270; stars, 273–4; sun, *243*, *244*, 271–3; UV light, 274
  environmental, 269–9; infrasound, 278; landmarks, 248, 270; magnetism, 244–5, 274–5; odors, 277–8
oriole, 302
Orizomys, 224
ornithologist, 98, 287
Ornithorhynchus anatinus, 317
ossicle
  auditory, 77
  Weberian, 72
Ostariophysi, 72
Osteichthyes, 14, 17–23, 113, 187, *196*, *229*, 231, 236
Osteodontornis, beak, 61
ostracoderm, 12, 13
ostrich, 58, *59*, 140, 224
otolith, 72
otter, 54
out-group, 111
oval window, 79
overkill, Pleistocene, 211
overlap, altitudinal, 151, 154
oviduct, 17
oviparity, 17, 47
ovoviviparity, 17, 18, 317
ovulation, 84
owl, 68, 76, 80, 319
oystercatcher, beak, 61

Pacific Northwest, 30, 44, 201, 241
Pacific Ocean, tropical islands, 215
Pacific salmon, 239
Pacific Tectonic Plate, 223
  tropical, 215
pack hunting, 325
paddlefish, *19*, 20, 219
paedomorph, 27
palate, paleognathus, 230
palatoquadrate, 16
paleontology, 41
pampas, 269
Panama, 53, 82
Pangaea, 7, *8*, *9*, 227
pangolin, 91, 301
panting, 50, 140
papilla, 74
parakeet, 76, *77*
parallelism, 51
paraphyletic group, 11, 113
parasite, 118, 186
  nest, 302, 318
  sexual, 114
parasitism, 205, 210
parental care, 35, 289, 316–19
parrot fish, 21
parsimony, 113
parthenogenesis, 113–15, 327
Parulidae, 99, 258, 266
Parus
  atricapillus, 125
  major, 294, 308
passenger pigeon, 210
Passerina cyanea, 264, 273, *275*, 282–3
passerine, 103
patch (food in), 295
pecking order, 326
pectoral girdle, 25
pekingese, 70
Pelecaniformes, 59
Pelecanus erythrorhynchos, 325
Pelomedusidae, 231
pelvic girdle, 25, *26*
penguin, 58, *59*, 93, 230
perilymph, 73
period, geologic, *3*
  Cambrian, 7, 12
  Carboniferous, 7, *9*, 11, 13, 26, 31
  Cretaceous, 18, 43, 51, 221, 222, 223, 227, 228, 229, 232, 252
  Devonian, 7, *8*, 12, 13, 15, 18, 20, 23, 25, *26*
  Jurassic, *9*, 27, 54, 222, 227
  Ordovician, 7, *8*, 12
  Permian, 21, 32
  Silurian, 7, 12, 13, 15, 23
  Tertiary, 2, 16, 43, 51, 221, 226, 229, 232
  Triassic, 7, 27, 33, 41, 43, 221, 222
  see also epoch, geologic; era, geologic
Perisorius canadensis, 134
Perissodactyla, 53, 54
Perognathus californicus, torpor in, 123
Peromyscus, 107
  maniculatus, 44, 125
perspective, 68
perturbation, accidental, 180
Peru, 238, 261
  desert, 199

Petromyzon marinus, 13
Phalacrocorax auritus, 325
Phalaenoptilus nuttalli, 123
phalanger, 51
Phalaropus tricolor, 261
pharyngeal cavity, 12, 15, 21
pharynx, 12
phascogale, 51
Phascolarctus cinereus, 212, 293
phellandrine, 213
phenogram, 108, *109*
pheromone, 310
Philippines, 309
Phoca
  hispida, 133
  vitulina, 248
phoebe, 252
Pholidota, 54
photophore, 22
photosynthesis, 117
Phyloscopus trochilus, 261
phylum, 94
physiological cause, 289
pied flycatcher, 285
Piedmont, 189, 202, 216, 237, 253
pigeon, *77*
  flightless, 58
  homing, 269, 272, 274–5, *276*, 277
pigment cell, 36
pileated woodpecker, *61*
Pine Subclimax, 202
Pinnipedia, 54
piñon pine, 125
pipefish, 301
Pipidae, 219, 230
Pipilo erythrophthalmus, 309
Piranga olivacea, 308
pitch (sound), 72
pitching motion, 15
pit viper, 87
Placodermi, 15
planetarium, 273
plankton, 20
plate tectonics, 219, 221
platypus, 51
Plectrophenax nivalis, 134, 260, 309
Plethodon, 85, 95, 96–7, *98*, *102*, 103, 219, 309
  cinereus, 85; aestivation, 129; pheromone use, 310
  clemsonae, 101
  cylindraceus, 151
  glutinosus, 29, 104, 108–11, 124, 151–8, 206, 314
  jordani, 29, 95, 101, 104, 106, 111, 124, 151–8, 206, 314
  metcalfi, 101, 151, *152*
  oconaluftee, 151, *157*
  shermani, 101
  yohahlossee, *98*, *152*
Plethodontidae, 85, 96, 190, 294
Poa pratensis, 180
plover, 257
Pluvialis dominica, 257, 287
pocket gopher, 54
pocket mouse, torpor in, 123
Podicepidiformes, 59
Poeciliopsis, parthenogenesis in, 114–15
polar bear, 54, 133, *134*, 185
Polynesia, *217*, 222
Polypteri, 20
pond, experimental, 167–8
poorwill, hibernation, 123
population
  dynamics, 145, 154
  ecology, 145–9
  history, 177, *178*
  limitation of, 287, 328
porcupine, 226, 301
postbreeding wandering, 128
Potamotrygon, 16
potassium-40, 5
prairie chicken, 330
precocial young, 303
predation, 167, 169, 170, 205, 210–12
  avoidance, 289, 327
  prediction, 173
predator, 118, 186, 293
  avoidance, 300–5, 321–4
  nest, 146
  organized group, 315–26
  relationship to prey, 180–1, 289
prediction, 152, 215, 290
  a posteriori, 295
  quantitative, 147–9, 177
premaxilla, 41, 78
premolar, *see* tooth
pressure sensing, 86–7
Pribilof Islands, 237, 287
primary (feather), 56
Primates, 53, *69*, 223
Proboscidia, 53
Procellariformes, 59, 277
producer, in ecological context, 117
productivity (ocean vs. stream), 237
profitability, 295
prolactin, 286
protein, 108–10
protection, of young, 319
Protopterus, 17–18
Prototheria, 51
proximate cause, 289
prussic acid, 213
Pseudacris crucifer, 169–71
ptarmigan, 44, 133, 303
Pteraspida, 12
Ptilonorhynchidae, 316
Ptilonorhynchus violaceus, 316
Puerto Rico, 223
puffin, beak, 61
pug, 70
pumpkinseed sunfish, 167–9
pupil, 36, 67
pygmy hippopotamus, 223
Python molurus, temperature regulation, 137
quadrate, 39, 78
Queensland, 213
quill, 55

rabbit, 54, 69, *138*, 300
radioactive decay, 4
radiation, heat, 124
radius, 25
rail, flightless, 58
rainfall, 198–201
Rajiformes, 16
Rana
  catesbiana, 74
  sylvatica, 132
  utricularia, 169, *170*, *171*
Rangifer tarandus, 248
rate of increase, intrinsic, 147, 155, 179
ratio of increase, 146
ratite, 230, 231
rat kangaroo, 51
rattlesnake, 80; *see also* sidewinder
Rattus, 210
ray, 15, 16
reaction, biochemical, 49
realm, zoogeographic, 193–4, *195–6*, 219, 225
rear-fanged snake, 39
recombination, genetic, 105, 326, 327
Recurvirostra americana, 322
red-backed salamander, 310
red beds, Devonian, 24
red-bellied newt, 85, 242
red crossbill, *61*, 293
red-eyed vireo, 198
red fox, 300
red-grey vole, *178*
redhead (duck), 254
red knot, 260
Red Sea, 256
red squirrel, 208–9
red-tailed hawk, *61*
red-winged blackbird, *77*, *81*, 309
reference system, 96
refractive index, 66, 70
removal, experimental, 152, 163–4
replication, 163
reproductive isolation, 100, 110
Reptilia (reptile), 11, 30–41, 74, *75*, 96, 188, *190*, *196*, 221, *229*, 231, 237
resource, 118
  common, 167
  limited, 118, 158
  necessary, 205
  species, 291
  utilization, 158
respiration, 117
  via oral cavity vs. skin, 172
result, negative, 163
retina, 66–8
rhea, 55, 224
Rheobatrachus, 58, 224
rhinoceros, 302

Rhyncocephalia, 222
rhythm, endogenous, 124
rib, 35
rice rat, 224
Rift Valley, *see* Great Rift Valley
ring-billed gull, 311, *313*, 318
risking life (altruistic), 327
river otter, 126
Riverside, Calif., 135
robin, 302
rock rabbit, 54
Rodentia (rodent), 46, 53, 122, *178*, 209
  desert, 80, 158–61
  hystricomorph, 226
rod (in vision), 68
rolling, 15
rose-breasted grosbeak, *61*
Ross's goose, 254
rostrum, 21
round window, 26
ruby-throated hummingbird, *61*, 288
ruff, 311
ruffed grouse, 319
rufous-sided towhee, 309
rusty lizard, 145–9

Sable Island, 248
saber-toothed cat, 212, *213*, 323
saccule, 72
sage grouse, 312, *313*, 321, 330
Sahara, 139, 256
salamander, 27–8, *29*, 73, 74, 85, 108, 124, 129, 151–8, 187, 190, 237, 309
  classification, 95–8
  food, 157
  lungless, 172
  response to magnetic field, 244–5
  streambank, 172–4
  terrestrial, 151
salmon, 14, *71*, 236
  fingerlings, 86
  homing of, 239–41, 242
sandpiper, 61, 237
San Francisco, 330
San Juan Capistrano, 265
San Juan River, Nicaragua, 15
Sarcopterygii, 17–18, *19*, 25
Sardinia, 185
Sargasso Sea, 201, 236
sargassum weed, 246
satin bowerbird, courtship display, 316
Sauromelus obesus, 143
savannah, 53
  west African, 38
scala tympani, 76
scala vestibuli, 76
scale, 91
scale tree, 23
Scaphiopus, 127
  hammondii, 142
  holbrooki, 169–71
scarlet tanager, 308
scat, 83
Sceloporus
  clarki,163
  merriami, 164
  olivaceous, 145–9
  undulatus, 38, 163
scent, 83
  offensive, 45
  see also odor
scent marking, 309
scincid, *75*
Sciurus, 208–10
  carolinensis, 125
sclera, 66
Scotia Arc, 228
Scotland, 242, 300
screech owl, 103
seabird, colonial, 311
sea horse, 301
sea lamprey, 13, 236
seal, 54, 54, 248
  blubber, 133
  carcasses, 287
  fur, 237, 311
  milk, 44
sea lion, 54
sea otter, 54
search image, 295
sea snake, 39
season, 121–31
sea turtle, 93, 245–7
  migration, 215–17, 237
secondary (feather), 56
seed, 158–9
  storage, 125
selection, *see* group selection; natural selection
semicircular canal, 71–2
sequence, physiological (migration), 281–5
Serengeti, 237, 248
Serpentes, 188
sex, 327–8
  determination, 33
shark, 15–16, *19*, 88
sharp-shinned hawk, 278
shearwater, 58
ship, sailing, 199
shivering, temperature control, 137
shorebird, 254, 322
shorttail weasel, *45*
shoveler, *61*
shrew, 83, 124
Shroud of Turin, 5
shrub, desert, 198
Siberia, 7, 201, 256
Sicily, 255
sidewinder, 142
sight, 65–70
significance, statistical, 154
similarity, degree of, 108
siphon sac, 17
Sirenia, 54
Sirenidae, 96
size, at metamorphosis, 171
skate, 16
skeleton, axial, 13, 25
skin
  gland, 310
  graft, 114
skitter (on surface film), 35
skull, 14
skunk, 45, 83, 302
slipstreaming, 325
sloth, 51
smell, sense of, 82–4; *see also* odor; scent
Smilodon, 212, *213*
snake, 39–41, 74, 84, 93, 221
  markings, *40*, 41
  secretion of skin lipid, 247
snow bunting, 133, 134, 260, 309
snow goose, 103, *105*, 262
snowshoe hare, 44, 133, 177, *178*
soaring, 56, 141, 256
sociobiology, 328–30
solicitation behavior, 316
Solnhofen Limestone, 55
sonar, 82
song, 76
songbird, 103
song sparrow, 77
Sonoran desert, 163
South Africa, 18, 256, 260
South America, 7, 17, 53, 58, 111, 211, 216, 219, 221, 224, 226, 228, 237
  northern, 269
  west coast, 201
South Atlantic, 220, 227, 237
South Carolina, 38, 101, 253
Southampton Island, 254
South Magnetic Pole, 7
spadefoot, 127, 142
Spain, 256, 278
sparrow, 44, 76, 302
sparrowhawk, 77
speciation, 100, 104
species, 92, 94, 99–116
  application of concept, problems, 107–15
  defined, 100
  other, importance of, 205–18
  redefinition, 111
  relationship to area, 213
  sibling, 103
spermatophore, 28
Spermophilus, 138–9
  lateralis, 122, *123*
Spheniscidae, 230, 231
Sphenodon punctatus, 25, 93, 222
Spilogale putorius, 302
spine, 301
spiny anteater, 317
spiracle, 15
spotted skunk, 302
spring, 126–8

spruce, 205, 236
  cone, 293
Squalus acanthias, 237
Squamata, 33, 35, 188, *221*
squirrel, 46, 137–8, 301
stapedius, 78
stapes, 78
starling, 324
sterovilli, 72
stingray, 16, *19*
stomach content, of fish, 168
stork, 256
storm petrel, 86
stream
  capture, 191
  habitat, 97
  mountain, 172
Strigiformes, 59
striped skunk, *45*
structure, feeding, size ratios in, 172
Struthio camelus, 140
sturgeon, *19*, 20, 86
Sturnella neglecta, 311
Sturnus vulgaris, 271, 321
subfamily, 95, 194
subjectivity, 304
subphylum, 95
succession, stage in, 216
suction feeding, 21
Sudan, 261
Sulawesi, 193
summer, activity, 128–9
sun clock, 244
sunfish, 167–9
superposition, law of, 1
surface
  film, 35
  grinding, 47
  ratio to volume, 173
  respiratory, 190
Surinam, 216
survival, 163, 171, 265
  age specific, 145
  on migration, 251
Swainson's thrush, 262
swallow, 56, 256, 261
  "hibernation" of, 235
swallowing, 39, 108
swamp rabbit, 54
swamp sparrow, 77
sweat, 50, 139
swift, 56
swimming, 21
Switzerland, 256, 278
swordfish, 21
Sylvia, 284–5
Synapsida, 33

Tachyglossus, 317
tadpole, 85, 169–72; *see also* frog
Tahiti, 257
taiga, 237, 256
tailed frog, 30, 222
tail wind, 261
Tamias striatus, 125
Tamiasciurus hudsonicus, 86
tank, experimental, 169
Tanzania, 211
tapetum lucidum, 66
tapir, 54, 219, 225
Taricha rivularis, 242
tarpon, 21
tarsal, 25
Tasmanian devil, 51
Tasmanian wolf, 1
taste, sense of, 84
taxonomist, 94
taxonomy, 101
  numerical, 107–11
  see also classification
tectonic plate, 6
Teiidae, 84
Teleostii, *19*, 21, *72*
temperature
  average, distributional correspondence, 195–8
  body, 49–50, 124, *138*
  and feeding, *125*
  lower critical, 135
  migration timed by, 281
Tennessee, 101, 206
tensor tympani, 78
Terrapene carolina, 127
territoriality, 29, 289, 327, 328
  behavioral, 307–14
  benefits vs. costs, 312
territory, 36, 289
  defined, 307
  kinds, 311
test, a priori, 296
testis length, premigratory, 284
Testudinae, 33
Testudinata, 33, *221*
Tethys Sea, 7
Tetrao urogallus, 303
tetrapod, 96, *221*
Texas, 38, 110, 145
Thalarctos maritimus, 133
theory, 287, 290–1
  assumptions of, 110, 113, 291–2
  of competition, 154–7
  of distribution, 219–33
  of evolution, 100
Therapsida, 43
thermoneutral zone, 135
Thomson's gazelle, 248
Thraupidae, 266
three-spined stickelback, 308
thrush nightingale, 308
thunder, 278
Tibet, highlands, 256
tibia, 25
tiger salamander, 170
time, and migration, *see* migration, timing
time scale, geologic, 1–5
Timor, 193
Tinamidae, 231
titmouse, 124
Tobago, 264
toe, lobed, 59
tokay gecko, 308
Tonga, 185
tongue, 108
tooth
  canine, 46, 212
  carnassial, 47
  heterodont, 42, 45
  incisor, 46, 323
  milk, 42
  molar, 46
  pedicellate, 27
  permanent, 42
  premolar, 45
  vomerine, 108
topography, 189
torpedo ray, 23
torpor, 50, 122–3
touch, 86–7
toucan, beak, 61
trade wind, 197, 258
trait, types, 91, 93
  single, 108
transfer time (feeding), 297
tree, broadleafed, 205
tree ring, 4
tree shrew, *81*
tree swallow, 287
Triadobatrachus, 27
tribe, 96
triploid, 115
Tristan da Cunha, 210
Tropic of Cancer, 199
Tropic of Capricorn, 199
trout, 21
Trypanosoma brucei, 212
tsetse, 212
Tubulidenta, 54
tuna, 21
tundra, 205, 237, 256, 323
tundra swan, 261
Turdus migratorius, 317
Turkey, 252
turkey, 303
turkey vulture, 82–3, 265, 277
turtle, 30, 33, *34*, 70, 75, 93, 221, 301
  longevity, 34
  see also sea turtle
tusk, 46
Tympanuchus cupido, 330
tympanum, 73, 76
type locality, 101

ulna, 25
ultimate cause, 290
ultrasound, 82
undulation, 21
United States, *198*
  eastern, 216, 219
  northeastern, 330
  southeastern, 202, 216

southwestern, 80, 114, 158, *159*, 164, 261
see also specific state
United States Forest Service, Coweeta Hydrologic Lab., 172
United States National Museum, 101
uranium-235, 5
urea, 32, 143
uric acid, 31, 143
urine, 83
Urochordata, 12
Urosaurus ornatus, 163
Utah, 38, 158, 163, 254
utricle, 72

Vanellus vanellus, 309
variation, geographic, 101
varve, 4
veery, 262
vegetation, 198
Venezuela, 217, 257
ventricle, 49
Vermont, 235
vertebrate
primitive, 224
terrestrial, origin, 23–7
vibrissa, 44, 86
Victoria (Australia), 213
vicuna, 226
vine snake, 42
viper, 40
vireo, 302
Vireonidae, 226
Virginia, 99, 107, 125, 216, 258, 270, 330
vision, binocular vs. stereoscopic, 68
visual field, 68–70
viviparity, 17, 47, 317
volcano, 6
vole, 177, 181
Vulpes vulpes, 300
vulture, 56, 117

Wales, 208
wallaby, 51
Wallops Island, Va., 270
walrus, 54, 185
wapiti, 311
warbler, 56, 60, 99, 302
warning signal, 327–8
Washington (state), 138
water
chemical composition, 239
loss of, 139, *141*
watershed, 191
water shrew, 54
wave, direction of, 246
weasel, 44, 51, 83
Welland Canal, 13
westerly, 201
Western Hemisphere, 87, 211, 216
western meadowlark, 311
West Indies, 222, 258
West Virginia, 216
whale, 54, 82, 93
migration, 247–8
wheatear, 260
whitefish, 13
white pelican, 325
white pine, 236
cone, 293
white-throated sparrow, 251, 265, 270
wildebeest, 237, 248
willow warbler, 260
Wilsonia citrina, 308, 324
Wilson's phalarope, 261
wind pattern, 199–201
wing, types, 56–8
wing covert, 56
winter, activity, 121–6
wolf, 51, 83, 132, 225, 309, 323, 325
wombat, 51, 83
woodchuck, 50
hibernation, 122
woodcock, beak, 61
wood frog, 132
woodpecker, beak, 61
wood warbler, 99, 103
worm, 12
Wyoming, 4, 247

Xenarthra, 223

Yangtze River, 20
yawing, 15
yellow-breasted chat, 302
yellow-headed lizard, 36, 309
yellow-rumped warbler, *61*, 130, 287
Yucatan Peninsula, 288

Zapodidae, 122
zebra, 248
Zenaida macroura, 142
Zermatt (Switzerland), 205
zonation, altitudinal, vs. latitudinal, 195, 205
Zonotrichia albicollis, 251, 265, 270
Zugunruhe, *see* migratory restlessness